Helium Cryogenics

THE INTERNATIONAL CRYOGENICS MONOGRAPH SERIES

Current Volumes in this series

APPLIED SUPERCONDUCTIVITY, METALLURGY, AND PHYSICS OF TITANIUM ALLOYS • *E. W. Collings*
Volume 1: Fundamentals
Volume 2: Applications

CRYOCOOLERS • *G. Walker*
Part 1: Fundamentals
Part 2: Applications

THE HALL EFFECT IN METALS AND ALLOYS • *C. M. Hurd*

HEAT TRANSFER AT LOW TEMPERATURE • *W. Frost*

HELIUM CRYOGENICS • *Steven W. Van Sciver*

MECHANICAL PROPERTIES OF MATERIALS AT LOW TEMPERATURE • *D. A. Wigley*

STABILIZATION OF SUPERCONDUCTING MAGNETIC SYSTEMS • *V. A. Al'tov, V. B. Zenkevich, M. G. Kremlev, and V. V. Sychev*

SUPERCONDUCTING ELECTRON-OPTIC DEVICES • *I. Dietrich*

SUPERCONDUCTING MATERIALS • *E. M. Savitskii, V. V. Baron, Yu. V. Efimov, M. I. Bychkova, and L. F. Myzenkova*

Helium Cryogenics

Steven W. Van Sciver
University of Wisconsin—Madison
Madison, Wisconsin

PLENUM PRESS • NEW YORK AND LONDON

Library of Congress Cataloging in Publication Data

Van Sciver, Steven W.
Helium cryogenics.

(International cryogenics monograph series)
Includes bibliographies and index.
1. Liquid helium. 2. Helium at low temperatures. I. Title. II. Series.
QC145.45.H4V36 1986 536′.56 86-20461
ISBN 0-306-42335-9

A Division of Plenum Publishing Corporation
233 Spring Street, New York, N.Y. 10013

Printed in the United States of America

To Mary, Travis, and Courtenay

Preface

At least 10 years have elapsed since a comprehensive monograph concerned with the broad subject of cryogenics has been published. During this time a considerable quantity of research and development has been carried out in the field of cryogenics. Furthermore, there has been a certain degree of redirection of effort within the field, mostly driven by the variety of new applications, ranging from superconductive magnet systems to microelectronics. Greater emphasis is now being placed on low-temperature cryogenics, particularly that of liquid helium. Until now cryogenic books have provided a broad survey of materials and fluid properties over the entire cryogenic regime, $T \lesssim 150$ K. This approach does not allow sufficient detail in any particular area to bring the reader to the current level of understanding in the subject. In addition, the behavior of helium has been lumped with that of other cryogenic fluids, although the properties of helium are quite unique. As a result, a clear relationship has not been established between the fundamental understanding of helium fluids and their potential applications.

The present book has been written to fill this void. The approach is to survey the field of cryogenics, specifically as it pertains to helium fluids. This approach is more specialized than that contained in previous cryogenics books. Furthermore, the level of treatment is more advanced and a certain knowledge of fundamental engineering and physics principles has been assumed. Unlike previous books on liquid helium, the present treatment contains both engineering and physical descriptions. The goal throughout the work is to bridge the gap between the physics and engineering aspects of helium fluids to encourage their use and enhance their usefulness in low-temperature systems.

The content of the book is based on a course offered at the University of Wisconsin–Madison. Students who register for this course are almost exclusively at the graduate level. As a result, a reasonable background knowledge of physics and engineering has been assumed. Recommended prerequisites include a working knowledge of thermodynamics and statistical physics, heat transfer, and elementary solid-state physics. Without this background, the reader may find it necessary to review one or more of these subjects. A number of useful references are given at the end of the book.

The material contained in this book is divided into nine chapters. Chapter 1 introduces the basic principles of cryogenics, including a discussion of applications. Chapter 2 describes the thermal properties of materials at low temperatures, concentating on solids. This is not only a useful background review but it also introduces some fundamental physics, which is used in later chapters. Chapter 3 introduces helium as a classical fluid, concentrating on its physical aspects as they can be described using classical models. Chapter 4 then discusses helium as a quantum fluid, emphasizing the theory and experimental evidence associated with superfluidity. Chapter 5 turns to the engineering problem of heat transfer in superfluid helium, and how the fundamental understanding of helium introduced in Chapter 4 can be used to describe its characteristics. Chapter 6 concentrates on the problem of heat transfer in pool boiling normal helium. Chapter 7 extends the discussion of helium to fluid flow, including heat transfer and pressure drop. Chapter 8 discusses the thermodynamic aspects of liquefaction and refrigeration systems, including a discussion of actual refrigeration systems in use today. Finally, Chapter 9 summarizes some special topics of interest to both helium cryogenics and related disciplines. The goal here is to survey a few very specific areas of helium cryogenics and related disciplines which, although slightly outside the main scope of the text, are still important in low-temperature applications.

Throughout the writing of this book, I have received considerable assistance and encouragement from collegues, students, and friends. Their support should not go unrecognized. I would like to give thanks to two of my students, D. Scott Holmes and John G. Weisend II, for their critical review of the partially completed manuscript and for assisting in developing problems. A number of colleagues read sections of the manuscript and made substantive suggestions on improvements to be made. They are Drs. A. F. Clark, F. R. Fickett, and V. Arp, all of the National Bureau of Standards; Dr. L. Dresner, Oak Ridge National Laboratory; Prof. O. E. Vilches, University of Washington; Prof. J. T. Tough, Ohio State University; and Prof. R. F. Barron, Louisiana Tech University. Their help is greatly appreciated. The conversion of my handwritten version to a readable typewritten text was due to the efforts of Ms. Kay Ewers. This task was certainly second only to the actual writing in terms of the amount of effort involved. Production of the graphics must be credited to Ms. Helga Fack and her staff. Finally, I would like to acknowledge the indirect help that my family has provided in terms of encouragement and willingness to forego some leisure activities so that time could be devoted to the effort of writing this book. In retrospect, it has been worthwhile.

S. W. Van Sciver

Contents

Symbols, SI Units

A — area, m^2
— Gorter–Mellink parameter, m · s/kg
— Schottky specific heat coefficient, J · K/kg · ion

a — absorptivity
— coefficient of turbulence buildup, $W^{3/2}/m^{3/2}$
— van der Waals gas coefficient, N/mole
— vortex core radius, m
— Zeehman ordering parameter

B — Brillouin function
— magnetic flux density, T
— second virial coefficient, m^3/kg
— vortex interaction parameter

b — coefficient of expansion
— impact parameter, m
— van der Waals gas coefficient, $m^3/mole$

C — circulation, m^2/s
— cost, $
— heat capacity, J/K, $J/m^3 \cdot K$, J/mole K
— specific heat, J/kg · K
— third virial coefficient, m^6/kg^2

c — sound velocity, m/s
— surface wave speed, m/s

D — density of states
— diameter, m
— thermal diffusivity, m/s
— fourth virial coefficient, m^9/kg^3

d — diameter, m
— film thickness, m
— number of atomic layers

E — binding energy, J
— energy flux, J/m^2
— electric field, V/m
— internal energy, J
— radiant energy flux, W/m^2
— Young's modulus, N/m^2

e — charge of electron, coul
— specific internal energy, J/kg
— spectral energy density, W/m^3

F — force, N

f — bubble detachment frequency, s^{-1}
— Fermi–Dirac distribution function
— force per unit length, N/m
— friction factor
— heat conductivity function, W^3/m^5
— Maxwell-Boltzmann distribution

G — specific mass flux, $kg/m^2 \cdot s$

Gr — Grashof number

g — degeneracy
— Landé g-factor
— specific Gibbs potential, J/kg

g — gravitational acceleration, m/s^2

H — hamiltonian
— height, m
— magnetic field, A/m

h — heat transfer coefficient, W/m^2K
— hydrostatic head, m
— Planck's constant, J/K
— specific enthalpy, J/kg

J — exchange interaction
— particle collision term
— total angular momentum quantum number

j — momentum flux, kg/m^2s

K — forced convection parameter
— numerical constant

k — expansion parameter
— thermal conductivity, W/m · K
— wave number

k_B — Boltzmann constant, J/K

L — length, m
— Lorenz number
— orbital angular momentum quantum number

l — mean free path, m

M — magnetization, A/m
— molecular weight, kg/mole

m — mass, kg
— electron spin quantum number

N — number of particles

N_0 — Avogadro's number, mole^{-1}

Nu — Nusselt number

n — number density, m^{-3}
— statistical distribution

P — momentum, kg · m/s
— perimeter, m
— polarization
— refrigeration power, W

Pr — Prandtl number

p — pressure, Pa

Q — heat rate, W

q — critical quality
— heat flux, W/m^2
— isosteric heat, K

R — gas constant, J/mole K

Ra — Rayleigh number

Re — Reynolds number

r — radius, m

S — entropy, J/K, J/m^3 · K, J/mole K
— slip ratio
— spin quantum number
— vortex line dimension, m

s — specific entropy, J/kg · K

T — temperature, K

t — reduced temperature
— time, s
— transmission coefficient

U — attractive potential, J

u — velocity component, m/s

V — volume, m^3

v — velocity, m/s

v — specific volume, kg/m^3
— velocity component, m/s

W — work, J

w — Bénard convection parameter
— specific work, J/kg
— width, m

x — concentration
— coverage
— expansion circuit flow fraction
— position coordinate, m

y — position coordinate, m
— yield in a liquefier

Z — collision number, m^{-3}s^{-1}
— integrated heat conductivity function, W/m$^{5/3}$
— impedance
— partition function

z — fugacity
— position coordinate, m

Greek Letters

α — accommodation coefficient
— linear thermal expansion coefficient, K^{-1}
— Kapitza conductance parameter, W/m^2K^n
— void fraction

β — bulk expansivity, K^{-1}
— condensation coefficient
— geometrical factor in Poiseville flow
— $1/k_BT$

η — efficiency
— phase shift
— viscosity, $N \cdot s/m^2$

θ — angle

Θ — reduced temperature

Θ_D — Debye temperature, K

κ — compressibility, Pa^{-1}
— transient heat transfer parameter, $W^4 \cdot s/m^8$

λ — latent heat, J/kg
— wavelength, m

μ — chemical potential, J
— expansion coefficient, K/Pa, $T \cdot m^3K/J$
— magnetic movement
— roton mass, kg

ν — frequency, s^{-1}
— kinematic viscosity, m^2/s

ξ — coherence length, m

π — reduced pressure

ρ — density, kg/m^3
— resistivity, $\Omega \cdot m$

σ — conductivity, $(\Omega \cdot m)^{-1}$
— Kapitza conductance parameter, W/m^2K^4
— Lennard–Jones potential parameter, J
— scattering cross section, m^2
— Stefan–Boltzmann constant, W/m^2K^4
— stress, Pa
— surface tension, J/m^2

γ — coefficient of the electronic specific heat, J/mole K^2
— Curie constant, K
— Grüneisen parameter
— reduced velocity
— specific heat ratio
— statistical parameter

Δ — roton minimum energy, K

δ — thickness, m
— level splitting in Stark effect, J
— vortex line spacing, m

ε — emissivity
— energy level, J
— Lennard–Jones potential parameter, J

τ — reduced temperature
— time, s

Φ — potential, J
— wave function
— reduced equation of state

ϕ — spreading pressure, N/m
— azimuthal angle

χ — flow quality
— magnetic susceptibility
— vortex interaction parameter

Ψ — wavefunction
— parameter in cylindrical heat transfer

Ω — collision integral
— thermodynamic probability

ω — angular frequency, s^{-1}

Subscripts

a — acceleration, absorbed
B — Bohr
B_0 — Boyle
b — bath, black body
bp — boiling point
C — cold, Curie

c — critical
c_1 — lower critical
c_2 — upper critical
CL — classical
D — Debye
e — electronic, expansion, emitted
eff — effective
exp — experimental
F — Fermi
f — final state, film
fb — film boiling
fc — forced convection
G — Grüneisen
g — gas
gen — generation
gr — gravitational
H — hot, constant field
h — hydraulic, constant enthalpy
i — initial state, incident
ic — internal convection
int — internal
inv — inversion
j — Joule–Thomson
K — Kapitza
k — index
L — vortex line, liquid
LJ — Lennard–Jones
l — liquid
m — maximum, mean
n — normal, normal fluid, index
nc — normal fluid critical
ns — normal-superfluid
p — constant pressure, proton
ph — phonon
pl — const pressure liquid
R — reversible, recovery
r — roton, reduced, radiant
s — superconducting, superfluid, sound, constant entropy, surface, substrate
sat — saturated
sc — superfluid critical
st — isosteric
T — constant temperature, thermal de Broglie

t — transmitted, total
tt — turbulent
u — ultimate
v — constant specific volume
y — yield
λ — lambda point
$\perp$ — perpendicular component
0 — ambient, ground state
1 — first excited state, first
2 — second
3 — of ^{3}He, third
4 — of ^{4}He, fourth

Superscripts

* — critical, normalized parameter, effective
$\overline{}$ — average
$\cdot$ — time derivative
A — acoustic mismatch
c — critical
m — empirical power law
n — empirical power law
p — phonon radiation, empirical power law
q — empirical power law
r — empirical power law
α — empirical power law
β — empirical power law

Chapter 1

Cryogenic Principles and Applications

The science and technology of producing a low-temperature environment is generally referred to as cryogenics. The word cryogenics has its origin in the Greek language where "cryos" means frost or cold and "gen" is a common root for the English verb to generate. Strictly speaking, cryogenics means to produce cold, yet the term has developed a more general connotation over years of usage by engineers and scientists. Today, the word cryogenics is associated with the production and study of low-temperature environments. Thus, a cryogenic engineer is a person who specializes in these areas. The expertise of a cryogenic engineer can vary considerably within this discipline. For example, he or she may be concerned mostly with heat transfer aspects of low-temperature fluids or alternatively, improved methods of producing low temperatures. Expertise in cryogenic engineering is in demand in a wide variety of technical fields including advanced energy production and storage technologies, transportation and space programs, and a wide variety of physics and engineering research efforts.

Over the years the word cryogenics has developed several common usages. A cryogenic fluid is one that is used in the production of cold, while cryogenic machinery is the hardware used in achieving low-temperature environments. At first it would appear that all machinery and fluids used in cooling would be identified as cryogenic. Generally, however, it is accepted that the word cryogenics is reserved for those processes that take place below about 100–150 K. This distinction is established because it represents the point where permanent gases such as N_2, O_2, and Ar begin to liquefy. Sometimes cryogenics is used in reference to liquid natural gas (LNG) and its liquefaction technology.

1.1. TEMPERATURE SCALE

Assuming it is possible to refer to a range of temperature as cryogenic and that the range spans from absolute zero to about 100 K, we might ask the following question: "Why all the excitement over a range of temperature spanning less than a few hundred kelvins?" The answer to this question lies in the thermodynamic description of the temperature scale. To produce low temperatures, it is necessary to reduce the entropy of the system since there is a direct relationship between absolute temperature and entropy. The third law of thermodynamics states that as absolute zero is approached, not only does the entropy of a system go to zero, but the entropy change associated with an adiabatic process must also go to zero. In other words, the lower the absolute temperature the more difficult it is to obtain a unit temperature decrease. Therefore, it is much more meaningful to identify temperature as having a logarithmic rather than linear scale.

On a logarithmic temperature scale, the cryogenic range occupies a very large portion of achievable temperatures. Plotted in Fig. 1.1 is the temperature scale with the range of physical phenomena superimposed. It

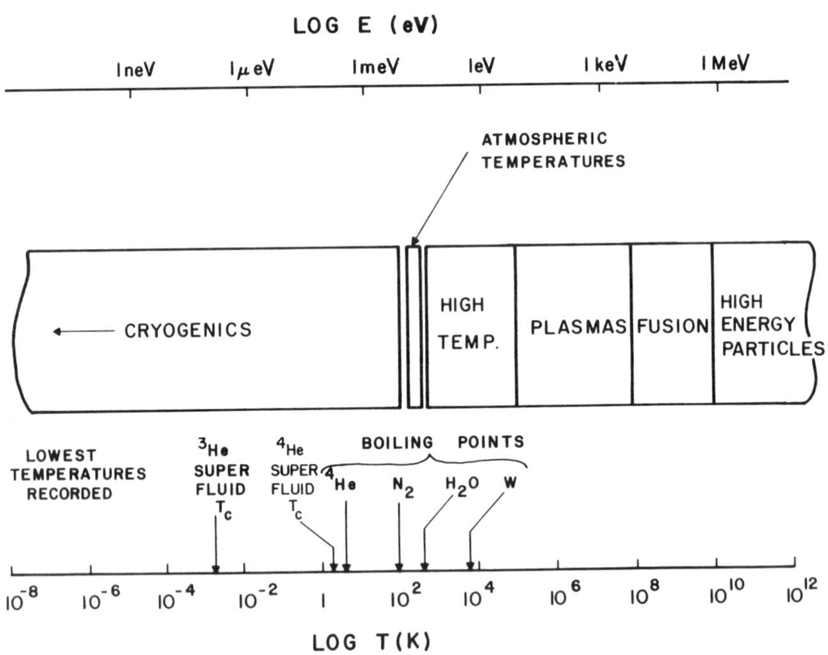

Fig. 1.1. Logarithmic temperature scale with superimposed physical phenomena. Note that $E = k_B T$ and 1 eV is equivalent to 11,609 K.

should be immediately clear that cryogenics applies to nearly half of the achievable temperature range. From the physical point of view, the range is interesting for the large number of phenomena that occur within it. A few examples of these phenomena are phase changes of common elements, magnetic ordering, solid-state transformations, and the quantum effects including superconductivity and superfluidity.

A subset of the field of cryogenics is identified with the most permanent of all gases, helium. There is probably more known about helium than any other fluid with the possible exception of water. Furthermore, the refrigeration and liquefaction of helium are very specialized, making it a separate subject in and of itself. The technological application of helium cryogenics is also unique because its use is in a temperature range inaccessible by other fluids and extremely exciting for its relationship to superconductivity. Helium cryogenics spans a smaller range of temperature, since the critical point of the most common isotope, ^{4}He, is 5.2 K. However, since helium does not solidify except under an external pressure exceeding 2.5 MPa (25 atm), fluid properties can be studied to as low temperatures as are physically possible. At the present time the minimum achievable bulk temperature is about 50 μK. The technology of achieving these ultralow temperatures is special and is relevant primarily to fundamental studies of condensed-matter physics. Since the emphasis of this book is on the technological applications of helium cryogenics, the phenomena associated with ultralow temperatures are not considered in much detail. Thus, helium cryogenics as described here emphasizes the range of temperature where the fluid has large scale potential applications as a coolant, $1 \lesssim T \lesssim 10$ K.

1.2. HISTORICAL BACKGROUND

Research and development into the field of cryogenics began slightly more than 100 years ago. Perhaps the most significant discovery in advance of actual liquefaction of a permanent gas was the realization that common gases such as O_2 and N_2 have similar behavior when cooled to temperatures near their critical points. In the early 1860s, substantial theoretical and experimental evidence was put forth identifying the concepts of phase separation and critical phenomena. These ideas, which are taken for granted today, form essential background to the understanding and application of the liquefaction and refrigeration processes.

To summarize phase separation and critical phenomena, reference is made to a useful general figure (Fig. 1.2) showing the states of matter when temperature T is plotted versus entropy S. Here the critical temperature T_c is at the maximum of the two-phase coexistence region. For obvious

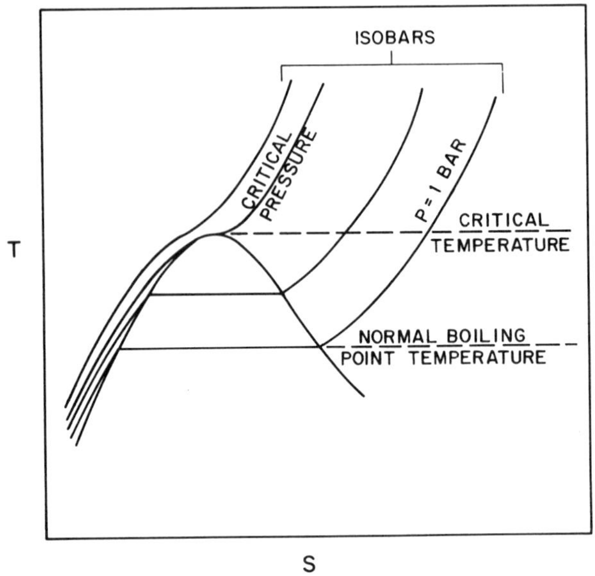

Fig. 1.2. Typical temperature–entropy diagram for a pure substance.

reasons, many engineers who work in refrigeration and liquefaction refer to the coexistence region as "the dome." The terms isobaric and isothermal processes are used to describe those events that occur at constant pressure or temperature. As a general rule of thumb, the normal boiling point T_b of cryogenic liquids is about half T_c, although there is considerable variability in this rule particularly for helium where $T_b/T_c \approx 0.8$.

Three major scientific developments in the late 19th century provided the essential framework for the liquefaction of helium and thus the beginning of helium cryogenics. The first of these developments occurred during the liquefaction of O_2 in 1877 by two separate investigators in Europe, Cailletet in France and Pictet in Switzerland. Each of these workers demonstrated the concepts of liquid–vapor coexistence in permanent gases near their normal boiling point. Furthermore, Pictet's method used the cascade principle where gases are successively cooled to lower temperature by exchanging heat with a higher normal boiling point liquid. Today, this method is often used in refrigeration systems that require the highest possible thermodynamic efficiency.

The second major development occurred in 1892 when Dewar introduced the vacuum insulated flask as a storage container for liquid cryogens. His concept consisted of a glass double-walled vacuum vessel with inner walls silvered to reduce radiation heat transfer. A drawing of an early "dewar" vessel used for liquid helium is shown in Fig. 1.3. Liquid

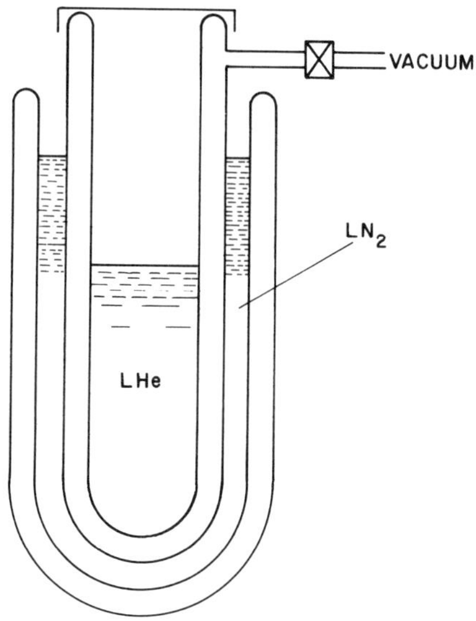

Fig. 1.3. Schematic of an early dewar vessel for containing liquid helium. The outer liquid nitrogen dewar provides a thermal radiation intercept.

nitrogen shielding is provided to reduce further the thermal radiation heat leak. The dewar is essential for the storage of liquid helium because of the fluid's extremely small latent heat.

Dewar made the final development that led the way to the liquefaction of helium. This involved the use of a Joule–Thomson valve to produce liquefaction of hydrogen, the last of the permanent gases with boiling points above that of helium. It is particularly significant that Dewar employed a Joule–Thomson valve, which is essentially isenthalpic expansion, because this method also is used in the production of liquid helium. With liquid hydrogen available (having a boiling point $T_b = 20.4$ K), it became possible to consider liquefying helium by a combination of the cascade principle and Joule–Thomson expansion. The chief remaining difficulty was that helium gas was a rare commodity around 1900.

It took 10 years for Kamerlingh Onnes at the University of Leiden to achieve the successful liquefaction of helium in 1908. At that time their total helium inventory was only 360 STP gas liters (equivalent to about 0.5 liquid liter), and yet Leiden rapidly became the center of liquid helium research. Among the achievements attributed to Onnes' group are the discoveries of superfluidity in liquid helium below 2.17 K and superconductivity in many elements.

For the next 30 years, production of liquid helium and its associated research and applications were limited primarily to a few specialized laboratories. Researchers actively investigated the properties of materials as well as those of liquid helium down to temperatures below 1 K. Included among their discoveries are: the properties of superfluid helium, identification of numerous superconducting materials, and the use of magnetic cooling to achieve temperatures below 1 K. At the same time, researchers envisioned large-scale applications of helium cryogenics but lacked the motivation and resources to develop the technologies.

In the 1940s commercial development of hydrogen and helium liquefaction equipment began, owing to a large extent to the efforts of S. Collins of MIT and the Arthur D. Little Company. For the first time, laboratories could purchase helium liquefaction plants and liquid helium become available on the open market. Liquid helium research became more widely practiced. This point in time marked the beginning of large-scale cryogenic engineering, which required individuals skilled at design and handling of cryogenic equipment. A first, military and space applications led the field. However, today a major thrust is in applications of superconductivity. As a result of this growth, helium cryogenic engineering has developed into a substantial discipline.

Since the development of commercial helium refrigerators, major progress has been made on the two fronts of research and development. In basic research, the ready access to large quantities of liquid helium has freed the researchers to push toward lower temperatures. Milestones in this progress must include: the development of ^{3}He–^{4}He dilution refrigeration technology permitting continuous cooling in the millikelvin temperature range; the approach to submillikelvin temperatures using techniques such as Pomaranchuk cooling and nuclear demagnetization; and the discovery of superfluidity in ^{3}He at about 2 mK.

Commercial development has also progressed substantially in the last 40 years. Larger and larger liquefiers and refrigerators are being produced. Presently, state-of-the-art may best be portrayed by the Fermi National Laboratory central liquefier, which has a 5000 L/h capacity at 4.5 K, and the Brookhaven refrigerator, which has a cooling capacity of 25 kW at 3.8 K. Besides larger-sized refrigerators, other major commercial developments include production of refrigerators that operate continuously at 1.8 K using superfluid helium, establishing standard laboratory-sized refrigerators that operate in the 10–100 W range, and the development of small cryocoolers in the 1 W range for cooling of detectors and other low-power applications.

In general, helium cryogenics has evolved into a well-established discipline providing the environment for a wide range of technologies. What is in store in the future? Trends include development of a wider range of standard products both from the very small refrigerators to the largest ones. New demands are being placed on the systems being produced: higher thermodynamic efficiency, greater reliability, and cleaner operation, to name a few. Considerable research is aiding this progress and new technologies are putting special demands on helium cryogenic systems.

1.3. APPLICATIONS FOR CRYOGENICS

As motivation to the discussion of the properties and production of low-temperature fluids, it is useful to identify the major applications for cryogenics technology today. Some of these are commercial enterprises, while others are still primarily in the stages of research and development. It is possible to separate these applications into at least five major categories:

1. Storage and transport of gases
2. Separation of gases
3. Biological and medical applications
4. Altering material properties by reduced temperature
5. Superconductivity

The last of these categories is particularly of current interest and relevance to helium cryogenics. It also represents a major thrust area for research and development.

Large-quantity storage and transport of gases are best achieved with the help of cryogenics. It is much more efficient from the standpoint of total weight, to transport cryogenic fluids in the liquid state rather than as a pressurized gas. Furthermore, the gases evolved from a storage dewar can maintain a lower impurity content than is common in high-pressure gas storage. There are a number of examples where cryogenic storage and transport are widely practiced. At relatively high temperatures, liquid natural gas (LNG) is transported on a large scale. Liquid oxygen is stored in large quantities for applications in steel production as well as providing high-purity gaseous oxygen supplies for hospitals. Another major application for cryogenic storage and transport is in liquid fuel rockets where LO_2 and LH_2 are common propellants. Even helium is often transported in the liquid state. This is not only because many users do not possess the necessary liquefaction equipment but also to save weight for transportation.

The separation of gas mixtures, particularly air, is another commercial enterprise in which cryogenics plays a major role. By using the physical properties of adsorption, that is, the tendency for gases to condense on cold surfaces, it is possible to separate gas mixtures by differences in their adsorption rates. This procedure is used commonly in extracting O_2 and N_2 from air but also in the purification of LNG. In fact, the major source of helium gas is as a by-product of natural gas production. Related to gas separation is the technology of cryopumping. The physical process of adsorption provides a mechanism for clean, high-speed pumping systems. For this process to be effective, the pumping surfaces must be well below the critical temperature of the gas to be pumped.

Biological and medical uses for cryogenics are extensive. In these applications the goal is to store, modify, or destroy a biological structure by reducing its temperature. Storage of cellular structures in liquid nitrogen is a common practice, the largest of these being the storage of blood plasma. Other examples of this technology include storing cattle semen for artificial insemination and the preservation of food. Apart from cellular storage, medicine is making increasing use of cryogenics. In a procedure known as cryosurgery, selected areas of tissue are frozen and removed with less difficulty than by conventional methods.

The basic properties of materials change with the temperature is reduced. This effect is being used in several engineering applications of cryogenics. A good application for material property variation is in the recycling industry. Cryogenic recycling uses low temperatures to separate materials. The approach takes advantage of differential thermal contraction and the increased brittle nature of materials at low temperatures. There are numerous examples of composite materials that can be recycled by this method. Thermal contraction can also be used in the construction of mechanical structures. The assembly of a close tolerance connection can be facilitated if one of the components is first cooled in a cryogenic flüid to make it slightly smaller. In these applications there is little need to reduce the temperature below the normal boiling point of liquid nitrogen because in general mechanical properties do not change significantly below this temperature.

Besides mechanical properties, low temperatures also are used to change electrical properties of materials. One of the major applications of this process is in the cooling of detectors and other electronic sensors. The low temperature reduces the thermal noise and provides an isothermal environment for the sensor. Examples of devices that use low temperatures include semiconducting particle detectors and large-scale infrared detectors for astrophysical studies. A recent demonstration of this concept is the infrared radio astronomy space telescope (IRAS) which had its detector cooled to liquid helium temperature.

The technology of superconductivity warrants special attention as an application of cryogenics. The largest-scale application of superconductivity is in magnet technology. At present, development programs are underway in support of high-energy physics accelerators, magnetic fusion confinement systems, energy storage, magnetohydrodynamics (MHD), magnetic levitation, and whole-body magnetic resonance imaging (MRI) scanners. In addition, large-scale superconducting power transmission is an application of superconductivity which has similar cryogenic problems. Each of these applications requires large and sophisticated cryogenic systems.

Small-scale applications of superconductivity are also being developed with a concern for their cryogenic systems. Two major areas of small-scale development are in superconducting electronics. The superconducting computer and the superconducting magnetometer both rely on special electronic circuitry cooled to liquid helium temperatures. The cryogenic system requirements for these applications are special mostly because of their being smaller than those currently available from conventional refrigeration equipment suppliers.

1.4. THERMODYNAMIC LAWS

Thermodynamic principles and concepts are of fundamental importance to the field of cryogenics. Thermodynamics is the basis for calculations of the properties of cryogenic fluids and in the performance of refrigeration and liquefaction systems. This requires a sound understanding of the principles of thermodynamics. There are three basic laws of thermodynamics that apply to all systems and are of particular interest to the field of cryogenics. It is assumed that the reader is familiar with these laws through a previous course in thermodynamics. However, for completeness and commonality of notation a review of these laws is presented here.

1.4.1. First and Second Laws of Thermodynamics

The first law of thermodynamics applies to a process that takes a thermodynamic system from an initial state to a final state. The law involves the definition of what is called an adiabatic path which is a process taking the system from one point to another without the production or absorption of heat. For an adiabatic path, the work necessary to bring the system from the initial to the final state is independent of the path chosen between the points. This statement can be considered a definition of work; that is, work done on the system is equal to the increase in the potential energy. Incidentally, if the process is not adiabatic, heat is produced and conservation of energy demands that the amount of extra heat be included in the total amount of work done. The mathematical formulation of the first law of thermodynamics can be written

$$Q = E_f - E_i + W \tag{1.1}$$

where E is the internal energy, a state function dependent only on the initial and final points of a path in thermodynamic space.

Now consider a cyclic process where a system in question is taken from the initial state through the final state and back to the initial state by some other path. This process, shown schematically in Fig. 1.4., may be, for example, an idealization of a cycle used to liquefy a cryogenic fluid.

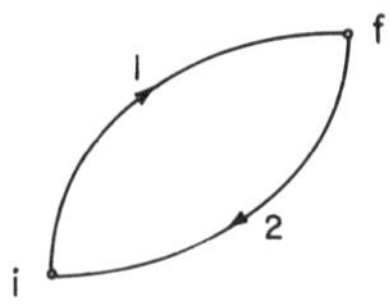

Fig. 1.4. Adiabatic paths between two thermodynamic states.

Application of the first law demands that the difference between the heat and work for the two paths must be equal in magnitude but opposite in sign. Thus, for the entire cycle, the sum of the heat and work must be identically zero.

To make the first law into a more useful form, the initial and final states are brought arbitrarily close together, resulting in a differential form for (1.1):

$$\delta Q = dE + \delta W \tag{1.2}$$

Here it is important to note that the only total differential is that of the internal energy. The differential work term in (1.2) can take on several forms, dependent on what type of system is of interest. In particular, three important systems are:

Liquid–gas system: $\delta W = p\,\delta V$

Magnetic system: $\delta W = -\mu_0 \mathbf{H} \cdot \delta \mathbf{M}$

Electric system: $\delta W = \mathbf{E} \cdot \delta \mathbf{P}$

Note that the latter two differentials are vector quantities dependent on the direction of the applied fields, while the pV product is directionally independent. Fortunately, it is a rare occurrence for more than one type of work to be important in a particular system or process.

The second law of thermodynamics is concerned with the conversion of heat into work and the efficiency with which this can take place. The second law works in concert with the first law to describe correctly the behavior of an ideal thermodynamic system. The second law is defined often in terms of heat engines and their performance of work:

> It is impossible to construct an engine that does work while exchanging heat with only one reservoir.

Thus, any engine that performs work must have at least two reservoirs. A refrigerator, which is simply an engine running in reverse, also requires two reservoirs, the low-temperature one from which heat is absorbed and the high-temperature one into which heat is ejeted. Schematic configurations of a heat engine and a refrigerator are shown in Fig. 1.5. The attached sub-

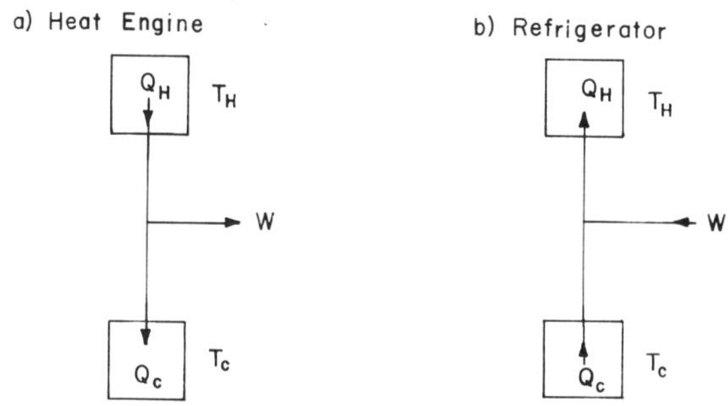

Fig. 1.5. Schematics of a heat engine and a refrigerator.

scripts refer to the hot (H) and cold (C) reservoirs. By definition, the thermodynamic efficiency η of an engine is given according to the relationship

$$\eta = \frac{W}{Q_H} \tag{1.3}$$

where Q_H is the heat absorbed from the high-temperature reservoir and W is the work done by the engine. It is apparent that an efficiency of unity can never be achieved because the second law requires two reservoirs and absolute zero is unattainable according to the third law of thermodynamics.

The entropy S is a state function that is particularly useful in cryogenic systems. A further aspect of entropy is that it has fundamental definitions in statistical mechanics, allowing a bridge between thermodynamic and statistical descriptions of thermodynamic systems. The thermodynamic defintion of entropy is couched in the form of a state change along a reversible path. If a thermodynamic system undergoes a differential change of state, the entropy is defined as

$$dS = \frac{dQ_R}{T} \tag{1.4}$$

where the subscript R refers to a reversible path. If the process takes the system from the initial to final state, a finite change in entropy occurs according to the integral of (1.4):

$$S_f - S_i = \int_i^f \frac{dQ_R}{T} \tag{1.5}$$

Note that (1.5) only defines the change in entropy ΔS and that it is necessary to set its value at the initial point. The third law of thermodynamics defines the entropy at absolute zero to be identically zero ($S_0 \equiv 0$).

A corollary to the definition of entropy is the Clausius theorem, which refers to the entropy associated with a closed cycle such as that shown in Fig. 1.4. If such a cycle follows reversible paths, the Clausius theorem states that the entropy change associated with this cycle is identically zero. It follows that a cycle that is not reversible must result in an increase in entropy.

A useful application of the entropy concept is obtained by considering a Carnot refrigerator. The Carnot cycle is shown schematically by the four-step process in Fig. 1.6. The first step in the cycle takes the thermodynamic system isothermally (at constant temperature) from (a) to (b), decreasing the entropy from S_2 to S_1. This process would be akin to isothermal compression of a gas at high temperature. The second step is an isentropic process (constant entropy) that reduces the temperature of the system from T_H to T_C, taking it from (b) to (c). The third step then heat exchanges with the low-temperature reservoir, a process (c) to (d) that takes place isothermally. Finally, the cycle is completed by an isentropic return to the original point (a). Since the Carnot cycle is reversible the work done is equal to the area enclosed by the cycle. The amount of heat absorbed from the low-temperature reservoir is $Q_C = T_C \Delta S$ and the heat ejected in the hot reservoir is $Q_H = T_H \Delta S$. For a Carnot cycle the ratio of Q_C to Q_H is simply the ratio of absolute temperatures,

$$\frac{Q_C}{Q_H} = \frac{T_C}{T_H} \tag{1.6}$$

which leads to a simple way of describing the efficiency of a Carnot engine.

Before leaving the topic of entropy it is worthwhile discussing its alternate definition, which is based on statistical mechanics. The statistical definition of entropy is associated with order in the system. The greater the order the lower the entropy. Presumably, if absolute zero were attainable, the system would be completely ordered. As part of the statistical picture, it is common to observe that any system will tend to evolve to a state of

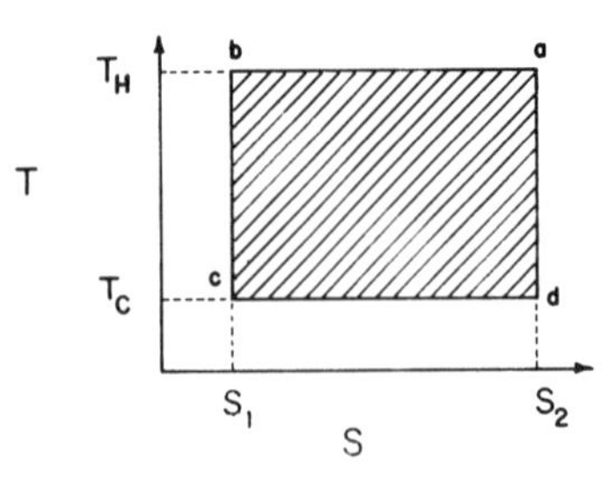

Fig. 1.6. Carnot cycle.

greater disorder and therefore higher entropy. For example, consider two vessels connected by a valve, one containing O_2 and the other N_2. If the valve is opened, irreversible mixing of the two gases occurs. The system will have greater disorder and higher entropy. The statistical definition of entropy evolves from introduction of a thermodynamic probability function Ω, which is a measure of the occupation of states in the thermodynamic system. Entropy is given mathematically by a function of this probability:

$$S = k_B \ln \Omega \tag{1.7}$$

where k_B is the Boltzmann constant equal to 1.38×10^{-23} J/K. Incidentally, other state functions of the system are also defined in terms of Ω. The statistical definition of entropy leads to a natural physical understanding of absolute zero temperature. At absolute zero, the system is in a ground state at which point the probability of a state being occupied is unity. Consequently, the probability function is equal to unity ($\Omega = 1$) and by the definition in (1.7) $S_0 \equiv 0$.

1.4.2. Third Law of Thermodynamics

The unattainability of absolute zero, the third law of thermodynamics, has fundamental significance in numerous aspects of helium cryogenics. Since the difficulty with which low temperatures are achieved increases as the temperature is decreased, cryogenics, by definition, is an attempt to fight this thermodynamic law. An additional aspect to the third law is that it describes the behavior of several thermodynamic functions as absolute zero is approached. For example, considering an isothermal change in pressure in a liquid–gas system, we can express the entropy change as

$$S(T, p) - S(T, 0) = -\int_0^p \left(\frac{\partial V}{\partial T}\right)_p dp \tag{1.8}$$

an expression obtained by integrating one of Maxwell's relations. Similarly, for a magnetic system, an isothermal change in magnetic field leads to the following expression:

$$S(T, H) - S(T, 0) = \mu_0 \int_0^H \left(\frac{\partial M}{\partial T}\right)_H dH \tag{1.9}$$

where M is the magnetization of the material. By combining (1.8) and (1.9) with the third law of thermodynamics a statement can be made about the behavior of V and M as absolute zero is approached. Specifically, since ΔS

approaches zero as $T \to 0$, both the volume and magnetization must approach constant values.

In addition to the above conventional statement of the third law and unattainability of absolute zero, the Nernst–Simon statement concerns itself specifically with entropy change:

> Entropy change associated with an isothermal reversible process in a condensed system approaches zero as $T \to 0$.

Since the Nernst–Simon statement deals only with the entropy change, the absolute entropy of a system at $T = 0$ must be a universal constant. It can be shown further that this constant can arbitrarily be set identical to zero.

The unattainability of absolute zero can be demonstrated by a number of examples. One useful example is based on the Nernst–Simon statement of the third law ($\Delta S \to 0$). Consider the schematic *T-S* diagram in Fig. 1.7 showing two isobars (constant pressure curves). For this type of system, it is fairly straightforward to achieve cooling and reduce the temperature by performing a constant entropy (isentropic) expansion from T_i to T_f. It would appear possible to extend this procedure toward absolute zero by a repeated application of isothermal compression from p_1 to p_2 and subsequently isentropic expansion. However, the third law states that $\Delta S_0 = 0$ so the two isobars must approach the same entropy at $T \to 0$. It should take an infinite number of steps to reach absolute zero.

As a conclusion, consider the application of the third law to a cryogenic system, a liquid–gas system that undergoes an isentropic expansion (Fig. 1.8). The mass of the fluid which transforms from liquid to gas is

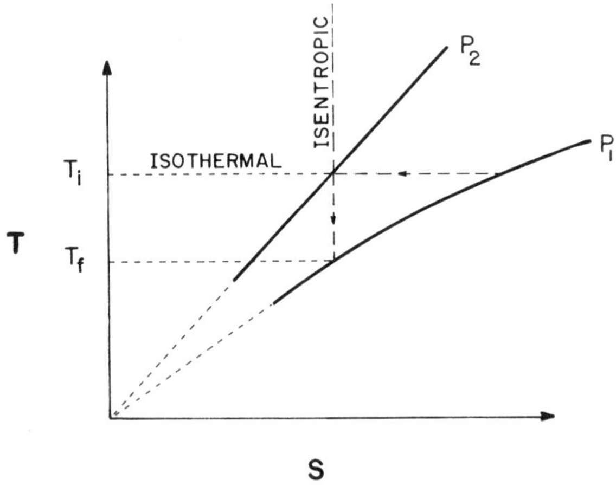

Fig. 1.7. Schematic *T-S* diagram showing isothermal and isentropic processes.

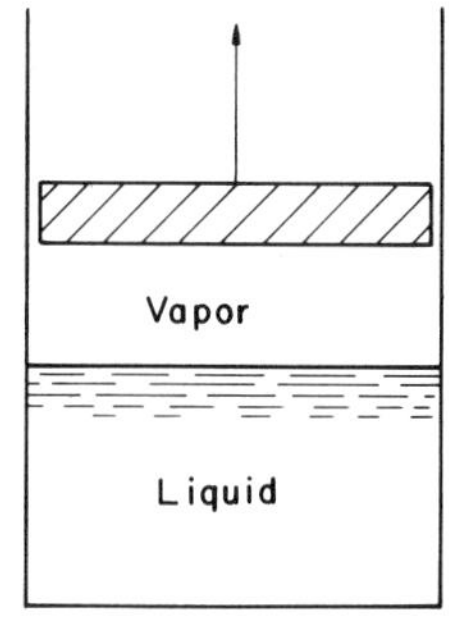

Fig. 1.8. Isentropic expansion of a liquid–gas mixture.

defined as dm. The heat removed from the liquid and taken away by the vapor is given the relationship

$$dQ = T\,\Delta S\,dm \tag{1.10}$$

The total heat capacity of the remaining liquid is simply the product of the liquid mass and specific heat. A finite change in mass will produce a finite temperature change, the details of which are governed by the equation

$$\int_{T_i}^{T_f} \frac{dT}{T}\frac{C}{\Delta S} = \ln \frac{m_f}{m_i} \tag{1.11}$$

If ΔS has a finite value at $T = 0$, (1.11) allows absolute zero to be reached with no contradiction in thermodynamics. However, if $\Delta S \to 0$ as is required by the third law of thermodynamics, it is necessary for $C \to 0$ as well. For the special case where $C/\Delta S$ has a limit equal to a constant as absolute zero is approached, integration of (1.11) yields an expression for the ratio of final to initial temperatures:

$$\frac{T_f}{T_i} = \left(\frac{m_f}{m_i}\right)^a \tag{1.12}$$

where a is a constant of order unity. Therefore, if absolute zero were achievable there would be no liquid remaining, a situation contrary to the actual goals of the cooling scheme.

Chapter 2

Low-Temperature Materials Properties

Before delving into the topic of helium cryogenics, it is useful to have a working knowledge of the properties of materials at low temperatures. This knowledge is useful because materials have behavior that must be taken into account when considering the problems of refrigeration, heat transfer, or storage of helium. Furthermore, many of the properties of helium are understood in terms of general models developed to treat the properties of different materials at low temperatures.

The study of material properties at low temperatures continues to be an active field of research. Current investigations include studies of the properties of materials at ultralow temperatures, $T \approx 1$ mK, new materials such as alloys and composites, and new types of investigation on traditional materials. Much of this work is fundamental in nature. However, since material properties play such an important role in the design and construction of low-temperature systems, it is essential to have a thorough knowledge of their behavior.

The present chapter is a survey of those properties which are of greatest importance to cryogenics. Included in the discussion are the behavior of the heat capacity, electrical and thermal conductivities, thermal contraction, and some special properties of materials—specifically magnetic spin systems and superconductors. Most of the descriptions are on the basis of thermodynamic or solid-state physics principles. More extensive discussions can be found in a number of references in the literature.[1–4]

2.1. HEAT CAPACITY

The heat capacity is a fundamental state property of matter. It represents the amount of energy needed to raise the temperature of a material one degree. The heat capacity per unit mass is called the specific heat. For cryogenics, heat capacity impacts numerous calculations including the dynamics of cooling, thermal energy stored, and transient heat transfer.

As with many physical properties, the heat capacity is defined in terms of thermodynamic state variables. It can be written as a derivative of either the entropy S or internal energy E. Because these state functions are described in a liquid–gas system in terms of pressure p, temperature, and specific volume v, one variable usually is held constant when defining the heat capacity. The constant volume heat capacity can therefore be written

$$C_v = T\left(\frac{\partial S}{\partial T}\right)_v = \left(\frac{\partial E}{\partial T}\right)_v \tag{2.1}$$

or the constant pressure heat capacity can be written

$$C_p = T\left(\frac{\partial S}{\partial T}\right)_p \tag{2.2}$$

It is also possible to define the heat capacity for other external variables held constant. For systems where magnetic properties are of significant importance, C_H or C_M may be used to designate the heat capacity at constant magnetic field or magnetization.

Numerous thermodynamic relationships exist to reduce physical properties to familiar form for analysis. One such relationship between C_p and C_v is given by

$$C_p - C_v = -T\left(\frac{\partial v}{\partial T}\right)_p^2 \left(\frac{\partial p}{\partial v}\right)_T = \frac{Tv\beta^2}{\kappa} \tag{2.3}$$

where $\beta = (1/v)(\partial v/\partial T)_p$ is the volume expansivity having units of inverse temperature and $\kappa = (1/v)(\partial v/\partial p)_T$ is the isothermal compressibility and has units of inverse pressure.

2.1.1. Heat Capacity of Solids

Experimental data exist for the heat capacity of many solids over the low-temperature range. For simple solid materials such as metals and crystalline insulators, there is a very good match between experimental measurements and theory. Experimental measurements near and above room temperature give close correspondence with the classical model of harmonic oscillators due to DuLong and Petit. The heat capacity at constant pressure is not much different from that at constant volume and both are closely equal to $3N_0k_B = 3R$, where N_0 is Avogadro's number = 6.023×10^{23} mol^{-1} and k_B is Boltzmann's constant = 1.38×10^{-23} J/mol · K. The gas constant $R = 8.31$ J/mol · K. The origin of this classical value is from assigning $\frac{1}{2}k_BT$ for each of the three kinetic and three potential degrees of freedom in the three-dimensional crystal.

At low temperatures, there is markedly different behavior according to the type of solid considered. For insulators, the dominant temperature dependence is proportional to T^3 while at very low temperatures, $T \lesssim 1$ K, metals have heat capacities that are linearly proportional to temperature. The distinction between C_p and C_v vanishes as $T \to 0$ for all solids, which can be used in conjunction with (2.3) to describe the limiting temperature dependence of other physical properties such as the volume expansivity.

The theory of solid-state heat capacity is reasonably well established. Two relatively simple theories are available to describe the behavior of metals and crystalline insulators over the entire temperature range of interest. The first such theory describes the specific heat of the quantized lattice vibrations or phonons that exist in a solid. For most solids, except metals at very low temperatures, the phonon contribution to the heat capacity dominates. To calculate the phonon contribution to the heat capacity, it is necessary to begin with an expression for the internal energy E_{ph} in terms of phonon frequencies ω:

$$E_{\text{ph}} = \frac{h}{2\pi} \int \omega \, d\omega \, D(\omega) \, n(\omega) \tag{2.4}$$

where $D(\omega)$ is the phonon density of states, $n(\omega)$ is the statistical distribution function, and h is Planck's constant having a value of 6.63×10^{-23} J · s. Since phonons obey Bose-Einstein statistics, the distribution function is given by

$$n(\omega) = \frac{1}{\hbar\omega/k_BT - 1} \tag{2.5}$$

where $\hbar = h/2\pi$. The model-dependent choice in the theory of phonon heat capacity is included in the selection of the proper density of states function $D(\omega)$. This problem can be quite complicated but fortunately many materials obey the simplifying assumptions inherent in the Debye theory. The Debye theory assumes the density of states is described by a continuum of levels up to the characteristic frequency ω_D, referred to as the Debye frequency. The density of states is then proportional to ω^2 and is shown schematically in Fig. 2.1.

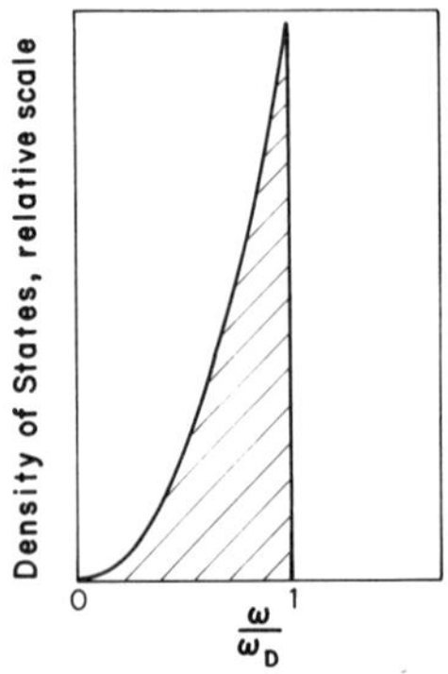

Fig. 2.1. Density of states $D(\omega)$ versus ω for the Debye model, assuming constant phonon velocity.

Inserting the Debye density of states into the equation for the internal energy (2.4), a nearly closed-form solution for this quantity is obtained:

$$E_{\text{ph}} = 9\,RT\left(\frac{T}{\Theta_D}\right)^3 \int_0^{x_D} dx\,\frac{x^3}{e^x - 1} \tag{2.6}$$

where $x = \hbar\omega/k_B T$, $x_D \equiv \Theta_D/T$, and Θ_D is the Debye temperature of the solid in question. The Debye temperature has a simple form from this model:

$$\Theta_D = \frac{hc}{2\pi k_B}\left(6\pi^2 \frac{N}{V}\right)^{1/3} \tag{2.7}$$

where c is the speed of sound in the material. In practice, the Debye temperature may be a function of more variables than just the number density as described in (2.7).

The constant volume heat capacity in the Debye model can be calculated directly from differentiation of (2.6) with respect to temperature

$$C_{\text{ph}} = 9\,R\left(\frac{T}{\Theta_D}\right)^3 \int_0^{x_D} dx\,\frac{x^4 e^x}{(e^x - 1)^2} \tag{2.8}$$

The definite integral within (2.8) can be analyzed numerically once Θ_D is known; however, the form of C_{ph} can be checked by evaluating (2.8) at either high or low temperatures. For high temperature, the exponentials within the integral may be expanded leading to the relationship

$$C_{\text{ph}} = 3\,R \quad \text{for } T \gg \Theta_D \tag{2.9}$$

which is the classical Dulong–Petit value. At low temperatures, the upper limit of the integral may be taken to be infinite, which leads to an exact solution,

$$C_{\text{ph}} = \frac{12\pi^2}{5} R \left(\frac{T}{\Theta_D}\right)^3 \quad \text{for } T \ll \Theta_D \tag{2.10}$$

The latter expression nicely reproduces the cubic temperature dependence of the heat capacity observed for many materials at low temperatures.

The simplicity of the Debye model and the dominance of the phonon contribution to the heat capacity make it a useful tool for calculations related to cryogenic problems. It is a fairly simple matter to tabulate the Debye values for C_{ph} and E_{ph} in terms of T/Θ_D as is shown graphically in Fig. 2.2. These results can be combined with known Debye temperatures which typically range from 100 to 1000 K and are listed in Table 2.1. It is therefore a straightforward calculation to estimate C_{ph} and E_{ph} of common solids. This analysis is usually accurate to within 10–20%.

For metals at low temperatures, $T \approx 1$ K, there is an additional contribution to the heat capacity due to the conduction electrons. Fortunately, as with the phonon contribution, the electron contribution can be approximately described by a simple theory known as the free-electron model. As before, the internal energy E_e is written in terms of the density of states $D(\varepsilon)$:

$$E_e = \int \varepsilon \, d\varepsilon \, D(\varepsilon) f(\varepsilon) \tag{2.11}$$

where ε is the electron energy used as a variable in this case instead of frequency in the Debye model. Electrons obey Fermi–Dirac statistics and have a different distribution function,

$$f(\varepsilon) = \frac{1}{e^{\beta(\varepsilon - \mu)} + 1} \tag{2.12}$$

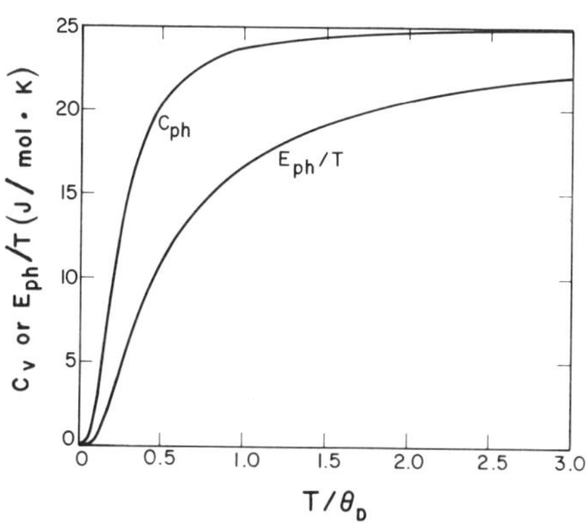

Fig. 2.2. Debye specific heat and internal energy functions.

Table 2.1. Debye Temperatures for Common Elements in Cryogenics[a]

Element	Θ_D(K)
Ag	225
Al	428
Au	165
Cd	209
Cr	630
Cu	343
Fe	470
Ga	320
Hf	252
Hg	71.9
In	108
Nb	275
Ni	450
Pb	105
Sn	200
Ti	420
V	380
Zn	327
Zr	291

[a] From Kittel.[1]

where μ is the chemical potential and $\beta = 1/k_B T$. The free-electron model further assumes that the Fermi surface is spherical, having an energy ε_f dependent on the total number of electrons per unit volume,

$$\varepsilon_F = \frac{\hbar^2}{2m}\left(3\pi^2 \frac{N}{V}\right)^{2/3} \tag{2.13}$$

where m is the mass of an electron equal to 9.11×10^{-31} kg. The density of states in the free-electron model can be written

$$D(\varepsilon) = \frac{V}{2\pi^2}\left(\frac{2m}{\hbar^2}\right)^{3/2} \varepsilon^{1/2} \tag{2.14}$$

Ordinarily, the electron temperature is far below the Fermi temperature $T_F = \varepsilon_{F/k_B}$ and only a small fraction of the electrons near the Fermi surface contribute to the thermal properties. Because $T \ll T_F$, the electrons in a metal generally are referred to as a degenerate Fermi gas. A degenerate Fermi gas has a simple limiting form for the heat capacity,

$$C_e = \gamma T \tag{2.15}$$

where $\gamma = \frac{1}{3}\pi^2 D(\varepsilon_F) k_B^2$ is sometimes called the Sommerfeld constant and $D(\varepsilon_F)$ is the electron density of states at the Fermi surface. The coefficient of the electronic heat capacity has been measured extensively and some typical values are listed in Table 2.2. For copper it is interesting to note that the electron and phonon contributions to the heat capacity are equal at about 3.8 K.

2.1.2. Heat Capacity of Gases

The thermal properties of gases, particularly those of interest to cryogenics, are well established, and theory exists to interpret their behavior. This subject will be considered in more detail in Chapter 5 for the case of helium gas. At this point it is sufficient to describe the heat capacity of most gases as their deviation from ideal behavior. For ideal monatomic gases, the internal energy is equal to $\frac{3}{2} RT$, for three translational degrees of freedom. This relationship works quite well for low-density monatomic gases at temperatures well above their boiling points. If the gas is more complex, for example, consisting of more than one atom per molecule, the

Table 2.2. Coefficient of the Electronic Specific Heat for Various Metallic Elements of Technical Interest[a]

Element	γ (mJ/mol · K^2)
Ag	0.646
Al	1.35
Au	0.729
Cd	0.688
Cr	1.40
Cu	0.695
Fe	4.98
Ga	0.596
Hf	2.16
Hg	1.79
In	1.69
Nb	7.79
Ni	7.02
Pb	2.98
Sn	1.78
Ti	3.35
V	9.26
Zn	0.64
Zr	2.80

[a] From Kittel.[1]

internal energy is larger. For a diatomic gas the internal energy has a high-temperature limit of $\frac{5}{2}RT$ because in addition to the translational motion two rotational degrees of freedom are excited.

The constant volume heat capacity C_v for most simple gases well above their condensation temperatures is normally between $\frac{3}{2}R$ and $\frac{5}{2}R$. If an equation of state exists between p, v, and T it is a relatively straightforward problem to analyze (2.3) to determine C_p, the heat capacity at constant pressure. C_p is always greater than C_v because it includes contributions due to expansion of the gas. For an ideal gas where the equation of state is simply $pv = RT$, $C_p - C_v = R$. Numerous analytic attempts at defining a good general equation of state for real gases have met with only modest success. For most gases it is easier to use tabulated values for such properties as the heat capacity. A good example exists for helium where the thermodynamic properties are tabulated for a wide range of temperature and pressure (see Appendix A.1).

2.1.3. Heat Capacity of Liquids

The liquid state is more complex than solids or gases and cannot be described using simple physical models. Some properties of liquids are more like those of solids because the densities of these two states are similar. However, liquids do not possess long-range order, so their excitation spectrum cannot be modeled easily. Other properties of liquids are described better in terms of gaslike behavior. These may include the transport properties such as viscosity and thermal conductivity. In general, the liquid state possesses more complex behavior than either solids or gases; thus, there are few simple models useful in interpreting its behavior.

The heat capacities of cryogenic liquids are established experimental quantities. Tabulated data exist for most ranges of interest. Of all liquids, the heat capacity of liquid helium is probably the most measured and understood. Yet it is also very complex. At low temperatures < 1 K, liquid helium exhibits a region that is dominated by phonon excitations and consequently has a heat capacity proportional to T^3. Above this temperature, the character of the heat capacity changes significantly with a higher-order excitation, termed "roton," taking over but eventually giving way to the λ transition at 2.172 K. At still higher temperatures, liquid helium has a heat capacity character similar to other simple liquids, being a gradually increasing function of temperature up to the boiling point.

2.2. CONDUCTIVITIES—ELECTRICAL AND THERMAL

The electrical and thermal conductivities are nonequilibrium transport properties that determine, among other things, the rate at which a disturbance approaches equilibrium. In the steady state, the electrical conductivity of a metal controls the resistance to flow of electricity while the thermal conductivity controls the heat transport. The thermal conductivities of insulators at low temperatures are generally much smaller than that of metals owing to the phonon transport being less effective than electron transport.

2.2.1. Electrical Resistivity of Metals

Near room temperature, the electrical resistivity of most pure metals decreases monotonically with temperature following an approximately linear relationship. This tendency is the result of electron–phonon scattering and is the dominant temperature-dependent contribution to the resistivity $\rho(T)$. At low temperatures, there is a trend to a constant value, which finally is approached when the metal is near liquid helium temperature. The constant value of low temperature is referred to as the residual resistivity ρ_0 and is strongly dependent on metal purity and amount of lattice imperfections. Generally, these two effects are additive, obeying what is known as Matthiessen's rule that the total resistivity is the sum of two contributions,

$$\rho = \rho_0 + \rho(T) \tag{2.16}$$

As an example of the behavior of electrical resistivity, consider Fig. 2.3,

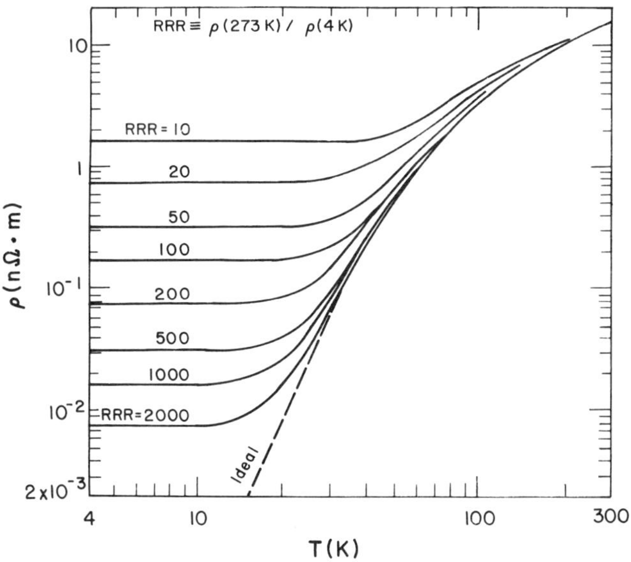

Fig. 2.3. Electrical resistivity versus temperature of differing purities of copper, $\rho(273\text{ K}) = 15.45\ \text{n}\Omega \cdot \text{m}$ (reprinted from Powell and Fickett[10]).

which is for various purities of copper. It should be noted that the temperature at which essentially constant resistivity is obtained decreases with increasing purity. In fact, the residual resistivity ratio [RRR = $\rho(273\text{ K})/\rho(4.2\text{ K})$] is used often as a measure of the purity of the metal. The other point to note in the figure is that the high-temperature ($T \approx 300\ K$) resistivity is essentially independent of purity, consistent with the dominance of electron–phonon scattering. Numerous references[4–8] are available which contain tabulated values for the electrical resistivities of elements and alloys.

The theoretical interpretation of electrical conductivity associates the loss mechanism with scattering processes between the electron and lattice. Considering a low-frequency transport of electrical current in a metal, we can relate the conductivity to the mean scattering time, $\tau \approx l/v_F$, where l is the mean free path between electron scattering events and v_F is the Fermi velocity. Elementary theory of electrical conductivity σ gives

$$\sigma = \frac{ne^2\tau}{m} \tag{2.17}$$

where n is the number of conduction electrons per unit volume and m is the electron mass.

There are two principal types of electron scattering which determine the magnitude of the electrical conductivity. For fairly high temperatures, $T \approx \Theta_D$, the dominant mechanism is due to electrons scattering off quantized lattice vibrations, that is, phonons. A simple way to see the temperature dependence of this effect is to identify the magnitude of the phonon scattering with the mean square displacement of the lattice, $\langle x^2 \rangle$. In a simple harmonic solid, this quantity is proportional to $k_B T$, the thermal energy of the lattice. Since the electrical resistivity should be proportional to the magnitude of phonon scattering, near room temperature the resistivity of metals should be proportional to T, a fact borne out by experiment.

For low temperatures, $T \ll \Theta_D$, the phonon scattering continues to decrease, while scattering from lattice imperfections begins to control the conductivity. In this region the resistivity approaches a temperature-independent value determined primarily by the impurities and imperfections in the lattice. For metallic elements, a few parts per million of impurity can have a profound effect on electron transport. At the lowest temperatures, with the purest samples, scattering off the surface of the metal can contribute substantially to the resistivity.

At intermediate temperatures, $T \approx \Theta_D/3$, the resistivity varies smoothly between the two regions. Many metals have a roughly T^5 dependence which can be attributed to the phonon scatterer population being proportional to T^3 and the probability of scattering through large angle having a T^2 dependence. The resistivity is therefore proportional to the product of these two factors.

2.2.2. Thermal Conductivity of Metals

Thermal conductivity is a property that determines the temperature gradient across a substance in the presence of a flow of heat. In all materials there are several contributions to the thermal conductivity k. For metals, the principal conduction mechanisms are electronic and lattice, the latter one being small for pure metals. The electronic thermal conductivity can be understood by a similar model as used for electrical conductivity.

The behavior of k can be understood in terms of a kinetic theory model for gases of electrons and phonons.[9] Such simple models work very effectively to explain the limiting behavior of the thermal conductivity. In particular, the thermal conductivity may be written

$$k = \tfrac{1}{3} C v l \tag{2.18}$$

where C is the heat capacity per unit volume, v is the velocity, and l is the mean free path. Using the free-electron model, the electronic contribution to the thermal conductivity can be calculated by inserting the electronic specific heat, (2.15), and the Fermi velocity, $v_F = (2\varepsilon_F/m)^{1/2}$, into (2.18). Thus,

$$k_e = \frac{\pi^2 n k_B^2 T \tau}{3m} \tag{2.19}$$

where τ is the mean scattering time identical to that considered for electrical conductivity. At high temperatures, $T > \Theta_D$, $\tau \approx 1/T$ and the thermal conductivity should approach a constant. At low temperatures, τ is approximately constant, implying the thermal conductivity should be linearly proportional to T. As an example, Fig. 2.4 shows plots of the thermal conductivities of copper analogous to Fig. 2.3 for electrical resistivity. The limiting behavior of this system is consistent with the simplified theoretical picture.

Since the electronic thermal and electrical conductivities have similar scattering processes, a correspondence exists between these two properties. The Wiedemann–Franz (W-F) law generally assumes that for a metal the ratio of the thermal conductivity and electrical conductivity is a function of temperature only. Furthermore, for the free-electron model, this ratio is a simple value,

$$\frac{k_e}{\sigma} = \frac{\pi^2}{3}\left(\frac{k_B}{e}\right)^2 T = L_0 T \tag{2.20}$$

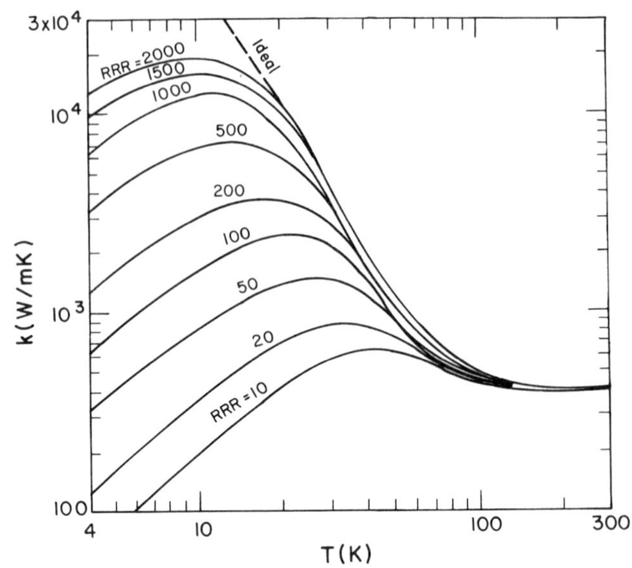

Fig. 2.4. Thermal conductivity versus temperature of differing purities of copper, $k(273\ \mathrm{K}) = 401\ \mathrm{W/m \cdot K}$ (reprinted from Powell and Fickett[10]).

where the quantity $L_0 = (\pi^2/3)(k_B/e)^2 = 2.45 \times 10^{-8}\ \mathrm{W \cdot \Omega/K^2}$ is referred to as the Lorenz number. In the model L_0 is totally independent of material properties and temperature.

Experimental evidence indicates that the *W-F* law works effectively only at temperatures near room temperature and at very low temperatures $(T \ll \Theta_D)$.[11] This fact has to do with the asymmetry imposed on the Fermi surface subjected to a thermal gradient rather than a transport current. At intermediate temperatures, the experimentally defined Lorenz ratio $(L_0 = k/\sigma T)$ is always less than the value predicted by the *W-F* law combined with the free-electron model. The amount of deviation is strongly dependent on the purity of the sample, with the less pure having a smaller deviation. The behavior of the Lorenz ratios with sample purity are plotted in Fig. 2.5. Considerable effort has been applied to understanding these effects, the details of which are beyond the scope of the present discussion.

2.2.3. Lattice Thermal Conductivity

The thermal conductivity of insulators and the lattice contribution to the thermal conductivity of metals is understood similarly to the electronic contribution although the thermal carrier is now a phonon gas. It is still possible to apply (2.18) although the heat capacity must be that due to the

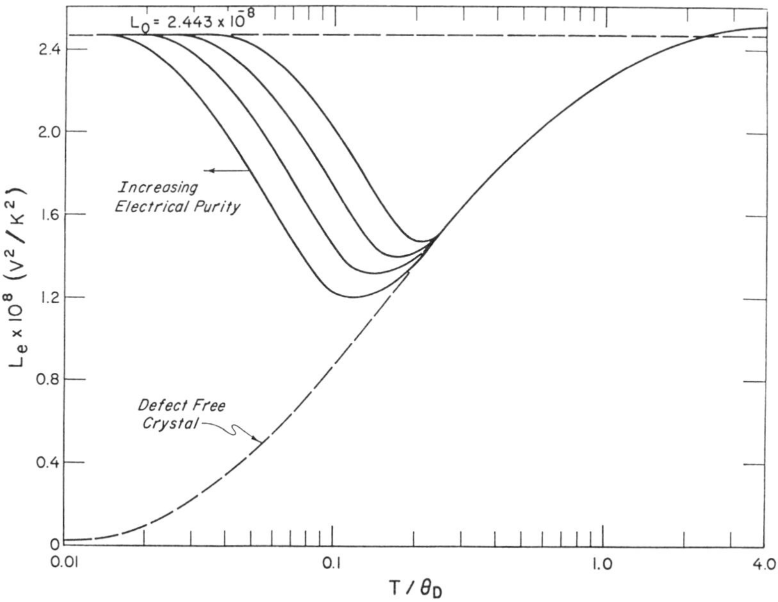

Fig. 2.5. Electronic Lorentz ratio for pure metals and defect-free metals (reprinted from Sparks and Hust[11]).

lattice, C_{ph}, which at low temperatures is proportional to T^3. The appropriate velocity is the speed of sound and l is the phonon mean free path. Most insulators have thermal conductivities that are several orders of magnitude smaller than that of common metals. At high temperatures their behavior is complicated by the details of the phonon density of states, but generally the thermal conductivity decreases monotonically with temperature. At low temperatures, where scattering times become approximately independent of temperature, the thermal conductivity decreases more rapidly, approaching zero as T^n where $n \approx 3$.

The thermal conductivity of some technical materials are listed in Table 2.3. Because the list includes a wide variety of alloys and amorphous insulators, a considerable range in values is displayed. The table only contains a limited number of technically interesting materials, indicating that a continuing area of research is the determination of thermal conductivities of new materials. This need is particularly evident with the growing use of composite materials in low-temperature applications.

Table 2.3. Thermal Conductivity (in W/m · K) for Some Technical Alloys, Glasses, and Plastics[a]

Material	0.1 K	0.4	1	4	10	40	80	150	300 K
Al 5083	—	—	0.7	3	8	34	56	80	120
Al 6063	—	—	8	35	87	270	230	200	200
Brass (70% Cu, 30% Zn)	0.06	0.2	0.7	3	10	37	65	85	120
Cu + 30 Ni	0.006	0.03	0.1	0.5	2	~12	~20	~25	30
Cu + 2 Be	—	—	—	1.9	5.1	20.5	35	—	—
Constantan	0.006	0.02	0.1	0.8	3	14	18	20	23
Iconel	—	—	0.05	0.4	1.5	7	10	12	14
Manganin	0.006	0.02	0.06	0.5	2	7	13	16	22
Silicon bronze (96% Cu, 3% Si, 1% Mn)	—	—	—	—	1.5	7	14	—	25
Soft solder (60% Sn, 40% Pb)	—	—	—	16	42	52	52	—	~50
Stainless steels (18/8)	0.008	0.03	0.08	0.3	0.7	5	8	11	15
Wood's metal	—	—	—	4.0	12	20	23	—	—
Nylon	—	0.0006	0.003	0.01	0.04	—	—	—	—
Polystyrene	—	—	0.01	0.03	—	—	—	—	—
Pyrex	0.0003	0.003	0.01	—	—	—	0.5	0.8	1.1
Soft glass	—	—	0.015	0.1	0.2	0.3	0.5	—	—
Teflon	0.00002	0.0004	0.004	0.05	0.1	0.2	0.2	—	—
Vitreous silica	—	—	—	0.1	0.1	0.3	0.5	0.8	1.4
Epoxy	0.00004	0.0007	0.007	0.06	0.06	—	—	—	—

[a] After White.[2]

2.3. THERMAL CONTRACTION

All materials experience a change in physical dimension when cooled to low temperatures. This effect, normally referred to as thermal contraction in the field of cryogenics, is of the order of a few tenths of a percent change in volume in most materials. It can have profound effect on the design of engineering devices operating in a low-temperature environment, largely because the thermal contractions of different materials vary considerably. Since most devices constructed to operate in cryogenic systems are fabricated out of a number of different materials, one of the major concerns is the effect of the differential thermal contraction and associated thermal stress due to material differences. Differential contraction is especially important to the design of vacuum seals, structural supports, and electrical insulation.

It is of considerable importance to understand the thermal contraction behavior of technical materials. There are a number of good reviews[10–14] in the literature on this particular subject. The thermal expansion or contraction is normally given a thermodynamic definition. In this manner, it can be combined with other thermodynamic state properties to make some prediction of the details of the properties of materials at low temperatures. The thermal expansion of liquid helium has some very interesting properties which are considered in detail in Chapter 5. For liquids and gases, the most meaningful factor to consider is the volume expansivity given by

$$\beta = \frac{1}{V}\left(\frac{\partial V}{\partial T}\right)_p \tag{2.21}$$

where β is in general a function of temperature. For solids, where the changes in individual dimensions may be different due to anisotropic effects, the linear thermal expansion coefficient,

$$\alpha = \frac{1}{L}\left(\frac{\partial L}{\partial T}\right)_p \tag{2.22}$$

is a more appropriate factor to consider. For isotropic materials, $\alpha = \frac{1}{3}\beta$, to first order. For many common solids near room temperature, the linear expansion coefficient is approximately constant.

It can be shown, by modeling a solid with oscillators, that thermal expansion is caused by anharmonic terms in the restoring potential. The Debye model assumes a solid is comprised of harmonic oscillators and therefore in its simplest form does not predict thermal expansion. An expression that effectively includes these anharmonic terms is the variation of the Debye temperature Θ_D with volume,

$$\gamma_G \equiv -\frac{d(\ln \Theta_D)}{d(\ln V)} \tag{2.23}$$

referred to as the Grüneisen relationship. A list of values for the Grüneisen constant of some elements is given in Table 2.4. The Grüneisen coefficient, which is nearly constant over a temperature range down to $T \approx \Theta_D/5$, can be used along with other thermodynamic properties to describe the thermal expansion coefficient,

$$\alpha = \frac{\gamma_G C_v \kappa}{3v} \tag{2.24}$$

where κ is the isothermal compressibility. For cubic metals at low temperatures, $T < \Theta_D/25$, the dominant temperature dependence of α is in the specific heat, $C_v \approx \gamma T + \beta T^3$.

At low temperatures, the expansion coefficient is far from linear and actually approaches absolute zero with zero slope. This fact can be seen based on thermodynamic relationships. The difference between the constant volume and constant pressure heat capacity, (2.3), is proportional to the square of the volume expansivity β. The quantity $C_p - C_v$ goes to zero as a strong function of temperature, indicating that β also must go to zero. This effect makes sense physically because the harmonic terms would be expected to dominate the interatomic potential at low temperatures.

Because of the nonlinear nature of the expansion of materials, it is often more useful to have the integrated thermal contraction for the purpose of design. Table 2.5 lists the integrated linear contraction of a number of common materials used in cryogenic applications. Note that metals typically have contractions in the range of 0.5% or less. Polymers such as

Table 2.4. Values for the Grüneisen Coefficient γ_G for Common Elements[a]

Element	γ_G
Ag	2.40
Al	2.17
Cu	1.96
Fe	1.6
Ni	1.88
Pt	2.54

[a] From Kittel.[1]

Table 2.5. Linear Thermal Contractions Relative to 293 K[a]

Substance	T(K): 0	20	40	60	80	100	150	200	250
Aluminum	41.4	41.4	41.2	40.5	39.0	36.9	29.4	20.1	9.6
Copper	32.6	32.6	32.3	31.6	30.2	28.3	22.1	14.9	7.1
Germanium	9.3	9.3	9.3	9.4	9.3	8.9	7.3	5.0	2.4
Iron	20.4	20.4	20.3	19.9	19.5	18.4	14.9	10.2	4.9
Lead	70.8	70.0	66.7	62.4	57.7	52.8	39.9	26.3	12.4
Nickel	23.1	23.0	22.9	22.6	21.8	20.8	16.5	11.4	5.4
Silicon	2.16	2.16	2.17	2.23	2.32	2.40	2.38	1.90	1.01
Silver	41.0	41.0	40.3	38.7	36.5	33.7	25.9	17.2	8.2
Titanium	15.1	15.1	15.0	14.8	14.2	13.4	10.7	7.3	3.5
Tungsten	8.6	8.6	8.5	8.4	8.1	7.6	5.9	4.0	1.9
Brass (65 % Cu, 35 % Zn)	38.4	38.3	38.0	36.8	35.0	32.6	25.3	16.9	8.0
Cu + 2 Be	32.4	32.4	31.9	31.6	30.0	28.3	22.0	16.0	7.0
Constantan	—	—	26.4	25.8	24.7	23.2	18.3	12.4	5.85
Invar[b]	4.5	4.6	4.8	4.9	4.8	4.5	3.0	2.0	1.0
304, 316 Stainless steel	—	29.7	29.6	29.0	27.8	26.0	20.3	13.8	6.6
Pyrex	5.6	5.6	5.7	5.6	5.4	5.0	3.95	2.7	0.8
Silica (1000° C)[c]	−0.1	−0.05	0.05	0.2	0.3	0.4	0.5	0.4	0.2
Silica (1400° C)[c]	−0.7	−0.65	−0.5	−0.3	−0.2	−0.05	0.2	0.2	0.1
Araldite	106	105	102	98	94	88	71	50	25
Nylon	139	138	135	131	125	117	95	67	34
Polystyrene	155	152	147	139	131	121	93	63	30
Teflon	214	211	206	200	193	185	160	124	75

[a] Units are $10^4 \times (L_{293} - L_T)/L_{293}$. Sources of data include *Thermophysical Properties of Matter* (1977), Corruccini and Gniewek (1961), and *American Institute of Physics Handbook* (1972). Compiled by White.[2]

[b] The expansion of Invar NiFe alloys containing ~36 % Ni is very sensitive to composition and heat treatment.

[c] These silicas were aged at 1000° C and 1400° C.

nylon or teflon contract three times as much as metals and can be as high as 2%. Some amorphous materials, paticularly glasses, have nearly zero or sometimes negative thermal contraction. Composite materials such as fiberglass epoxy often can have their thermal contraction predicted based on a linear combination of the two individual materials, taking into account the elastic modulus of each.[15]

2.4. MECHANICAL PROPERTIES

The mechanical properties of materials are of considerable importance in the design of cryogenic systems. The two properties that are most often of concern are the level of stress σ within the material and the modulus of elasticity E. These material properties enter into calculations of failure modes in mechanical structures.[16]

The uniaxial stress limits are also of interest in structural materials. The yield stress σ_y is that associated with a permanent deformation of about 0.2%. Sometimes yield refers to the stress level that produces a distinct change in the slope of the stress–strain curve. For structural components of systems, it is advisable to stay well below the yield stress of the material in use. Typically, a structural component may be designed so that the maximum stress does not exceed $\frac{2}{3}\sigma_y$. However, this value must be derated considerably if a cyclic stress is applied. The ultimate stress σ_u represents that level of stress necessary to fail a particular material under tensile load. In ductile materials, the ultimate stress is considerably greater than σ_y and can be associated with substantial deformation. On the other hand, brittle materials have $\sigma_y \simeq \sigma_u$. Composite structural materials such as fiberglass epoxy have even more complex behavior.

Table 2.6. Yield Stress σ_y of Several Materials as a Function of Temperature (MN/m^2)[a]

Material	σ_y(0 K)	σ_y(80 K)	σ_y(300 K)
304L-SS	547	460	406
6061-T6 Al	345	332	282
OFHC-Cu (Annealed)	90	88	75
Cu + 2 Be	752	690	552
Brass (70% Cu, 30% Zn)	506	473	420
Inconel X-750	940	905	815
G10-CR	758	703	414
Teflon	130	65	20

[a] See Refs. 17 and 18.

Table 2.7. Ultimate Stress σ_u of Several Materials as a Function of Temperature (MN/m^2)[a]

Material	σ_u(0 K)	σ_u(80 K)	σ_u(300 K)
304L-SS	1660	1509	659
6061-T6 A1	497	422	312
OFHC Cu (annealed)	418	360	222
Cu + 2 Be	945	807	620
Brass (70% Cu, 30% Zn)	914	804	656
Inconel X-750	1620	1496	1222
G10-CR	758	703	414
Teflon	194	86	21

[a] See Refs. 17 and 18.

For most common structural materials, the yield and ultimate stresses increase with decreasing temperature. The magnitude of this increase varies from around 10% in some metallic alloys to several 100% in polymeric materials. The increase in strength is seen to result from the reduced thermal excitations within the lattice, which inhibits the spread of dislocations. Listed in Tables 2.6 and 2.7 are the yield and ultimate stress limits for several materials of interest in cryogenic applications. Values listed in the tables are typical and considerable variation can occur depending on the treatment and form of the particular materials. More detailed tabulations can be found from several sources in the literature.[4,17]

The modulus of elasticity, or Young's modulus E, represents the change in stress level needed to cause a unit change in strain while the material is in the elastic region. Thus, Young's modulus is simply the slope of the stress–strain curve for small values of strain. As with the yield and ultimate stresses, Young's modulus also increases with decreasing tem-

Table 2.8. Young's Modulus E of Several Materials as a Function of Temperature (GN/m^2)[a]

Material	E(0 K)	E(80 K)	E(300 K)
304L-SS	201	205	190
6061-T6 A1	87	79	68
OFHC-Cu (annealed)	139	139	128
Cu + 2 Be	134	130	121
Brass (70% Cu, 30% Zn)	—	145	110
Inconel (X-750)	—	223	210
G10-CR	36	34	28

[a] See Refs. 4 and 17.

perature. A list of typical values for technical materials is shown in Table 2.8. Unlike the limiting stress values, Young's modulus is not as strongly affected by material treatment and form.

2.5. SPIN SYSTEMS

The low-temperature behavior of magnetic spin systems occupies a special place in the field of cryogenics. Use of spin entropy for cooling was first suggested in 1925. It has since been developed into a well-established technique. Cooling methods associated with the technique are reviewed in Section 9.4. However, in this chapter concerning low-temperature material properties, it is appropriate to introduce the solid-state aspects of paramagnetic spin systems. The present treatment is intended to introduce the microscopic aspects of paramagnetic spin systems.[18] Dealt with in detail, the subject can be quite complex and is therefore beyond the scope of this book.

The most important magnetic spin system to cryogenics is that of the paramagnetic salt. These materials have a number of characteristics in common. They are comprised of multicomponent ionic molecules, one ion of which is magnetic. The magnetic ion is sufficiently dilute within the material that at high temperature it approximates a free noninteracting spin system. The strength of interaction determines the ordering temperature, below which the materials usually become diamagnetic. There are numerous examples of paramagnetic salts, many of which have been used as coolants in demagnetization experiments. A partial list is given in Table 9.7. The ordering temperature of most paramagnetic salts is in the range of 1 K and below.

The other important spin system of interest in cryogenics is that which occurs from unpaired nuclear magnetic moments. Since the nuclear moment is smaller than the paramagnetic moment by the ratio of the electron to proton mass, $m_e/m_p = 1836$, nuclear magnetic properties are usually important only at very low temperatures. However, the strength of the interaction goes as the square of the magnetic moment, allowing nuclear demagnetization to be used even into the microkelvin range. Since the microkelvin temperature range is below that of interest in this book, nuclear spin systems are not considered here. Their behavior is analogous to paramagnetic salts although their application is at much lower temperature. In Chapter 9, nuclear demagnetization is discussed further in its connection with the cooling of ^{3}He.

The theory of paramagnetic spin systems is developed around the statistical behavior of free magnetic ion systems. Any free magnetic molecule has two types of magnetic movement associated with its electrical structure. The first, due to orbital motion of the electron, is given the quantum number L. The other, resulting from uncompensated electron spins, has a quantum number S. The total angular momentum number J is the resultant of these two individual magnetic moments. In general, atoms have nonzero values of L and S. However, certain ions, particularly magnetic Fe^{3+} and Cr^{3+} as well as the rare earths such as Gd^{3+}, have effectively an inert-gas orbital structure, so that $L=0$, but uncompensated electron spins, such that $J=S$.

Because there are different quantum numbers m_i associated with each spin state of the atom, there are $2J+1$ individual energy levels. Without the application of an external magnetic field, these levels are degenerate. However, in an external field the degeneracy is lifted, shifting the energy levels by an amount

$$\varepsilon_i = -g\mu_B B m_i \tag{2.25}$$

where μ_B is the Bohr magneton given by a ratio of fundamental constants, $\mu_B = eh/4\pi m = 0.927 \times 10^{-23}$ J/T. The magnetic field B is that which is applied locally to the atom and can be substantially different from the applied magnetic field due to external sources. The magnetic quantum number m_i takes on values in integral steps from J to $-J$. The quantity g in (2.25) is referred to as the Landé g factor, given by the relationship

$$g = 1 + \frac{J(J+1)+S(S+1)-L(L+1)}{2J(J+1)} \tag{2.26}$$

In the materials that are good paramagnetic coolants, $L=0$ so $g=2$.

For a material like gadolinium sulfate with $g=2$, individual energy levels within the magnetic system have a level separation that can be written in terms of absolute temperature,

$$\frac{\Delta\varepsilon}{k_B} = 1.4\, BT \quad \text{(K)} \tag{2.27}$$

where B is in tesla. This relationship establishes the boundary between the two regimes of ordering in a magnetic ion subsystem. The low-field, high-temperature regime occurs when the product $BT \ll 1$. In this regime, the magnetic ions essentially are disordered with the level spacing being small compared to $k_B T$. On the other end of the spectrum, the high-field, low-temperature regime, $BT \gg 1$, corresponds to the majority of the spins occupying the lowest energy level. Population numbers of higher levels are given by the Boltzmann distribution function.

It is appropriate to describe the behavior of a magnetic ion system using Boltzmann statistics. The starting point for statistical analysis is the definition of a partition function. For a magnetic ion system, the assumption is made that the total partition function is a product of the internal nonmagnetic part and the magnetic contribution,

$$Z_{\text{total}} = Z_{\text{int}} Z_B \tag{2.28}$$

In a magnetic ion system Z_B essentially is due to the Zeeman effect. The partition function may be written in the form

$$Z_B = \sum_{i=-J}^{J} \exp(g\mu_B B m_i / k_B T) \tag{2.29}$$

where the summation occurs over all magnetic quantum numbers m_i. The summation (2.29) has a closed form,

$$Z_B = \frac{\sinh(J+\frac{1}{2})\, a}{\sinh(a/2)} \tag{2.30}$$

where the parameter $a = g\mu_B B/k_B T$ is a measure of the ordering.

To obtain the magnetization it is necessary to differentiate the partition function with respect to the magnetic field,

$$M = Nk_B T \left(\frac{\partial \ln Z_B}{\partial B} \right)_T \tag{2.31}$$

where N is the total number of magnetic ions. Substituting (2.30) into (2.31), the magnetization takes the form

$$M = Ng\mu_B J \left\{ \frac{1}{J} \left[(J+\tfrac{1}{2}) \coth(J+\tfrac{1}{2})\, a - \tfrac{1}{2} \coth(a/2) \right] \right\} \tag{2.32}$$

The quantity in brackets is termed the Brillouin function $B_J(a)$. It is tabulated in a number of references on magnetic systems. Graphically, the Brillouin function is displayed in Fig. 2.6.

The temperature dependence of the magnetization in a magnetic ion subsystem is obtained by considering the limiting behavior of the Brillouin function. The high-field, low-temperature limit corresponds to the case where $a \gg 1$. For this case, the Brillouin function approaches unity and the magnetization approaches a constant value

$$M = Ng\mu_B J \quad \text{for } a \gg 1 \tag{2.33}$$

which represents a fully ordered magnetic system. In the low-field, high-temperature limit, it is necessary to expand the Brillouin function in the

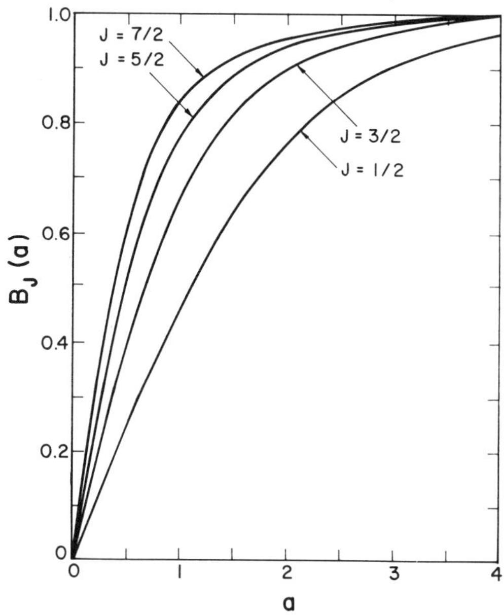

Fig. 2.6. The Brillouin function for different total spin quantum numbers J.

limit of small a. Thus, the magnetization is inversely proportional to the absolute temperature,

$$M = \frac{\gamma_C B}{T} \tag{2.34}$$

where γ_C, known as the Curie constant, is given by

$$\gamma_C = \frac{N g^2 \mu_B^2 J(J+1)}{3 k_B} \tag{2.35}$$

The magnetic susceptibility is then a derivative of the magnetization,

$$\chi = \left(\frac{\partial M}{\partial B}\right)_T = \frac{\gamma_C}{T} \tag{2.36}$$

This latter expression is the well-known Curie equation for noninteracting magnetic systems.

The above analysis has been successfully compared with experiment. Plotted in Fig. 2.7 are variations of the magnetization of several magnetic ion systems compared to the appropriate Brillouin function. The agreement is quite satisfactory. It should be possible to calculate with reasonable con-

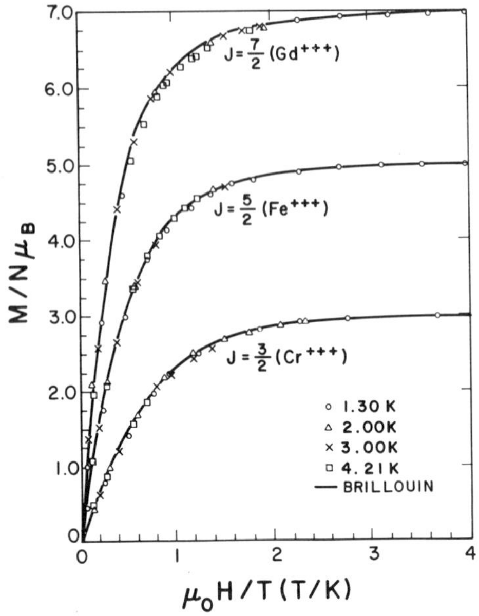

Fig. 2.7. Plot of magnetization per magnetic dipole, expressed in Bohr magnetons, against H/T for chromium potassium alum ($J=\frac{3}{2}$), iron ammonium alum ($J=\frac{5}{2}$), and gadolinium sulfate ($J=\frac{7}{2}$). The points are experimental results and the solid curves are graphs of the Brillouin equations (from Zemansky[18]).

fidence the magnetization of a particular paramagnetic ion system using the above simple model.

The low-temperature thermal properties of a magnetic ion system are considerably different from other more ordinary materials. These systems do possess phonon excitations, but a more dominant mechanism contributing to their thermal behavior at low temperature is due to crystal field splitting of the ionic energy levels within the lattice. This behavior, which is termed the Stark effect, is caused by the ionic electric fields establishing different energy levels analogous to those from the magnetic Zeeman effect. The level splitting is small, typically on the order of 10–100 mK, and consequently only becomes important at very low temperatures.

To derive an expression that adequately describes the Stark effect, it is necessary to make some assumptions concerning the structure of the energy levels. Often, crystal field splitting has associated with it two levels separated by an energy δ, with each level having a degeneracy g_n that is not lifted by the crystal field. With this model, it is possible to write the partition function associated with the internal system,

$$Z_{\text{int}} = g_0 + g_1 e^{-\delta/k_B T} \tag{2.37}$$

where g_0 and g_1 are the degeneracies of the ground and first excited states.

The internal energy of the magnetic ion system is a function of the total partition function,

$$E = Nk_B T^2 \left(\frac{\partial \ln Z}{\partial T}\right)_H + \mu_0 HM \tag{2.38}$$

By taking the derivative of the total partition function as defined by (2.28), the term associated with the Zeeman effect exactly cancels the second term in (2.38). The internal energy simplifies to

$$E = Nk_B T^2 \left(\frac{\partial \ln Z_{\text{int}}}{\partial T}\right)_H \tag{2.39}$$

which when combined with (2.37) leads to the expression

$$E = \frac{N\delta}{1 + (g_0/g_1)\, e^{\delta/k_B T}} \tag{2.40}$$

The heat capacity at constant magnetization is just the temperature derivative of the internal energy, $C_M = dE/dT$. It follows from (2.40) that the constant magnetization heat capacity is

$$\frac{C_M}{Nk_B} = \frac{\delta^2}{k_B^2 T^2} \frac{(g_0/g_1)\, e^{\delta/k_B T}}{[1 + (g_0/g_1)\, e^{\delta/k_B T}]^2} \tag{2.41}$$

This expression, known as the Schottky heat capacity, is plotted in Fig. 2.8 for different ratios of g_0/g_1. Note that the maximum in the Schottky heat capacity is approximately $\frac{1}{2}Nk_B$, which makes it very large compared to the other solid-state contributions at low temperatures.

It is important to bear in mind that the above analysis applies to a system of N magnetic ions. Typically, in paramagnetic salts the magnetic ion number is much smaller than the total number of particles in the

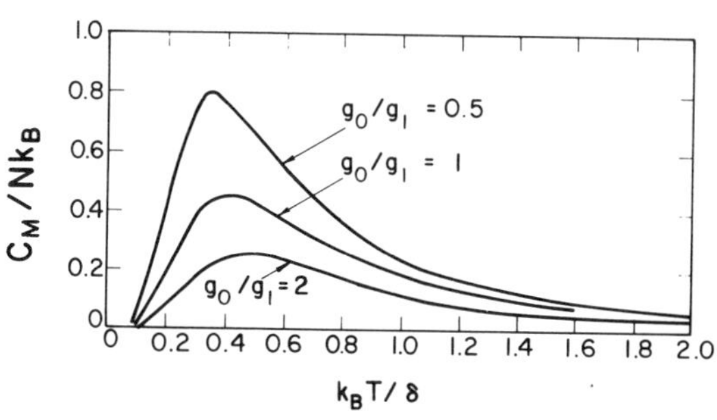

Fig. 2.8. Schottky heat capacity of a magnetic ion subsystem whose lowest energy level, in the absence of any external magnetic fields, is split into two states with degeneracies g_0 and g_1.

crystal. Therefore, if the heat capacity per unit mass or unit volume is desired, it is necessary to use the total molecular weight of the ionic salt.

Demagnetization of magnetic ion systems is a useful method of achieving low temperatures. The practical aspects of this method are discussed in Section 9.4 on magnetic refrigeration. Working materials used in magnetic refrigerators are generally of the type described above.

2.6. SUPERCONDUCTIVITY

Many metallic elements and a larger number of alloys transform into a state with essentially zero electrical resistance when cooled to temperatures close to that of the boiling point of liquid helium. This effect was first observed by Kamerlingh Onnes in 1911 as part of an investigation of the electrical resistance of pure metals at low temperatures. Onnes observed a sharp transition from the normal resistance state to one which had immeasurably small resistance. This new state, termed "superconductive," has spawned a great deal of fundamental research of both a theoretical and experimental nature. At the present time, a number of review monographs exist which survey the current understanding of the field of superconductivity.[19–21]

More recently, with the discovery of high-field superconductive materials, technical applications have come to fruition. In particular, large-scale superconductive magnets and superconductive electronics are now fairly well-established applied subfields of superconductivity. There are several reference books that cover this material in considerable depth.[22–25] The present discussion represents only a brief review of the properties of superconductors. However, it is important to note that much of the interest in helium cryogenics is brought about by these applications.

There are two distinct types of superconductors. Type I superconductors, which comprise most of the pure-element superconductors, have a sharp transition to the zero resistance state and a total screening of magnetic flux within the bulk—the Meissner effect. Unfortunately, with the exception of some special electronic devices, Type I superconductors have little practical application because of their rather low critical field H_C. Type II superconductors, sometimes referred to as "hard" superconductors, are much more practically useful because of their higher critical field and transition temperature. The behavior of a Type II superconductor is more complex owing to the existence of a mixed state which allows magnetic flux to penetrate the bulk material.

The regime of existence for a Type I superconductor is determined by two parameters, the critical temperature T_c and the critical field $H_c(T)$. For temperatures and magnetic fields below these critical values, the material exists in the superconducting state. There exists an empirical relationship between the critical temperature and field,

$$H_c(T) = H_0\left[1 - \left(\frac{T}{T_c}\right)^2\right] \tag{2.42}$$

where H_0 is the critical field at $T = 0$ K. Listed in Table 2.9 are the parameters for known Type I superconductors. Note the range of T_c varying from 7.2 K for Pb to 325 μK for Rh. Similar wide variations in the critical field are evident.

The thermodynamic properties of superconductors are described best in terms of their variation from that of the normal state. The difference in the Gibbs free energy between the superconducting and normal states is

Table 2.9. Critical Temperature and Critical Field of Type I Superconductors

Material	T_c (K)	$\mu_0 H_0$ (mT)
Aluminum	1.2	9.9
Cadmium	0.52	3.0
Gallium	1.1	5.1
Indium	3.4	27.6
Iridium	0.11	1.6
Lanthanum α	4.8	
β	4.9	
Lead	7.2	80.3
Lutecium	0.1	35.0
Mercury α	4.2	41.3
β	4.0	34.0
Molybdenum	0.9	
Osmium	0.7	~6.3
Rhenium	1.7	20.1
Rhodium	0.0003	4.9
Ruthenium	0.5	6.6
Tantalum	4.5	83.0
Thalium	2.4	17.1
Thorium	1.4	16.2
Tin	3.7	30.6
Titanium	0.4	
Tungsten	0.016	0.12
Uranium α	0.6	
β	1.8	
Zinc	0.9	5.3
Zirconium	0.8	4.7

written in terms of a condensation energy, which is a function of the critical field,

$$g_n - g_s = \tfrac{1}{2}\mu_0 H_c^2 \tag{2.43}$$

Combining (2.43) with the empirical relationship for the temperature dependence of the critical field (2.42) yields a relationship for the entropy difference between the two states,

$$S_n - S_s = \frac{2\mu_0 H_0^2}{T_c}\left[1 - \left(\frac{T}{T_c}\right)^2\right]\frac{T}{T_c} \tag{2.44}$$

where $S = \rho s$ is the entropy per unit volume of the superconducting material. The heat capacity difference is written similarly,

$$C_n - C_s = \frac{2\mu_0 H_0^2}{T_c}\left[1 - 3\left(\frac{T}{T_c}\right)^2\right]\frac{T}{T_c} \tag{2.45}$$

Experiments have shown an approximately cubic temperature dependence of the specific heat for $T \lesssim T_c$. However, at lower temperatures, $T < 0.5\ T_c$, an exponential temperature dependence is observed, indicating an energy gap in the electron density of states.

A Type I superconductor exposed to an external magnetic field $H < H_c$ will exclude the flux from penetrating into its bulk. This behavior, known as the Meissner effect, is shown schematically in Fig. 2.9. There are essentially two equivalent models for describing the Meissner state. The first is to assume that because the superconductor has no electrical resistance, screening currents are established on the surface opposing the changing flux. These currents flow in a layer at the surface of thickness $\lambda \approx 50$ nm, known as the London penetration depth. λ is one of the fundamental characteristic lengths used to define the behavior of a superconductor. The alternate description is to consider the superconductor as a perfectly diagmagnetic body such that its magnetization always equals the negative of the applied field, $M = -H$. Describing a superconductor as being a perfect diamagnet has the advantage of always predicting the flux exclusion

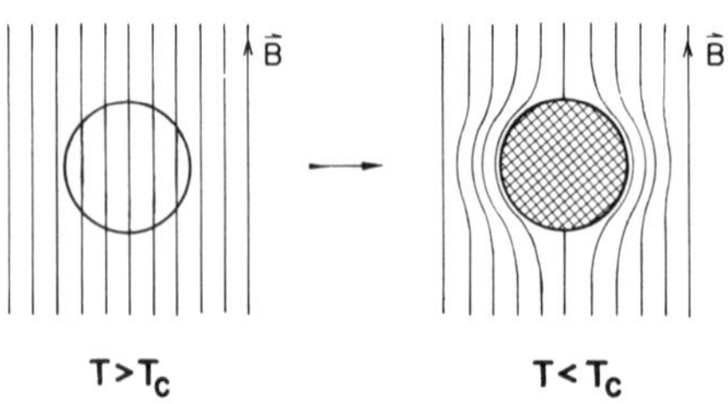

Fig. 2.9. Meissner effect in a superconducting sphere cooled in a constant applied field. Below T_c, $B = 0$ within the superconductor.

condition independent of the order of the applied field and low-temperature environment.

The complete theory of superconductivity is based on microscopic interactions between the electrons and phonons within the lattice, leading to correlated behavior of the electrons, known as Cooper pairing. The mathematical treatment of this model is quite complex, requiring a thorough knowledge of advanced quantum mechanics, and is well beyond the scope of this brief survey. Nevertheless, there are some important results of the microscopic theory which are helpful in understanding the general behavior of superconductors.

Most advanced theories of superconductivity introduce a second characteristic length, the coherence length ξ. In the microscopic theory, the coherence length is roughly the size of a Cooper pair, while in macroscopic theory it represents the spatial distance over which the superconducting to normal transition can occur, that is, variation of the order paramater. The coherence length is a strong function of the crystal imperfections. Superconductors with large values of $\xi \gtrsim \lambda$ are Type I, while Type II superconductivity occurs for those cases where $\xi \lesssim \lambda$.

The other major success of the microscopic theory is its ability to predict the superconducting properties of a metal based on knowledge of the electron and phonon energy distributions. The electron–phonon interaction which produces Cooper pairing causes a gap in the density of electron states. This gap is the origin of the exponential specific heat at low temperatures. The width of the gap is directly proportional to the superconducting transition temperature. In the microscopic theory due to Bardeen, Cooper, and Schrieffer,[26] an exact formula is derived for the critical temperature T_c,

$$T_c = 1.14\,\Theta_D \exp\left(\frac{-1}{UD(\varepsilon_F)}\right) \tag{2.46}$$

where Θ_D is the Debye temperature, and $D(\varepsilon_F)$ is the electron density of states a the Fermi surface. U is an attractive potential brought about by the correlated behavior of the superconducting electrons. This attractive potential is related to the electron–phonon interaction which leads to Cooper pairing of the superconducting electrons. Two interesting conclusions can be made based on (2.46). First, metals with high resistances near room temperature and possessing large electron–phonon interactions will also be more likely to be superconductors. This result is approximately borne out by experiment. Second, metals with even numbers of valence electrons and having a smaller $D(\varepsilon_F)$ are less likely to be superconductors. Empirically, it is found that the transition temperatures of superconductors peak with odd numbers of valence electrons in support of this theoretical result. In its fully

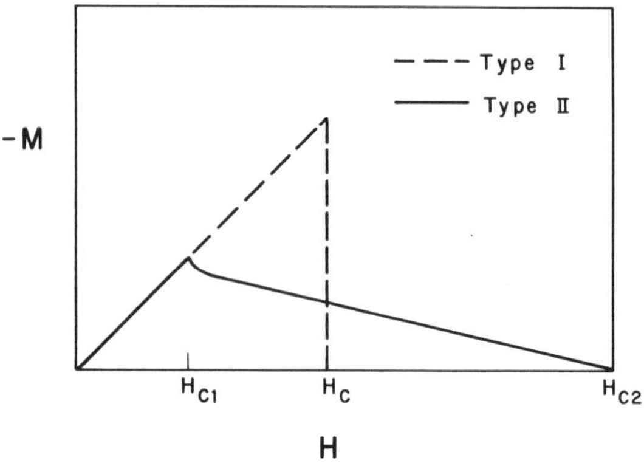

Fig. 2.10. Magnetization curves for Type I and Type II superconductors.

developed formalism, the microscopic theory of superconductivity is considered to be one of the major triumphs of theoretical solid-state physics.

Now consider the behavior of Type II superconductors. The distinction between Type II and Type I superconductors is in the magnetic behavior. In a Type I superconductor magnetic flux is totally excluded provided $H < H_c(T)$. A Type II superconductor can allow penetration of magnetic flux under certain circumstances. This penetration in an annealed Type II superconductor forms a regular array, effectively bringing small regions of the superconductor into the normal state. Such behavior, which is a deviation from the Meissner state, is referred to as the mixed state in Type II superconductivity.

Because of flux penetration, the Type II superconductor in the mixed state is no longer a pefect diamagnet. A typical magnetization curve of a Type II superconductor is shown in Fig. 2.10. Also indicated is a magnetization curve for the same material if it were Type I. In Type II superconductors there are two critical fields. The lower critical field H_{c1} represents the transition from the Meissner state to the mixed state. The upper critical field H_{c2} marks the maximum field for which any superconductive behavior is present. Listed in Table 2.10 are some typical Type II superconductors. Note that although the T_c may be only slightly higher than that of some Type I materials, H_{c2} is often orders of magnitude larger than H_c.

The mixed state in a Type II superconductor has an equilibrium condition in which the flux penetrates the material in a uniform array of individual flux lines called a fluxon lattice. In a pure annealed Type II superconductor, these flux lines are free to move about within the crystal. If the superconductor is subjected simultaneously to an external field and

Table 2.10. Critical Temperature and Upper Critical Field of Common Type II Superconductors

Material	T_c(K)	$\mu_0 H_{c2}(T)$
Nb	9.3	0.29
V	5.4	0.7
NbTi	10	15
Nb_3Sn	18	28
V_3Ga	15	26
Nb_3Ge	23	38

transport current, the flux lines will move under the influence of the Lorentz force, $\mathbf{F}_L = \boldsymbol{J} \times \boldsymbol{B}$. This flux motion will cause dissipation and the superconductor will not carry much current. Fortunately, in real superconductors it is possible to pin the flux against imperfections and crystal defects and thus flux flow. Flux pinning is very important in practical superconductors, allowing substantial transport currents to be carried to fields approaching H_{c2}.

For high-field applications there are a limited number of superconductors that are available commercially. The two materials that are employed most often in devices are NbTi and Nb_3Sn. NbTi is a binary alloy of approximately equal weight percent of each component. It has good mechanical properties and reasonably high $\mu_0 H_{c2} \approx 15$ T and $T_c \approx 10$ K. NbTi is the most commonly applied superconductor in all magnetic devices except those requiring particularly high fields, $\mu_0 H > 10$ T. Nb_3Sn, the other common practical superconductor, is an intermetallic compound of a general class of materials called the A-15's. Its mechanical properties are poor but Nb_3Sn has superior high-field characteristics, $\mu_0 H_{c2} \approx 28$ T and $T_c \simeq 18$ K, making it particularly well suited for high-field magnetic devices. Both these materials can be made with sufficient flux pinning sites to permit good transport current characteristics. Plotted in Fig. 2.11 are the critical current densities as functions of field and temperature of these two materials. For materials to be practically useful they need to carry currents in excess of 500 A/mm^2 which defines the region of application for these two materials. Also shown in the figure are the range of temperatures obtainable by the two lowest-temperature cryogenic fluids, liquid helium and liquid hydrogen. For reasonable current densities in high magentic fields, it is apparent that liquid helium provides the only practical coolant.

In addition to the above mentioned materials, there are several other superconductors which are being developed. One approach is to modify metallurgically the superconducting properties of the commercial materials. This approach is most effective with NbTi containing a small percentage of

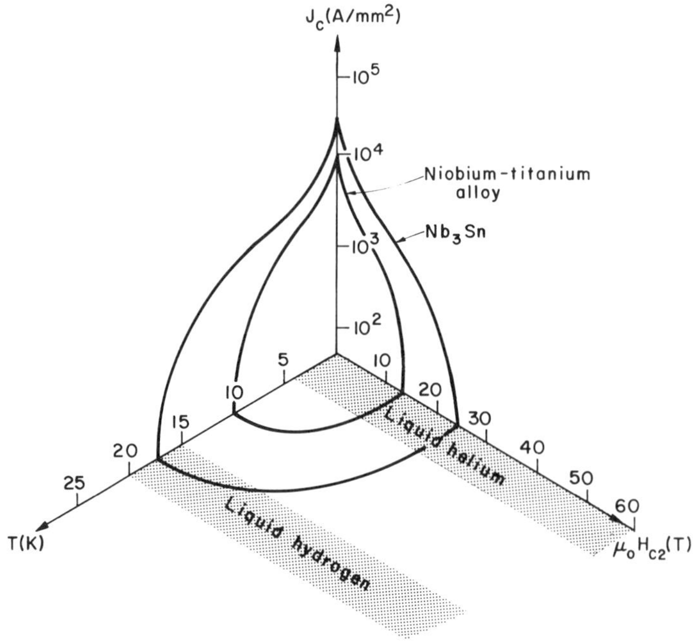

Fig. 2.11. Upper ciritcal field, temperature, and current density for commercial superconducting materials NbTi and Nb_3Sn (from Schwartz and Foner[27]).

Ta. In this case, the upper critical field and current density at low temperatures, $T \approx 2$ K, are enhanced over that of the binary alloy.[28] For the A-15 compounds, apart from Nb_3Sn, small quantities of V_3Ga have also been produced having higher current densities than Nb_3Sn, particularly at reduced temperature.[29] There are many other high-field superconducting materials emerging. Plotted in Fig. 2.12 are the critical field–temperature characteristics of the best high–field superconductors. Some of these materials have extremely high critical parameters; however, they have yet to be fabricated in large enough quantities to prove useful in devices.

The problems of superconducting materials are of great concern to helium cryogenics. Superconducting materials require helium temperature environments to achieve their properties, but more importantly, the behavior of superconductive devices are governed largely by problems of heat transfer, efficient cooling, and safety. For example, the properties of a superconductor are conducive to carrying electric current provided the material remains below the local critical temperature and field. Thermal equilibrium is not always possible so superconductors must be fabricated in a low-resistance matrix material like copper or aluminum to provide the

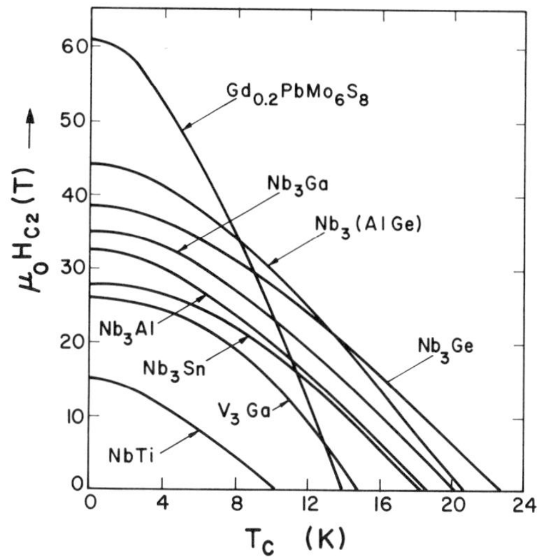

Fig. 2.12. Upper critical field and critical temperature relationships for the best high-field superconductors (compiled by M. N. Wilson[25]).

current-carrying capacity should the superconductor enter the normal state. Proper analysis of this problem requires knowledge of the heat tranfer and fluid flow conditions present in the particular magnetic device.

PROBLEMS

2.1. Calculate the difference between the constant pressure and constant volume heat capacities of aluminum at 300 K.

2.2. Derive an expression for the temperature at which the electron and phonon contributions to the heat capacity are equal. Of the elements listed in Tables 2.1 and 2.2, which has the highest value of this temperature?

2.3. a. Calculate the temperature at which the maximum in the Schottky heat capacity occurs for $g_0/g_1 = 2$.

b. Derive a form for C_M at the limit where $k_B T \gg \delta$.

2.4. a. Assuming the empirical form for the critical field of a Type I superconductor,

$$H_c = H_0 \left[1 - \left(\frac{T}{T_c} \right)^2 \right]$$

derive an expression for the entropy change associated with the superconducting transition.

b. *Adiabatic magnetization* is the constant entropy application of the magnetic field that brings the superconductor into the normal state. Derive an expression for the final temperature T_f resulting from this process.

c. Choose a Type I superconductor and calculate the value of $\Delta T_{\max}$, the maximum value for the temperature difference occurring from *adiabatic magnetization* (neglect the phonon heat capacity).

2.5. Beginning with the expression for a free magnetic ion system,

$$M = Ng\mu_B JB_J(a)$$

show that in the low-field, high-temperature limit the magnetic susceptibility obeys a Curie law.

2.6. a. Derive Eq. (2.19) using the relationship for the Fermi energy (2.13) and $k = \frac{1}{3} Cvl$.

b. For solids with negligible electronic specific heat in the temperature range below $\Theta_D/10$, show that the thermal conductivity may be written

$$k_{\text{ph}} = 78\, nk_B \left(\frac{T}{\Theta_D}\right)^3 c_s^2 \tau$$

where c_s is the speed of sound in the material.

c. Show the temperature dependence of k_e and k_{ph} at $T < \Theta_D/10$. Assume that the mean free path is a constant due to scattering from impurities, crystal defects, or sample boundaries.

d. Determine the high-temperature ($T > \Theta_D$) dependence of k_e and k_{ph} using $l_e \propto 1/k_B T$ and $l_{\text{ph}} \propto e^{-\Theta_D/2T}$.

2.7. a. Determine the mean free path in a copper sample with a residual resistivity ratio RRR = 50 for temperatures below 10 K. Use the following values for copper: Fermi velocity $v_F = 1.57 \times 10^6$ m/s and the electron concentration $n = 8.45 \times 10^{28}\ m^{-3}$.

b. Estimate the thermal conductivity of the same copper at 2 K.

2.8. The maximum electrical current that can be carried by a Type I superconductor wire is limited to the current that produces the critical field H_c at the conductor surface.

a. Derive an expression for the critical current of a cylindrical superconducting wire of radius R as a function of temperature.

b. How much current can a 1 mm diameter indium wire carry at 1.8 K? [Hint: Use Ampère's Law to determine the magnetic field at the surface of the wire.]

2.9. A normal metal wire carrying an electrical current will produce heat, raising the temperature of the wire. Estimate the time required for a copper wire carrying a current density of 100 A/mm^2 to heat up from 4.2 to 300 K. Assume $C_p = C_v$ and use approximations for both the resistivity and the specific heat as functions of temperature.

2.10. A sample of wire is made by codrawring a copper tube over an aluminum rod. The aluminum has a cross-sectional area A_1 and the copper A_2. The wire is stress free at ambient temperature, $T = 293$ K. Derive a relationship for the stress in either component of the wire as a function of A_1, A_2, and material properties when the wire is cooled to 4.2 K. For equal cross-sectional areas ($A_1 = A_2$), calculate the stress in the copper and compare it with the yield of annealed OFHC copper.

REFERENCES

1. C. Kittel, *Introduction to Solid State Physics*, 5th ed., Wiley, New York, 1976.
2. G. K. White, *Experimental Techniques in Low Temperature Physics*, 3rd ed., Monographs on Physics and Chemistry of Materials, Clarendon Press, Oxford, 1979.
3. O. V. Lounasmaa, *Experimental Principles and Methods Below 1 K*, Academic Press, New York, 1974.
4. R. P. Reed and A. F. Clark, *Materials at Low Temperatures*, American Society of Metals, Metals Park, Ohio, 1983.

5. L. A. Hall, *Survey of Electrical Resistivity Measurements on 16 Pure Metals in the Temperature Range* 0–273 K, NBS Technical Note 365, U.S. Government Printing Office, Washington, DC, 1968.
6. R. A. Matula, Electrical Resistivity of Cu, Au, Pd, and Ag, *J. Phys. Chem. Phys. Ref. Data* **8**, 1147 (1979).
7. G. T. Meaden, *Electrical Resistance of Metals*, Plenum Press, New York, 1965.
8. A. F. Clark, G. E. Childs, and G. H. Wallace, Electrical Resistivity of Some Engineering Alloys at Low Temperatures, *Cryogenics* **10**, 295 (1970).
9. R. Berman, *Thermal Conduction in Solids*, Clarendon Press, Oxford, 1976.
10. R. L. Powell and F. R. Fickett, Cryogenic Properties of Copper, International Copper Research Association, Dec. 1979.
11. J. G. Hust and L. L. Sparks, *Lorenz Ratios of Technically Important Metals and Alloys*, NBS Technical Note 634, U.S. Government Printing Office, Washington, DC, Feb. 1973.
12. R. J. Corruccini and J. J. Gniewek, *Thermal Expansion of Technical Solids at Low Temperatures*, NBS Monograph 29, U.S. Government Printing Office, Washington, DC, 1961.
13. T. H. K. Barron, J. G. Collins, and G. K. White, Thermal Expansion of Solids at Low Temperatures, *Adv. Phys.* **29**, 609 (1980).
14. R. S. Krishnan, R. Srinivasan, and S. Devanarayanan, *Thermal Expansion of Crystals*, Pergamon, Oxford, 1979.
15. G. K. White, Metals and Alloys: Expansion and Contraction, *Adv. Cryog. Eng.* **30**, 407 (1984).
16. R. J. Roark and W. C. Young, *Formulas for Stress and Strain*, 5th ed., McGraw–Hill, New York, 1975.
17. *Handbook on Materials for Superconducting Machinery*, Metals and Ceramics Information Center, Batelle, Columbus, Pub. #MCIC-HB-04 (Jan. 1977).
18. The present treatment is based primarily on material available in M. W. Zemansky, *Heat and Thermodynamics*, 5th ed., McGraw–Hill, New York, 1968; or M. W. Zemansky and R. H. Dittman, *Heat and Thermodynamics*, 6th ed., McGraw–Hill, New York, 1981.
19. A. C. Rose–Innes and E. H. Rhoderick, *Introduction to Superconductivity*, 2nd ed., International Series in Solid State Physics, Vol. 6, Pergamon Press, New York, 1978.
20. M. Tinkham, *Introduction to Superconductivity*, McGraw–Hill, New York, 1975.
21. E. A. Lynton, *Superconductivity*, 3rd ed., Chapman Hall Ltd. Science Paperbacks, London, 1969.
22. E. W. Collings, *Applied Superconductivity, Metallurgy, and Physics of Titanium Alloys*, Vols. 1 and 2, Plenum Press, New York, 1985.
23. J. E. C. Williams, *Superconductivity and Its Applications*, Pion Ltd., London, 1970.
24. H. A. Brechna, *Superconducting Magnet Systems*, Springer–Verlag, Berlin, 1973.
25. M. N. Wilson, *Superconducting Magnets Monographs on Cryogenics*, Clarendon Press, Oxford, 1983.
26. J. Bardeen, L. N. Cooper, and J. R. Schrieffer, Theory of Superconductivity, *Phys. Rev.* **108**, 1175 (1957).
27. B. B. Schwartz and S. Foner, Large Scale Applications of Superconductivity, *Phys. Today* **30**, 34 (1977).
28. D. C. Larbalestier, Niobium–Titanium Superconducting Materials, in *Superconducting Materials*, S. Foner and B. B. Schwartz (Eds.), Plenum Press, New York, 1981.
29. D. U. Gubser, T. L. Francavilla, D. G. Howe, R. A. Muessner, and F. T. Ormand, Multifilamentary V_3Ga Wire: A critical Review, *IEEE Trans. Magn.* **Mag-15**, 385 (1979).

Chapter 3

Helium as a Classical Fluid

Of all the cryogenic fluids, helium exhibits behavior that most nearly approximates that of an ideal fluid. This fact is caused mostly by the weak intermolecular potential that helium enjoys. It further manifests itself in the fact that helium has the lowest critical point of all fluids, $T_c = 5.2$ K, $p_c = 0.226$ MPa. As a result of this near ideality, much of the behavior of gaseous and liquid helium above the superfluid transition can be treated in terms of classical models. This is not to say that quantum effects do not contribute to the behavior. Rather, certain features of helium in this temperature and pressure range are controlled by a combination of physical phenomena which can be qualitatively if not quantitatively described in terms of classical models. Conversely, certain characteristics of helium, most notably that of the liquid state below the superfluid transition and also the solid state, have properties which are so determined by quantum mechanics that classical physics cannot be used in a meaningful way to interpret their behavior. The quantum aspects of helium are discussed separately in Chapter 4.

3.1. HELIUM PHASE DIAGRAM

As a beginning point to the description of liquid and gaseous helium, it is useful to make reference to a phase diagram, Fig. 3.1. In addition to the conventional characteristics such as the critical point and two-phase coexistence, there are several unique features to the phase diagram which should be noted. First, unlike all other substances, the solid state is not obtainable in helium at any temperature unless an external pressure in excess of 2.5 MPa (25 bars) is applied. This characteristic, an artifact of the quantum nature of helium, is caused by the large zero point energy of the helium molecule. The lowest energy state of helium is that of the liquid, albeit the superfluid state, He II.

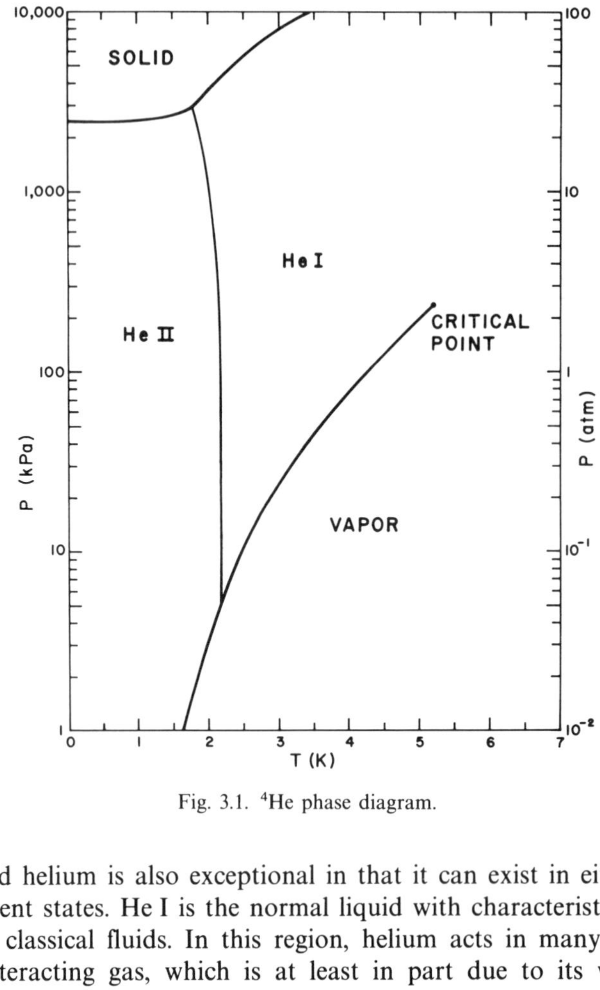

Fig. 3.1. ^{4}He phase diagram.

Liquid helium is also exceptional in that it can exist in either of two very different states. He I is the normal liquid with characteristics that are typical of classical fluids. In this region, helium acts in many ways as a weakly interacting gas, which is at least in part due to its weak intermolecular interaction and low viscosity. The viscosity of He I is comparable to that of air at room temperature.

The superfluid state, which is more correctly referred to as He II, has physical features that are truly exceptional. Most notable of these features are the transport properties, with a vanishingly small viscosity and an apparent thermal conductivity many orders of magnitude larger than liquids or even high-conductivity solids. The line that separates the two liquid states is termed the λ-line. This designation was adopted because the specific heat near the transition has the shape of the Greek letter λ. The λ-transition temperature is 2.172 K at saturated vapor pressure and decreases gradually with increasing pressure until it intersects the solid coexistence boundary at 1.763 K, $p = 29.74$ bars.[1]

The phase diagram of helium is also unique in that it lacks a triple point of coexistence between liquid, vapor, and solid, because the solid state can exist only under an external pressure. The intersection of the λ-line with the two phase coexistence regions of liquid and vapor or solid and liquid is not a triple point, although it occasionally is identified incorrectly as such. As is discussed in the next chapter, the λ-transition represents a second-order phase transition, which means it has a discontinuous slope in the temperature dependence of the entropy. There is no latent heat of formation of the He II state and coexistence of He II and He I in equilibrium conditions is not possible.

In the present chapter the properties of helium which are best described in terms of classical models are reviewed. Thus, the subject matter is dominated by the gaseous and normal liquid state, He I. Even with this somewhat arbitrary separation, it is not possible to analyze fully the properties of these "classical" fluids without some reference to the effect of quantum statistics on their behavior. Chapter 4 is reserved for the description of helium as a quantum fluid. Therefore, the bulk of the quantum mechanical analysis of helium is presented there with comparison to the quantum fluid, He II.

3.2. GASEOUS HELIUM

Helium at temperatures substantially above its normal boiling point behaves more like an ideal gas than any other commonly known fluid. This fact is brought on by the weak interatomic potential and the spherically symmetric molecular configuration. Of course, some deviation from ideality must occur for gaseous helium, otherwise physical phenomena such as liquefaction would not occur. Because of its similarity to an ideal gas, it is often most beneficial to consider properties of gaseous helium in terms of extensions from the ideal gas model.

The first goal of a physical model used to describe helium gas is the development of an understanding of its state properties. From an experimental viewpoint, the most common measurements are those of specific heat, C_p and C_v, and the state properties of pressure, temperature, and volume. The latter measurements lead to the equation of state for the gas, which in its complete form can be combined with the thermodynamic principles to determine the relevant state functions including entropy and enthalpy.

The equation of state, which is a functional relationship among p, V, and T, cannot be determined exactly for a particular real gas. However, there exist a number of approximate relationships that also give considerable physical insight into the processes involved. The most notable approximation to the equation of state is known as the virial expansion, which is an expansion involving intermolecular interactions. Furthermore, in developing a model to describe the interactions between helium gas molecules, particularly near the critical point $T_c = 5.2$ K, a concern for quantum mechanical effects is needed. The most common isotope of helium, ^{4}He, obeys Bose–Einstein statistics, which must be taken into account in the complete description of the gas.

3.2.1. Intermolecular Interactions

Any microscopic model for the behavior of helium has as its basis an understanding of the intermolecular potential.[2,3] This interaction is exactly that which is ignored in the ideal gas model. If helium were a classical gas and the intermolecular potential were known, it would be possible to compute the state properties directly. Unfortunately, this approach is not possible, making it necessary to use empirical potentials for describing the molecular interactions. These potentials must contain the physics of the interaction, which is composed primarily of two terms: (1) the hard core repulsive interaction and (2) the weaker attractive interaction due to the London-dispersion mechanism.

The hard core repulsive interaction can be understood in rather simple terms. Molecules occupy volume, which is defined by the steepness of the repulsive potential. This contribution would naturally be expected to cause deviations from ideal gas behavior because the latter model assumes the molecules are noninteracting point particles. At high temperatures and low densities the corrections to the equation of state due to the hard core potential are usually small.

The attractive term in the molecular interactions is brought on by a mechanism known as the London-dispersion interaction, so termed because it is the same process responsible for light dispersion by molecules. To understand this interaction, it is important to note that a helium molecule is not truly inert. Actually, it is composed of a nucleus containing the positive charge and a "sea" of negative electrical charge surrounding the nucleus. At finite temperature the negative charge has a tendency to oscillate about the nuclear center of the molecule causing a dipole moment. Because of the oscillating dipole, there is a momentary electrostatic potential set up in the vicinity of the molecule which varies as $V_d(r) \approx 1/r^3$. When two molecules are in close proximity and have oscillating dipole moments that are out of phase, there will be net attractive interaction. The interaction scales as the product of the two oscillating dipole moments, $\phi_a(r) \approx 1/r^6$. Detailed quantum mechanical analysis by London has suggested that the attractive interaction is a sum of terms, the first of which is the dipole contribution. As with the hard core potential, the attractive interaction causes deviations from the ideal gas equation of state. It has the largest effect at low temperatures and high densities, resulting in a tendency for the molecules to exhibit collective behavior and eventually leading to liquefaction.

Numerous examples of empirical potentials have been proposed to describe the total interaction of inert spherically symmetric molecules such as helium. Probably the most successful, because of its mathematical simplicity and physical basis, is the Lennard–Jones 12-6 potential. This interaction has a mathematical form,

$$\Phi_{\mathrm{LJ}}(r) = 4\varepsilon\left[\left(\frac{r_0}{r}\right)^{12} - \left(\frac{r_0}{r}\right)^{6}\right] \qquad (3.1)$$

that depends on the determination of two parameters: ε, which is a measure of the depth of the potential well, and r_0, which is a dimensional scale of the molecular radius. Plotted in Fig. 3.2 is the Lennard–Jones potential for a helium molecule with the quantities ε, r_0, and r_m, the radius of the potential minimum, all indicated. Listed in Table 3.1 are these parameters for some common cryogenic fluids.[2] Other examples of

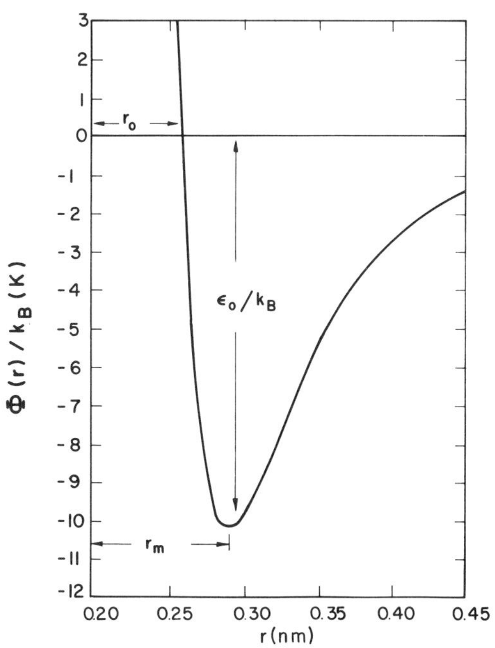

Fig. 3.2. Lennard–Jones 12-6 potential for ^{4}He.

Table 3.1. Coefficients of the Lennard–Jones 12-6 Potential for Common Cryogenic Fluids

	r_0 (nm)	r_m (nm)	ε_0/k_B (K)
He	0.2556	0.2869	10.22
Ne	0.2789	0.3131	35.7
Ar	0.3418	0.3837	124.0
N_2	0.3681	0.4132	91.5
O_2	0.3433	0.3853	113.0

empirical potentials with greater or lesser degree of mathematical simplicity and physical basis are the hard sphere potential, square well potential, and the Buckingham exp-6.[4]

3.2.2. Virial Expansion

Suitable selection of the intermolecular potential is insufficient in itself for approximation of the equation of state. It is necessary to adopt a form for the equation of state which incorporates the intermolecular potential. Such a relationship has numerous forms, but the best for physical interpretations is the virial expansion, which can be written as an expansion in the specific volume v:

$$pv = RT\left(1 + \frac{B}{v} + \frac{C}{v^2} + \frac{D}{v^3} + \cdots\right) \tag{3.2}$$

The quantities B, C, D, and so on are the second, third, fourth, and higher-order virial coefficients. Note that a special case of the virial expansion is that of the van der Waal equation of state.

The physical basis for the virial expansion in the form shown in (3.2) is for each coefficient to describe a particular type of interaction. Thus, the second virial coefficient B should incorporate corrections to the equation of state due to two-body interactions. The third virial coefficient C incorporates corrections due to three-body interactions, and higher-order terms follow the same sequence. The requirement for any theoretical description of the virial coefficients is that it must include the many-body interactions and use a physically realistic potential.

The second virial coefficient is the largest term in the expansion and therefore makes the largest correction to the ideal gas law. From classical mechanics, the second virial coefficient can be written in closed form,

$$B_{\text{cl}} = -\frac{N}{2}\int_0^\infty (e^{-\beta\phi(r)} - 1)\, 4\pi r^2\, dr \tag{3.3}$$

where $\phi(r)$ is the interaction potential assumed to be spherically symmetric. The quantity $\beta = 1/k_B T$ incorporates the thermal energy of the system. By considering (3.3) it is easy to see that the problem of analytical determination of B becomes quite difficult when complex potentials are employed.

The virial expansion becomes even more difficult to analyze when considering gaseous helium at relatively low temperatures because of the need to include quantum mechanical effects. A complete description of the quantum virial expansion problem is beyond the scope of this book. However, it is instructive to survey the problem to understand better some of the difficulties. To determine the quantum virial expansion, the equation of state is expanded in terms of the fugacity $z = e^{-\beta\mu}$, where μ is the chemical potential. This is a fairly standard approach in quantum statistical mechanics.[4,5] The result is an expression for the pressure in terms of this quantity,

$$\frac{p}{k_B T} = \sum_k b_k z^k \tag{3.4}$$

where b_k contains the physics of the problem. These coefficients must be determined by substitution in the time-independent Schrödinger equation,

$$-\frac{h^2}{2m_r}\left[\nabla^2 r\psi + \frac{l(l+1)}{r^2} r\psi\right] + \Phi(r)\, r\psi = Er\psi \tag{3.5}$$

where $m_r = m_1 m_2/(m_1 + m_2)$ is the reduced mass, l is the angular momentum quantum number, and ψ is the spherically symmetric wave function. The solution to the above problem leads to a set of energy levels, some of which are bound states and some of which are in the continuum. Applying quantum mechanical scattering methods, we calculate the phase shift $\eta_l(q)$, where q is the energy parameter. The phase shift serves the purpose of incorporating interactions into one parameter. The wave function solution to the Schrödinger equation includes the phase shift as part of its argument,

$$\psi \approx \sin(qr - \tfrac{1}{2}\pi l + \eta_l) \tag{3.6}$$

The virial coefficients are then found by summing over all states, bound and continuum. For ^{4}He this exercise leads to a computed virial coefficient shown in Fig. 3.3. Also shown in the figure is the classical second virial coefficient for comparison. Clearly, the two quantities are different, particularly below the Boyle temperature defined by the virial coefficient going to zero; for ^{4}He $T_{Bo} = 23.2$ K. At temperatures below around 10 K, there is about a factor of 2 difference between the classical and quantum virial coefficients. Also displayed in Fig. 3.3 are the experimentally determined virial coefficients, which show close agreement with the calculated quantum coefficient using the Lennard–Jones potential.[6,7]

A number of empirical representations of the second virial coefficient have been developed. A simple general form for the function is described by McCarty,[8]

$$B(T) = \sum_i b_i T^{(3-i)/2} \tag{3.7}$$

where on the order of nine terms in the expansion are required to fit adequately the temperature dependence over the whole range. A less cumbersome empirical fit good below the Boyle temperature was suggested by Keller,[4]

$$B(T) = a - \frac{b}{T} \tag{3.8}$$

where, for ^{4}He, $a = 23.05$ cm^3/mol and $b = 421.77$ cm$^3 \cdot$ K/mol $\cdot$ K.

Before leaving the subject of the Boyle temperature, it is useful to consider its physical significance. Since T_{Bo} marks the transition between a positive and a negative virial coefficient, it also marks a change in dominant interaction potential. Below T_{Bo}, B is negative, implying that the pressure is less than would occur for a truly ideal gas. Since the molecules must be attractive to cause this, the dominant interaction is the London

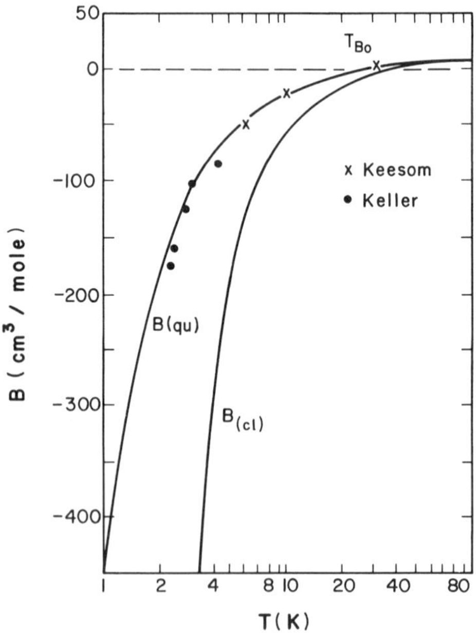

Fig. 3.3. Second virial coefficient for ^{4}He.

dispersion. On the other hand, above T_{Bo}, B is positive, implying the pressure is greater than that of an ideal gas. To have this effect requires an excess repulsion and therefore the hard core potential must be the major contribution. Thus, the value of the Boyle temperature defines the range of dominance of the various contributions to the intermolecular interaction.

The third virial coefficient C is more difficult to calculate because it involves three-body interactions. For the classical picture and the hard sphere potential, C can be shown to vary as the square of the molecular volume.[4] Thus, this coefficient is of little importance except at relatively high densities and low temperatures. Efforts to calculate the third and higher virial coefficients for helium have been hampered by the lack of experimental data. Furthermore, the exact form of these coefficients is less important to application than the development of an empirical equation of state suitable for modeling the helium properties over a wide range of temperatures and pressures.

3.2.3. Empirical Equations of State

Development of the theory of quatum virial expansions is helpful for understanding the complexities of gaseous helium at low temperatures. However, for applications, the above format is far too cumbersome to allow rapid calculation of properties. For this reason, empirical equations of state have been developed which make no attempt to understand the physical problem, but rather to provide a calculational tool. With a closed-form solution for the equation of state, it is a rather straightforward process to derive the other important thermodynamic properties.

The simplest empirical equation of state takes into account the deviations from ideal behavior in terms of a compressibility factor z. As defined, the compressibility factor may be written

$$z = \frac{pv}{RT} \tag{3.9}$$

which obviously is equal to unity for an ideal gas. The value of z is given empirically in either graphical or tabular forms since it depends on state variables. The compressibilitiy factor for helium is displayed in Appendix 2.

Several empirical equations of state have been developed for gaseous helium.[8,9] In general, it is not possible to describe adequately the entire p-v-T surface by one equation. The forms of the expansions have pressure as a dependent variable:

$$pv = RT\left[1 + \frac{B(T)}{v}\right] + \sum_k \left(\frac{g_k(T)}{v^{k+2}}\right) \tag{3.10}$$

where $g_k(T)$ is a complex summation function over temperature with empirically determined coefficients. By this method, correlations to existing data have been found. Properties derived from these correlations are readily available in tabular form.[1,8,9] A portion of these derived properties are reproduced in Appendix 1.

3.3. LIQUID HE I

The thermodynamic properties of normal liquid helium (He I for $T > T_\lambda$) are not radically different from those of the gaseous state. This fact is brought on by the liquid having a weak intermolecular interaction and a relatively low density. The specific gravity of liquid helium at saturated vapor pressure is between 0.145 and 0.0675 at T_c, making it—next to hydrogen—the lowest-density condensed fluid. The thermodynamic properties of liquid helium have been measured extensively and reported in the literature.[8–14] Tabulated in Appendix 1 are a number of these properties as given by McCarty.[8] Because of the complexity of the physics of normal liquids, it is necessary to resort to empirical fits to data on the thermodynamic properties. First principle calculations are quite difficult and beyond the scope of this treatment. A review of the experimentally determined properties of liquid helium is presented here with appropriate comparison to the gaseous state.

3.3.1. Density

The saturation density ρ is perhaps the most important thermodynamic property of He I because it is needed to derive the equation of state for the liquid. Plotted in Fig. 3.4 is the saturated density of liquid helium from T_c to below T_λ. Also shown for scale in the graph is the density of the saturated vapor over the same temperature range. A useful quantity to keep in mind is that the saturated liquid is about 7.4 times more dense than its vapor at the normal boiling point. The maximum density occurs at 2.178 K, 6 mK above T_λ in saturated liquid helium. Below T_λ the density is only weakly temperature dependent, becoming essentially constant at $T \lesssim 2.0$ K.

The slope of the density profile is a direct measure of the volume expansivity β. Under saturated vapor pressure, this quantity is plotted in Fig. 3.5 from results of Kerr and Taylor.[12] By definition the volume expansivity at saturated vapor pressure is given by

$$\beta_{\text{sat}} = -\frac{1}{\rho}\left(\frac{\partial \rho}{\partial T}\right)_{\text{sat}} \tag{3.11}$$

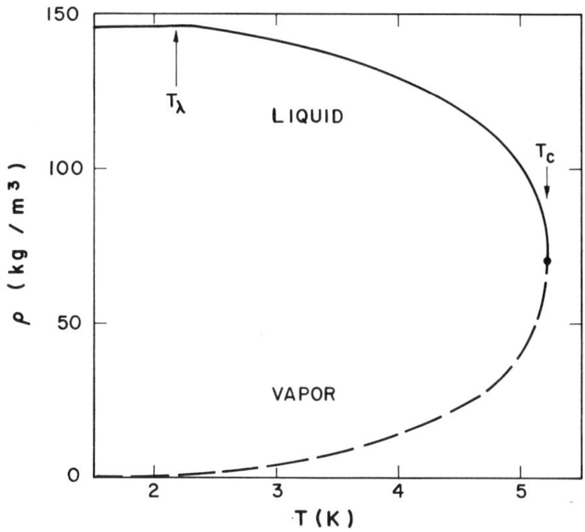

Fig. 3.4. Density of saturated liquid helium.

where the magnitude and slope of the density along the saturated vapor pressure curve are important. Note that not only does β go through zero at T_λ, but it has a discontinuity in slope.

There exists a thermodynamic relationship between the expansion coefficient along the saturated vapor pressure curve, β_{sat}, and the same quantity at constant pressure,

$$\beta_p = \beta_{\text{sat}} + \kappa \left(\frac{\partial p}{\partial T}\right)_{\text{sat}} \tag{3.12}$$

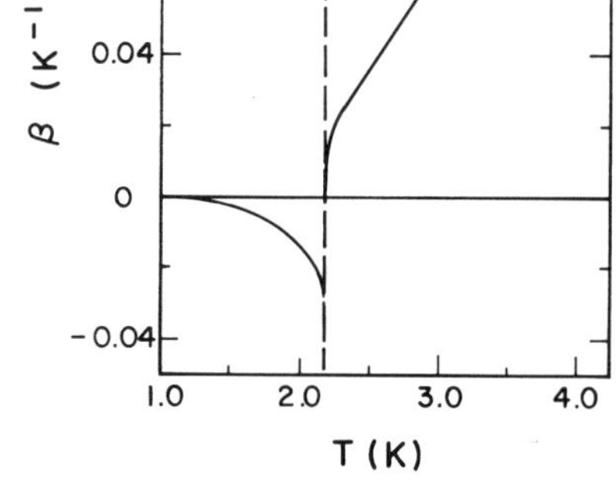

Fig. 3.5. Bulk thermal expansivity of ^{4}He (from Kerr and Taylor[12]).

where the bulk isothermal compressibility,

$$\kappa = -\frac{1}{V}\left(\frac{\partial V}{\partial p}\right)_T$$

is an additional parameter that must be known to compute β_p. The bulk isothermal compressibility is not strongly temperature dependent except near the critical temperature. Typical values for κ are about 120 Pa^{-1} below 1 K, increasing to 140 Pa^{-1} at 3 K. However, the compressibility does have a discontinuity at the λ point which is consistent with the signature of a second-order phase transition.

3.3.2. Thermal Properties

Of the thermodynamic properties of liquid helium, the entropy, specific heat, and latent heat of vaporization have some of the most unique behavior. Shown in Fig. 3.6a–c are these three quantities as determined along the saturated vapor pressure curve. The most pronounced changes in the behavior of the thermal properties occur below T_λ. A complete physical description of this region is given in Chapter 4, which is concerned specifically with He II.

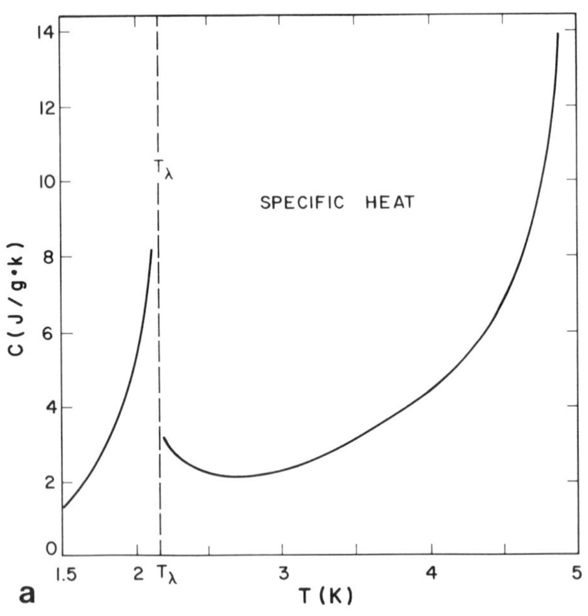

Fig. 3.6a. Specific heat of liquid helium at saturated vapor pressure.

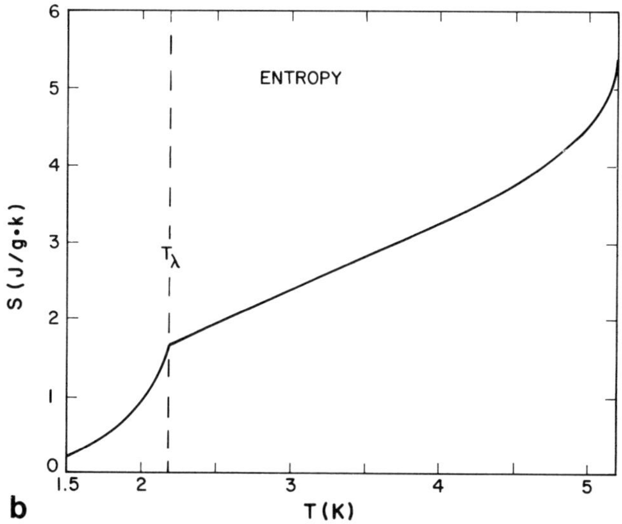

Fig. 3.6b. Entropy of liquid helium at saturated vapor pressure.

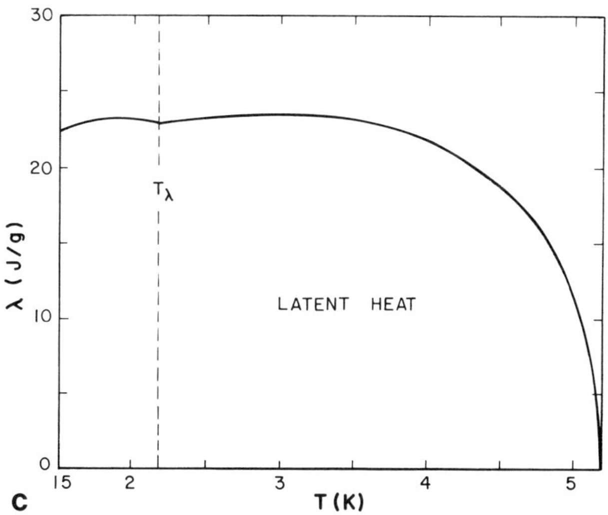

Fig. 3.6c. Latent heat of vaporization of liquid helium.

The entropy and specific heat above T_λ are not expected to have thermal behavior strongly different from that of a gas. Similarity to an ideal gas is expected because liquid helium has a low density and weak interaction. For example, the specific heat of an ideal monatomic gas at constant volume is $C_v = \frac{3}{2}R$ and at constant pressure is $C_p = \frac{5}{2}R$. An ideal gas of mass 4 would therefore have a $C_v = 3.12$ J/g · K and a $C_p = 5.19$ J/g · K. As can be seen in Fig. 3.6a, the specific heat of liquid helium under saturated vapor pressure varies from 2.3 J/g · K just above T_λ to 5.2 J/g · K at 4.2 K. At temperatures much below T_c, the volume expansivity is a small contribution, making C_{sat} approach that of C_v. However, near T_c the volume expansion makes a substantial contribution, and the saturated values are more similar to constant pressure, $C_{\text{sat}} \approx C_p$. That the specific heat of liquid He I brackets that of an ideal monatomic gas indicates that, in many instances, He I is treatable in terms of modified gas laws. The importance of this statement will become evident in the next section on transport properties.

The latent heat of vaporization λ of liquid helium is displayed in Fig. 3.6c. This quantity, which is only defined along the saturated vapor pressure curve, is given by the Clausius–Clapeyron equation,

$$\left(\frac{dp}{dT}\right)_{\text{sat}} = \frac{\lambda}{T(v_g - v_l)} \tag{3.13}$$

where v_g and v_l are the specific volumes of the gas and liquid, respectively. The heat of vaporization represents the energy required to take a unit of helium from the liquid to the vapor state. It is therefore an indirect measure of the strength of the intermolecular bonds associated with the formation of the liquid state. It is possible to compute λ from intermolecular empirical potentials. This problem is beyond the scope of the present treatment; however, on the experimental side there appears to be a useful relationship between the heat of vaporization and the normal boiling point of many liquids. Trouton's law[15] suggests that the ratio of λ to T_b should be approximately constant,

$$\frac{\lambda}{T_b} \approx \text{constant} \tag{3.14}$$

where the constant has a value of about 75 J/mol · K. This law, which is based primarily on experimental evidence, is obeyed to within 20% for most common liquids. However, it does not predict the behavior of cryogenic liquids. The largest deviation from Trouton's law occurs for helium where the ratio of $\lambda/T_b \approx 20$ J/mol · K.

3.3.3. Surface Tension

Before leaving the discussion of He I properties, it is useful to touch briefly on the subject of surface tension. The surface tension σ has units of energy per area and is associated with the increased energy of the liquid state caused by it having a free surface. Microscopically, the surface tension results from the molecules at the interface being at a higher energy because they have fewer nearest neighbors than those in the bulk. The surface tension, as with the heat of vaporization, goes to zero at T_c where the distinction between the liquid and vapor vanishes.

For liquid helium, the surface tension is quite small owing to its weak intermolecular binding. For classical liquids, the temperature-dependent relationship for σ should take the form

$$\sigma(T)=\sigma_0\left(1-\frac{T}{T_c}\right) \tag{3.15}$$

This expression when applied to He I gives a value for $\sigma_0 \simeq 0.5308\ \mathrm{mJ/m^2}$. An alternative empirical fit has been suggested by Keller[4] to agree better with the data over the temperature range including He II. In this case, the form for σ is shown to follow a nonlinear power law, such that

$$\sigma(T) \approx \sigma_0 - AT^{7/3} \tag{3.16}$$

where $\sigma_0 = 0.37\ \mathrm{mJ/m^2}$ and $A = 7.9 \times 10^{-3}\ \mathrm{mJ/m^2 \cdot K^{7/3}}$. Plotted in Fig. 3.7 are these two empirical forms for σ. Experimental evidence[16] agrees most

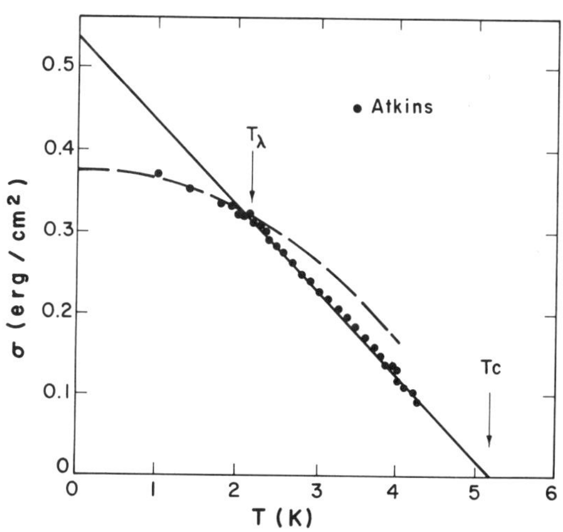

Fig. 3.7. Surface tension of liquid ^{4}He with classical form (—) and He II behavior (- - -).

closely with the classical formulation (3.15), except below the λ-point where quantum effects become important. Below T_λ, the fit described by (3.16) agrees most closely with the data.

3.4. TRANSPORT PROPERTIES OF GASEOUS AND LIQUID HE I

To derive a model that describes the transport properties of helium, account must be taken of the same effects introduced in the development of the equation of state. More specifically, any complete transport theory must include particle–particle interactions and quantum effects at low temperatures. This approach involves the computation of transport coefficient, similar to virial coefficients in the equation of state, which contain the physics of the interaction. Once these transport coefficients are determined, they can be used to describe the deviation from ideal classical behavior of a fluid such as helium. Several accounts of the theory of the quantum mechanical transport coefficients exist in the literature. Since a fundamental understanding of the derivation of this theory is not particularly necessary for the use of helium transport properties in subsequent chapters, only a cursory review of the principles is included here. For a more complete description of the theory of transport coefficients the reader is encouraged to review the appropriate references.[4,5]

Transport properties such as thermal conductivity and viscosity are set aside from the state properties because they are associated with nonequilibrium phenomena. In fact, these properties are fundamental to the understanding of how a system returns to equilibrium through collision mechanics. Thus, the viscosity and thermal conductivity are applied directly to the interpretation of the behavior of helium under fluid flow and heat transfer conditions.

3.4.1. Maxwell–Boltzmann Distribution

The beginning point for a model used to describe nonequilibrium phenomena is usually the Maxwell–Boltzmann distribution function. This function, $f_0(\mathbf{v})$, is a measure of the probability of finding a molecule with velocity $\mathbf{v}$ in a volume at constant temperature T and number density n:

$$f_0(\mathbf{v}) = n\left(\frac{m}{2\pi k_B T}\right)^{3/2} e^{-m(\mathbf{v}-\mathbf{v}_0)^2/k_B T} \tag{3.17}$$

where $\mathbf{v}_0$ is the average velocity of the gas. The distribution is peaked at the most probable velocity, $\bar{\mathbf{v}} = (2k_B T/m)^{1/2}$. Furthermore, the root mean square velocity is defined by the integral of this function over velocity space, $\mathbf{v}_{rms} = (3k_B T/m)^{1/2}$. These quantities are plotted on the distribution function displayed in Fig. 3.8. The Maxwell–Boltzmann distribution function represents the equilibrium velocity profile. Consequently, a good measure of the degree to which a particular system deviates from equilibrium is through the difference between its true distribution and the Maxwell–Boltzmann distribution.

Consider a system containing a fluid which is slightly out of equilibrium. The important quantity to determine for this system is the rate at which it returns to equilibrium through interparticle collisions. To find this rate, it is first necessary to evaluate the number of collisions that occur between molecules in the system. The number of collisions per unit volume and time at a given position r is defined as

$$Z = \iint \sigma_{tot} |\bar{\mathbf{v}}_1 - \bar{\mathbf{v}}_2| \, f(r, \mathbf{v}_1, t) \, f(r, \mathbf{v}_2, t) \, d^3\mathbf{v}_1 \, d^3\mathbf{v}_2 \tag{3.18}$$

where a new quantity has been introduced, σ_{tot}, the total scattering cross section, which is an averaged quantity evaluated at the difference in the mean velocities of the two distribution functions.

Before proceeding with the development of the transport coefficients, it is useful to make some numerical estimates of quantities defined above. The number of collisions per unit time can be estimated by assuming the scattering cross section $\sigma_{tot} \approx \text{constant} \approx \pi a^2$, the cross-sectional area of the molecule. It is then possible to take σ_{tot} out of the integral and the total number of collisions becomes

$$Z = \left(\frac{8}{\pi}\right)^{1/2} n^2 \sigma_{tot} \bar{\mathbf{v}} \tag{3.19}$$

Considering helium gas at 4.2 K, 1 bar, the most probable velocity $\bar{\mathbf{v}} = (2k_B T/m)^{1/2} = 1.32 \times 10^4$ cm/s and $n = 2.51 \times 10^{21}$ cm^{-3}, which is approximately inversely proportional to T at constant p. To estimate the cross section, it is necessary to know the molecular radius. This quantity is not well defined for a quantum mechanical system such as helium at

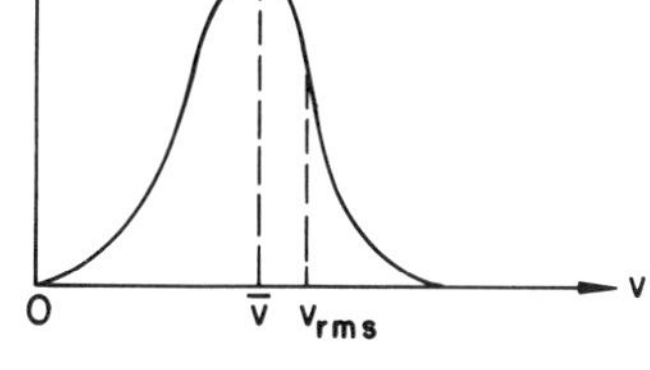

Fig. 3.8. Maxwell–Boltzmann distribution function.

low temperatures. For the present example, the hard core radius defined by the zero of the Lennard–Jones potential gives a value of $\sigma_{\text{tot}} \simeq 20.6 \times 10^{-16}\,\text{cm}^2$. This value leads to a particle collision rate $Z = 2.73 \times 10^{32}\,\text{cm}^{-3} \cdot \text{s}^{-1}$.

The number of collisions Z is not the most useful parameter to evaluate for numerical estimate purposes. A more physically satisfying quantity to estimate is that of the mean free path l. Consider a unit volume in space position $\mathbf{r}$ as shown in Fig. 3.9, containing n molecules per unit volume. The total number of collision paths is equal to the total number of collisions divided by 2 because it requires two molecules to produce a collision. The mean free path can be written in terms of these quantities, such that

$$l = \frac{n}{2Z}\bar{\mathbf{v}} \tag{3.20}$$

For the example considered above, that is, helium gas at 4.2 K, 1 bar, the mean free path works out to be 0.6 nm, which is only slightly larger than the molecular diameter. The mean scattering time τ is simply related to the mean free path by $\tau = l/\bar{\mathbf{v}}$. For helium gas at 4.2 K, $\tau \approx 4.6 \times 10^{-12}$ s.

To develop expressions for the transport coefficients, it is necessary to introduce nonequilibrium features into the distribution function and then apply these to a relationship between the dynamic variables, such as mass flux, momentum, and energy. The latter relationship is established in terms of the classical Boltzmann integral differential equation,

$$\frac{\partial f}{\partial t} = -\mathbf{p}\frac{\partial f}{\partial r} + J(ff) \tag{3.21}$$

where f is the actual distribution function of the system, $\mathbf{p}$ is the momentum vector, and $J(ff)$ is a collision term. The collision term introduces some simplicity into the problem because its argument restricts the dynamics of binary collisions.

Following the established correspondence between variables as given by (3.21), the next step is to define a relationship between the actual dis-

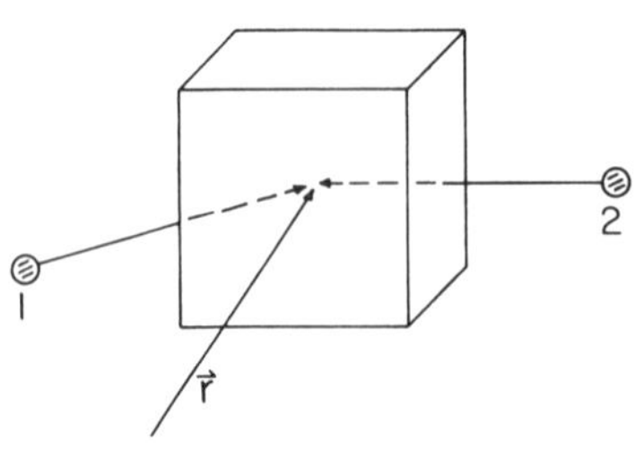

Fig. 3.9. Unit volume in particle space.

tribution function and the equilibrium form. One solution to the problem is to write the distribution function as a power series such that

$$f = f_0 + \alpha f_1 + \cdots \tag{3.22}$$

where f_0 is the Maxwell–Boltzmann distribution. Inherent in this assumed form is the requirement that the correction terms, αf_1 and so on, be small compared to f_0 so the series will converge with a finite number of terms. If only the first term in (3.22) is applied to the relationship between dynamic variables, the result is Euler's equation for ideal fluids. If the first two terms in the expansion are used, the ensuing differential equation is the Navier–Stokes equation. The function f_1 is comprised of the gradients in temperature and velocity.

To determine the transport coefficients and ultimately the transport properties, such as thermal conductivity and viscosity, it is necessary to introduce a set of temperature-dependent integrals $\Omega^{(n,s)}$ that describe the dynamics of the two-particle collisions:

$$\Omega^{(n,s)} = \left(\frac{2\pi k_B T}{m_r}\right)^{1/2} \int_0^\infty \int_0^\infty e^{-\gamma^2} \gamma^{2s+3} (1 - \cos^n \chi)\, b\, db\, d\gamma \tag{3.23}$$

where b is the impact parameter of the collision and m_r is the reduced mass [$m_r = m_1 m_2/(m_1 + m_2) = m/2$ for identical particles]. The other terms contained in the integral include γ, the reduced velocity, and χ the angle of deflection after collision, which is a function of the interparticle potential and impact parameter b. Obiously, the evaluation of (3.23) is complicated and varies considerably between systems.

3.4.2. Transport Properties

The transport properties of a pure gas such as helium are given in terms of the integrals of the form in (3.23). In particular, the viscosity η can be determined by

$$\eta = \frac{5}{8} \frac{k_B T}{\Omega^{(2,2)}} \tag{3.24}$$

and the thermal conductivity k as

$$k = \frac{25}{16} \frac{C_v k_B T}{\Omega^{(2,2)}} \tag{3.25}$$

Note that the ratio k/η is independent of the behavior of $\Omega^{(2,2)}$, having a value equal to $\frac{5}{2}C_v$. At high temperature and low density, this ratio should approach the value of $\frac{15}{4}R = 7.79$ J/g · K, consistent with that of an ideal gas.

Before considering how these models apply to the example of helium, it is worthwhile to discuss briefly the changes brought on by introducing quantum mechanics into the treatment. The approach has a similar impact to that of the quantum virial expansion. In particular, the integral expressions introduced in (3.23) now must include the full quantum mechanical scattering process. As before, this requires the development of a phase shift analysis for two-particle scattering. This approach is beyond the scope of the present treatment. Ultimately, the full quantum mechanical treatment can rely on the correspondence principle which states that at high temperature and low density the classical limit should be reached, giving way to agreement with the above described model.

For the specific case of helium in the gaseous or liquid state, the transport properties are reasonably well-established experimental quantities. In Fig. 3.10a and b are plotted the thermal conductivity and viscosity of helium for their applied pressures, 1.0, 0.1, and 0.01 MPa. At low temperatures, the three cases represent supercritical helium, normal helium at atmospheric pressure with associated change of state at 4.2 K, and subatmospheric helium with no change of state near 2.5 K. These two orders of magnitude in applied pressure represent the most interesting region for applications involving gaseous or liquid He I.

The first feature of these properties to note is their apparent insensitivity to pressure above about 30 K. Below this temperature, the attractive part of the potential contributes substantially to the interaction. At high temperatures, it is mostly hard core potential which is important. The continuously increasing values for k and η at high temperatures suggest a power law dependence that varies like T^n for both quantities. In fact, the data above 30 K approximately follow this law with $n \approx 0.6$. Referring to the temperature-dependent integrals (3.23). the leading terms suggest that $\Omega \approx T^{1/2}$, depending on the terms within the integral being weak functions of temperature. Assuming this to be the case, it would be expected that $k \approx \eta \approx T^{1/2}$, which is not far from the experimental result for $T > 30$ K, where C_v is temperature independent.

Apart from the temperature dependence, it is important to consider the magnitude of the transport properties relative to theoretical expressions. Considering the ratio of k/η as a function of temperature above 30 K, we see that this ratio makes a gradual increase from 7.4 to 7.8 J/g · K at 300 K. According to theory, $k/\eta = \frac{5}{2}C_v = 7.79$ J/g · K for an ideal monatomic gas. Thus, helium at high temperatures is a very good example of near ideality in a real gas. At low temperature, several factors cause a deviation from ideal behavior. The two main ones are the effect of

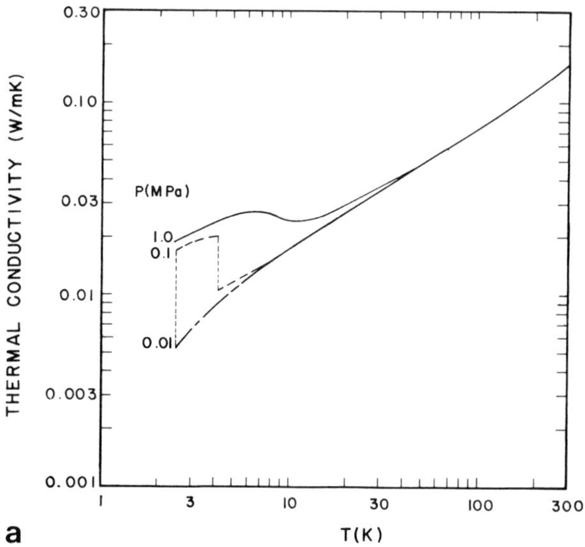

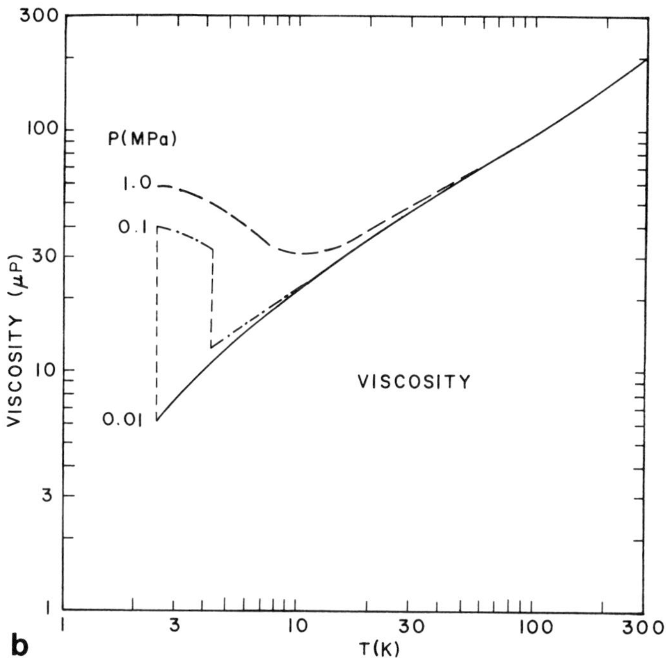

Fig. 3.10. (a) Thermal conductivity and (b) viscosity of ^{4}He as functions of T and p (data from McCarty[1]).

quantum statistics and interparticle interactions. Both these have increasing importance as the temperature is decreased.

As can be seen from the above theoretical review of transport properties, first principle calculations of the thermal conductivity and viscosity of a real fluid are very difficult and must be conducted by various expansion techniques. For the purposes of the user of helium gas or liquid, the fundamental approach is beyond interest. Of much greater concern is the behavior of a given fluid, which may be in the liquid or gaseaous state, under certain sets of conditions. The approach employed, which will be considered in greater detail in Chapters 6 and 7, is to develop sets of dimensionless numbers. These dimensionless numbers provide a fully general method of analyzing experimental data.

The concept of dimensionless numbers is introduced here because one of them, particularly the Prandtl number Pr, is determined by the transport properties. Based on dimensional analysis, combining the three properties of viscosity, heat capacity, and thermal conductivity yields the form

$$\mathrm{Pr} = \frac{\eta C_p}{k} \tag{3.26}$$

where the heat capacity is taken at constant pressure. The Prandtl number is a measure of the relative importance of thermal diffusion and mass flow for the transport of heat. For $\mathrm{Pr} \approx 1$, the velocity and temperature profiles are similar under conditions of a small pressure gradient.

For an ideal gas, the analyses developed in this section can be used to

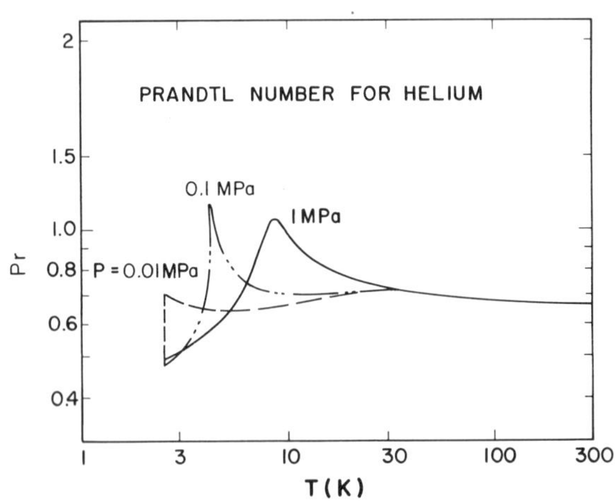

Fig. 3.11. Prandtl number for ^{4}He as a function of T and p.

Table 3.2. Experimental Value of the Prandtl Number for Some Common Liquids under Saturated Conditions

Liquid	T (K)	Pr
N_2	77	2.20
O_2	90	2.21
H_2	20.4	1.17
H_2O	311	4.52
Hg	750	0.004
He	4.2	1.15

predict the Prandtl number. Inverting the ratio of transport properties a numerical quantity is obtained,

$$\frac{\eta C_v}{k} = \frac{2}{5} \tag{3.27}$$

which is close to the form of the Prandtl number in (3.25) with the exception that the specific heat is measured at constant volume. For an ideal monatomic gas $C_p/C_v = \frac{5}{3}$, where this ratio sometimes enters computations of gas expansion. The above terms give a theoretical prediction of the Prandtl number for an ideal monatomic gas, $\mathrm{Pr} \approx \frac{2}{3}$. Plotted in Fig. 3.11 is the Prandtl number for helium gas and liquid at 1.0, 0.1, and 0.01 MPa. At high temperature the experimental values asymptotically approach 0.67 in agreement with prediction. For completeness, listed in Table 3.2 are the Prandtl numbers for a variety of liquids at both cryogenic and room temperatures. It is important to note that helium exhibits behavior nearest to that of an ideal gas. Liquids that exhibit anomalously large thermal conductivity, such as liquid metals, have correspondingly small Prandtl numbers.

PROBLEMS

3.1. a. Calculate the classical second virial coefficient using the square well potential

$$\begin{aligned} \phi(r)/k_B &= \infty && r < 0.25 \text{ nm} \\ &= -10 \text{ K} && 0.25 < r < 0.5 \text{ nm} \\ &= 0 && 0.5 \text{ nm} > r \end{aligned}$$

b. Find the Boyle temperature.

c. Using the virial expansion truncated after the second term, calculate the maximum inversion temperature.

3.2. Given a closed 1-liter volume containing 10 g of helium gas at 7 K:
   a. Calculate the pressure, assuming the gas obeys the ideal gas law.
   b. Calculate the pressure in the volume, for the equation of state based on the virial expansion truncated at the second coefficient.

3.3. Calculate the mean free path and collision frequency for nitrogen gas at saturated vapor pressure and 77 K (ρ_{N_2} at 77 K = 0.0046 g/cm^3).

3.4. Estimate the Prandtl number of nitrogen gas at room temperature, 300 K. The published value is Pr(N_2) = 0.716 at 300 K. Explain any difference between the estimated and published value.

3.5. Estimate the magnitude of the third virial coefficient for helium at the critical point.

REFERENCES

1. R. D. McCarty, *Thermophysical Properties of Helium-4 from 2 to 1500 K with Pressures to 1000 Atmospheres*, NBS Technical Note 631, U. S. Government Printing Office, Washington, DC, 1972.
2. J. O. Hirschfelder, C. F. Curtiss, and R. B. Bird, *Molecular Theory of Gases and Liquids*, Wiley, New York, 1954.
3. J. S. Rowlinson, *Liquids and Liquid Mixtures*, 2nd ed., Plenum Press, New York, 1969.
4. W. E. Keller, *Helium-3 and Helium-4*, Plenum Press, New York, 1969.
5. K. Hwang, *Statistical Mechanics*, Wiley, New York, 1963.
6. W. E. Keller, Pressure–Volume Isotherms of ^{4}He Below 4.2 K, *Phys. Rev.* **100**, 1790 (1955).
7. W. H. Keesom, *Helium*, Elsevier, Amsterdam, 1942.
8. R. D. McCarty, Thermodynamic Properties of Helium-4 from 2 to 1500 K at Pressures to 10^8 Pa, *J. Phys. Chem. Ref. Data* **2**, 923 (1973).
9. D. B. Mann, *The Thermodynamic Properties of Helium from 3 to 300 K between 0.5 and 100 Atmospheres*, NBS Technical Note 154, U.S. Government Printing Office, Washington, DC, 1962.
10. R. W. Hill and O. V. Lounasmaa, The Specific Heat of Liquid Helium, *Philos. Mag.* **8**, 2A, 143 (1957).
11. R. W. Hill and O. V. Lounasmaa, The Thermodynamic Properties of Fluid Helium, *Philos. Trans. R. Soc. London* **252A**, 357 (1960).
12. E. C. Kerr and R. D. Taylor, The Molar Volume and Expansion Coefficient of Liquid ^{4}He, *Ann. Phys.* **26**, 292 (1964).
13. C. T. Van Degrift, Ph.D. Thesis, University of California, Irvine, 1974.
14. C. F. Barenghi, P. G. J. Lucas, and R. J. Donnelly, Cubic Spline Fits to Thermodynamic and Transport Parameters of Liquid ^{4}He above the λ-Transition, *J. Low Temp. Phys.* **44**, 491 (1981).
15. M. W. Zemansky, *Heat and Thermodynamics*, 5th ed., McGraw–Hill, New York, 1968.
16. K. R. Atkins, *Liquid Helium*, Cambridge University Press, Cambridge, England, 1959.

Chapter 4

Helium as a Quantum Fluid

Some of the quantum mechanical aspects of gaseous and liquid helium have been introduced in Chapter 3 as deviations from the classical models. In particular, it was shown that the equation of state and transport models can be treated in terms of quantum mechanical scattering theory. These are not the most dramatic quantum features of helium. In fact, liquid helium below $T_\lambda = 2.172$ K at saturated vapor pressure behaves in such a way that its physical properties can be understood only by using a fully quantum mechanical model with a sizable fraction of the particles in the "condensed" ground state. This condensed state is envisioned to have a number of unique characteristics including zero entropy and viscosity.

In advance of discussing the quantum features of helium, it is worthwhile to ask two questions: What makes helium behave as a quantum fluid and why don't other fluids possess similar characteristics? To answer these questions, it is helpful to consider the most obvious quantum mechanical characteristic of liquid helium. This characteristic is not the existence of superfluidity, but rather the fact that helium does not solidify unless subjected to a sizable external pressure ($p \approx 25$ bars) even at absolute zero! The explanation for this phenomena can only originate in the quantum mechanical interpretation of matter. In this interpretation the position of a helium molecule cannot be defined absolutely but rather is limited by the spread of its wave function. Furthermore, the Heisenberg uncertainty principle describes the extent to which the position x of a molecule and its momentum p can be defined simultaneously:

$$\Delta p \, \Delta x \approx \hbar \tag{4.1}$$

where $\hbar = h/2\pi$ and $h =$ Planck's constant $= 6.63 \times 10^{-34}$ J · s. The Heisenberg uncertainty principle can be used to estimate the zero point energy of liquids.

An elementary quantum mechanical discussion can suggest why the liquid is apparently the lowest energy state of helium at absolute zero. For the solid state to exist it must possess some degree of long-range order. To have this long-range order, the lattice structure must be periodic with fixed positions for the helium molecules. The extent to which this localization is required is somewhat ill defined, but for the sake of the present discussion assume that the position of a given molecule must be defined to within 10% of the interatomic spacing in the lattice. For the case of solid helium the interatomic spacing is of the order of two atomic radii or 0.5 nm. Therefore, the position of a molecule must be defined to within ± 0.05 nm. Uncertainty in helium molecule position leads to an uncertainty in the momentum which is related to the zero point energy. Substituting for the momentum in terms of kinetic energy, $\Delta E \approx (\Delta p)^2/2m$, a relationship for the zero point energy is derived:

$$\frac{\Delta E}{k_B} \approx \frac{\hbar^2}{2mk_B(\Delta x)^2} \tag{4.2}$$

For the case of the common helium isotopes ^{4}He, with a $\Delta x \approx 0.05$ nm, the zero point energy in temperature units turns out to be $\Delta E/k_B \approx 24$ K. This energy is large relative to the potential energy well depth $\phi/k_B \approx 10$ K. Therefore, helium would not be expected to solidify.

For comparison consider the case of solid hydrogen. Since hydrogen is a diatomic molecule with a total mass half that of helium, it is reasonable to assume that its zero point energy would be about twice that of helium or $\Delta E/k_B \approx 50$ K. Hydrogen would not be expected to solidify if not for its very large intermolecular potential, $\phi/k_B \approx 100$ K, brought on by the non-spherical symmetry of the molecule. Hydrogen is observed to solidify with a triple point occuring around 14 K.

Helium is the only element that exists in the liquid state at absolute zero. It is therefore an ideal system to study as a quantum fluid. Since the common isotope of helium is ^{4}He, the molecule has integer spin and obeys Bose–Einstein statistics. A fundamental premise of this statistical model is to allow a large fraction of the population to be in the ground state, thereby affecting the physical behavior. It is worth noting that the rare isotope of helium ^{3}He obeys Fermi–Dirac statistics and would be expected to behave very differently from ^{4}He. In particular, Fermi–Dirac statistics denies the existence of a highly populated ground state by the Pauli exclusion principle. As can be seen in a more detailed discussion of ^{3}He in Chapter 9, the lighter isotope has physical properties that are quite different from those of ^{4}He.

In the present chapter, the concept of quantum fluids is introduced with helium as an example. Where possible, the bridge between experimental evidence and theory is described. The discussion emphasizes physical features including transport properties; however, the bulk of the heat transport description is reserved for Chapter 5.

4.1. IDEAL QUANTUM GASES

A system of noninteracting particles obeying quantum statictics is considered to be an ideal quantum gas. There are a number of physical systems in nature that approximate ideal quantum gas behavior. Common examples include the free electron and phonon gas models as they are used to describe the properties of the solid state. Helium is also a good example of a quantum gas; however, its gaseous state does not exist at sufficiently low temperature or high density to show pronounced quantum effects. Liquid helium, on the other hand, has a number of unique quantum features although it can hardly be considered noninteracting. However, understanding the behavior of an ideal quantum gas in the same temperature and density regime can be a helpful tool in interpreting the behavior of the liquid. To appreciate more fully the physics of quantum gases, particularly helium, the present section develops a theoretical description of an ideal Bose gas. A number of physical properties are calculated including entropy, specific heat, and phase transition thermodynamics. Several of these features are remarkably similar to those of liquid He II.

The theory of an ideal quantum gas begins with the correct statistical picture. For details on how to go about this development, the reader is referred to one of numerous monographs on statistical mechanics.[1,2] Since it is beneficial for appreciation of different statistical systems, the general concepts of quantum statistical mechanics as they apply to the helium problem are surveyed here.

Consider an ideal quantum gas consisting of N noninteracting point particles where N is a large number on the order of 10^{23}. It is possible to write down explicitly the Hamiltonian H for this system:

$$H = \sum_{i=1}^{N} \frac{p_i^2}{2m} \tag{4.3}$$

where p_i is the momentum of the ith particle and m is the particle mass. The Hamiltonian is independent of particle position. In its present form (4.3) is rather cumbersome since it is a sum over all particles. To simplify the analysis it is necessary to introduce a complete set of eigenfunctions to the Hamiltonian which are also eigenfunctions of the particular statistical picture. These eigenfunctions obey Bose statistics and are therefore symmetric under the interchange of particles. By contrast, the eigenfunctions of a Fermi system are antisymmetric under particle interchange.

For the ideal noninteracting system, energy eigenvalues are defined according to single-particle energy levels. It is possible to divide these single-particle levels into groups containing g_i levels, with an assigned average energy ε_i. The occupation number of the ith cell is then given by n_i where this value is assumed to be small compared to g_i. It follows that the total energy E and total particle number N are sums over these levels,

$$N = \sum_i n_i \tag{4.4a}$$

and

$$E = \sum_i n_i \varepsilon_i \tag{4.4b}$$

The difference between statistical pictures becomes evident through the allowable population density of each level within a given cell. For the Bose system any number of particles may occupy a given state. This leads to a different statistical distribution function than for the Fermi system, since the latter requires that the occupation number of a given state be either 0 or 1.

The classical limit of either statistical picture must lead to the Boltzmann statistical distribution. The correspondence principle is a formal statement of this limit. For quantum gases, it is required that both Fermi and Bose gases obey Boltzmann statistics in the limit of high temperature. As with a Bose gas, the Boltzmann gas does not restrict the occupation number of a particular level. However, the state that results from the interchange of particles is considered to be new for the Boltzmann gas but is indistinguishable for the Bose gas.

For each statistical picture presented above, there is a distribution function that describes the probability that a particle occupies a particular energy level. Since the number of particles in a given state is large, the difference between the actual and the most probable distribution functions is insignificant. The distribution function for each of the statistical pictures can be written in terms of the energy of the ith level:

$$n_i = \frac{1}{z^{-1} e^{\beta \varepsilon_i} + \gamma} \tag{4.5}$$

where the particular statistical pictures determine the value of γ. For Bose–Einstein statistics $\gamma = -1$, for Fermi–Dirac $\gamma = +1$, and for Boltzmann $\gamma = 0$. In the above equations z is referred to as the fugacity and equals $e^{\beta\mu}$, where μ is the chemical potential and $\beta = 1/k_B T$. Given the above expressions and description it is possible to solve for the properties of any particular system once the eigenvalues and boundary conditions are known.

Consider an ideal quantum gas confined to a cubical volume of linear dimension L. Since the particles are noninteracting, potential energy can be neglected. The problem then reduces to solving the time-independent Schrödinger equation,

$$-\frac{\hbar^2}{2m}\nabla^2\psi_n = \varepsilon_n \psi_n \tag{4.6}$$

where ψ_n are eigenfunctions of the system. Application of proper boundary conditions including continuity of the wave function and its derivative leads to a set of eigenvalues,

$$\varepsilon_n = \frac{h^2 n^2}{2mL^2} \tag{4.7}$$

where the momentum is also quantized such that $p = hn/L = n(\delta p)$. The quantity $\delta p = h/L$ can be considered to be the dimensions of the momentum space lattice.

The ideal gas problem deals with such a large number of particles that the discreteness of the momentum lattice becomes unimportant. It is therefore possible to make what is termed the continuum approximation which is equivalent to making the lattice dimension go to zero, $\delta p \to 0$. The approximation replaces the summation by an integral over momentum space. The total particle number may then be written

$$N = \sum_i n_i \tag{4.8a}$$

$$\simeq \frac{V}{h^3}\int n\, d^3p \tag{4.8b}$$

As is discussed below, care must be exercised when making this continuum approximation for the Bose system. Because of the potentially large number of particles in the zero momentum state, the discreteness of the lattice must be maintained for this state.

The integral in (4.8) is written explicitly by including the distribution function in the equation for the particle number density:

$$\frac{N}{V} = \frac{2\pi}{h^3}(2m)^{3/2}\int_0^\infty \frac{\varepsilon^{1/2}\, d\varepsilon}{z^{-1}e^{\beta\varepsilon} + \gamma} \tag{4.9}$$

Often the density of energy states $D(\varepsilon)$ in the particular statistical picture is defined in terms of the proportionality factor in (4.9) such that

$$D(\varepsilon)=\frac{2\pi V}{h^3}(2m)^{3/2}\varepsilon^{1/2} \tag{4.10}$$

A similar expression was introduced in Chapter 2 as part of the description of electron and phonon gas models in solids.

As an example, consider the statistical expression for the particle density (4.9) for a Boltzmann gas. For this special case, (4.9) can be simplified because $\gamma=0$:

$$\frac{N}{V}=\frac{2\pi(2m)^{3/2}}{h^3}z\int_0^\infty \varepsilon^{1/2}e^{-\beta\varepsilon}\,d\varepsilon \tag{4.11a}$$

$$=\frac{1}{\lambda_T^3}z \tag{4.11b}$$

where $\lambda_T=(h^2/2\pi mk_BT)^{1/2}$ is referred to as the thermal de Broglie wavelength. At the very least, (4.11) explicitly defines the fugacity z as a function of temperature and density. This quantity is used in the computation of the total energy, entropy, or other thermodynamic quantities of the Boltzmann gas.

The thermal de Broglie wavelength is a characteristic quantity that determines when quantum mechanics becomes an important factor in the behavior of a system. Explicitly, in the wave nature of matter, this quantity is the characteristic wavelength. An estimate can be made for the temperature at which quantum effects become important by allowing λ_T to be of the order of the particle size. For helium, the thermal de Broglie wavelength reduces to a value given by

$$\lambda_T(^4\mathrm{He})\approx\frac{0.87\ \mathrm{nm}}{T^{1/2}} \tag{4.12}$$

Equating λ_T with the particle diameter in liquid helium ($d\simeq 2r_m\simeq$ 0.51 nm), we find that the temperature at which quantum effects should become important is around 2.9 K. Although this temperature is rather close to the superfluid transition in liquid helium, $T_\lambda=2.17$ K, it should not be construed as a prediction of this effect. The thermal de Broglie wavelength being comparable to atomic dimensions makes no prediction of a condensation mechanism as is required for He II. Rather, it indicates approximately where to expect to observe the onset of pronounced quantum effects. Because of its smaller mass, ^{3}He would have a slightly higher transition temperature based on the above argument.

4.1.1. Density

Now consider a Bose gas contained in a volume V. In a similar fashion to the above discussion, an expression can be written for the number density subject to Bose–Einstein statistics:

$$\frac{N}{V} = \left(\frac{1}{\lambda_T}\right)^3 \frac{2}{\sqrt{\pi}} \int_0^\infty \frac{x^{1/2}\,dx}{z^{-1}e^x - 1} + \frac{1}{V}\left(\frac{z}{1-z}\right) \tag{4.13}$$

Here it has been necessary to separate off the population of the zero momentum state because of an anomaly that results when the discrete sum is taken to the integral form. The second term in (4.13) is therefore introduced to take care of the discontinuity when $z \to 1$. The above expression is a complete representation of the number density of an ideal Bose gas. It has the proper high-temperature limit since $e^{\beta(\varepsilon-\mu)} \gg 1$ leads to the exact Boltzmann form of (4.11).

The integral contained in the first term of (4.13) is a general function,

$$g_{3/2}(z) = \frac{2}{\sqrt{\pi}} \int_0^\infty \frac{x^{1/2}\,dx}{z^{-1}e^x - 1} \tag{4.14}$$

that is a specific example of a Riemann zeta function. This function actually is defined as a summation

$$g_n(z) = \sum_{l=1}^{\infty} \frac{z^l}{l^n} \tag{4.15}$$

where it is only bounded for values of $0 \leqslant z \leqslant 1$. The Riemann zeta function has been studied extensively and is tabulated in the literature.[3] For the special case in (4.14), Fig. 4.1 shows the form of this function. It is relatively well behaved, monotonically increasing from zero to $g_{3/2}(1) = 2.612$.

Returning to the expression for the number density of a Bose gas, (4.13), the first term on the right-hand side represents the number of par-

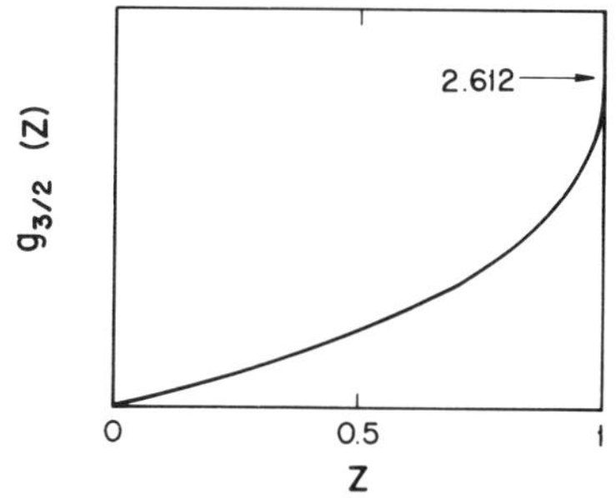

Fig. 4.1. Riemann zeta function for $n = 3/2$.

ticles in the noncondensed state. The second term is the density of particles in the ground state, N_0/V. Rearranging (4.13) to give an explicit expression for this quantity,

$$\frac{N_0}{V} = \frac{N}{V} - \frac{1}{\lambda_T^3} g_{3/2}(z) \tag{4.16}$$

In the above form, it is apparent that for the population of the zero momentum state to be finite, the right-hand side must be greater than zero. This requirement establishes the occurrence of a transition in a Bose gas. The transition temperature T_c is defined as the highest temperature that can sustain nonzero N_0/V, $g_{3/2}(1)$. Substituting this value and rearranging an expression for T_c, we obtain

$$T_c = \frac{h^2}{2\pi mk} \left(\frac{1}{v_c \, g_{3/2}(1)} \right)^{2/3} \tag{4.17}$$

where $v_c = (V/N)_{T_c}$ is the critical specific volume. It is important to note that the population of the zero momentum state is a continuous function and not until the system is at absolute zero do all the particles occupy this state. By combining (4.16) and (4.17) it is easy to show that the ground-state population is a temperature-dependent quantity normalized only by the critical temperature:

$$\frac{N_0}{N} = 1 - \left(\frac{T}{T_c} \right)^{3/2} \quad \text{for} \quad T \leqslant T_c \tag{4.18a}$$

$$= 0 \quad \text{for} \quad T > T_c \tag{4.18b}$$

This result is shown graphically in Fig. 4.2. Note that although the actual

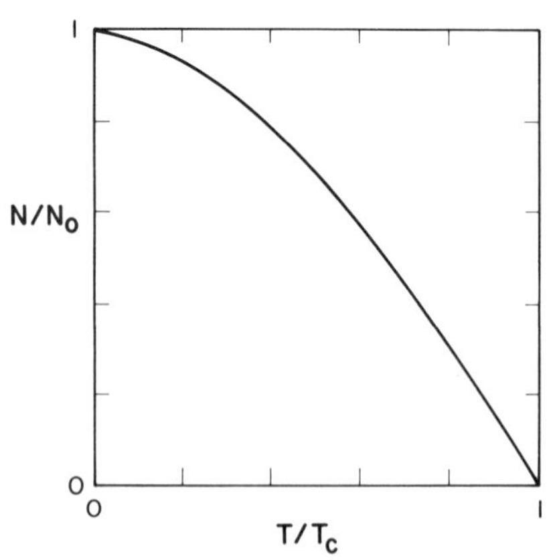

Fig. 4.2. Fractions of particles in the ground state as a function of reduced temperature for an ideal Bose gas.

value for N_0 varies continuously through the transition, there is a discontinuity in dN_0/dT at T_c.

4.1.2. Internal Energy

The internal energy E of an ideal Bose gas can be computed similarly. In terms of the discrete system, the expression for E is given in (4.4b) as a sum over the product $n_i\varepsilon_i$ for each of the particle states. This form is converted by means of the continuum approximation to an integral over the density of states function $D(\varepsilon)$:

$$E=\frac{2\pi}{h^3}(2m)^{3/2}V\int_0^\infty \frac{e^{3/2}\,d\varepsilon}{z^{-1}e^{\beta\varepsilon}-1} \tag{4.19}$$

It is important to note here that there is no longer a problem with the zero momentum state because although the population of the $p=0$ state may be finite the energy of that state is identically zero by definition.

Because the internal energy has a different form dependent on whether it is below or above T_c, it is useful to evaluate its behavior as a function of temperature. Substituting the quantities introduced as part of the density expression into (4.19), the internal energy may be written

$$\frac{E}{N}=\frac{3}{2}\left(\frac{1}{\lambda_T}\right)^3 vk_BTg_{5/2}(1) \qquad \text{for} \quad T\leqslant T_c \tag{4.20a}$$

$$=\frac{3}{2}\left(\frac{1}{\lambda_T}\right)^3 vk_BTg_{5/2}(z) \qquad \text{for} \quad T>T_c \tag{4.20b}$$

where $g_{5/2}(z)$ is the same Riemann zeta function defined in (4.15). In integral form,

$$g_{5/2}(z)=\frac{4}{3\sqrt{\pi}}\int_0^\infty \frac{x^{3/2}\,dx}{z^{-1}e^x-1} \tag{4.21}$$

As with $g_{3/2}(z)$, the above quantity is a smoothly varying function from 0 to its maximum value at $z\to 1$ (Fig. 4.3).

At this point, it is instructive to check the correspondence principle as it applies to the high-temperature limit of a Bose gas. Referring to the discrete form of the Riemann zeta function (4.15), we see that at high temperatures $g_n(z)\approx z+\cdots$ for all n. Substituting the leading term into the equations for the density and internal energy of a Bose gas, we obtain the classical limit $E/N=\frac{3}{2}k_BT$. Therefore, the theory as developed to this point has the proper high-temperature limit.

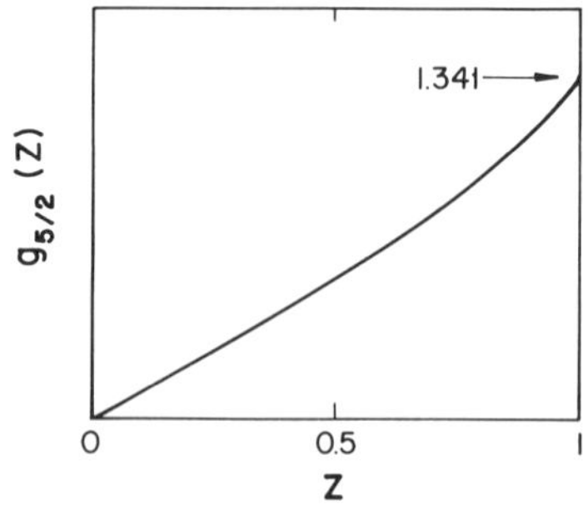

Fig. 4.3. Riemann zeta function for $n = \frac{5}{2}$.

4.1.3. Specific Heat

It is not possible to measure directly the internal energy of gas; however, the above theory can be used to predict measurable properties. One such property is the specific heat. The constant volume specific heat C_v is related simply to the internal energy through its temperature derivative, $C_v = (\partial E/\partial T)_v$. Therefore, taking derivatives of (4.20) with respect to temperature yields

$$\frac{C_v}{Nk_B} = \frac{15}{4} g_{5/2}(1) \frac{v}{\lambda_T^3} \qquad \text{for} \quad T \leqslant T_c \tag{4.22a}$$

$$= \frac{15}{4} g_{5/2}(z) \frac{v}{\lambda_T^3} - \frac{9}{4} \frac{g_{3/2}(z)}{g_{1/2}(z)} \qquad \text{for} \quad T > T_c \tag{4.22b}$$

The shape of the constant volume specific heat is shown graphically in Fig. 4.4. It is important to note three features in the form of the specific heat of an ideal Bose gas. First, at high temperatures $T \gg T_c$, $g_n(z) \to z \approx \lambda_T^3/v$, yielding the classical result that $C_v/Nk_B \to \frac{3}{2}$. This is to be expected based on similar arguments about the behavior of the internal energy. Second, below T_c the Riemann zeta function takes on a constant value and the specific heat is determined by the temperature dependence of the thermal de Broglie wavelength. Therefore, C_v goes to zero as $T^{3/2}$. It is possible to think of this behavior in a different way. Since the ground state has no energy of excitation, C_v is only a measure of the fraction of molecules in the excited states. Since this fraction decreases as $T^{3/2}$, it follows directly that the specific heat should have a similar dependence. The third aspect to the ideal Bose gas specific heat is that of its transition. The form as developed above does not predict a discontinuity in value of C_v but does show a cusp characteristic of a discontinuity of slope. A similar set of calculations for the constant pressure heat capacity C_p yields results that are not altogether different. As expected at high temperatures, the form of C_p approaches

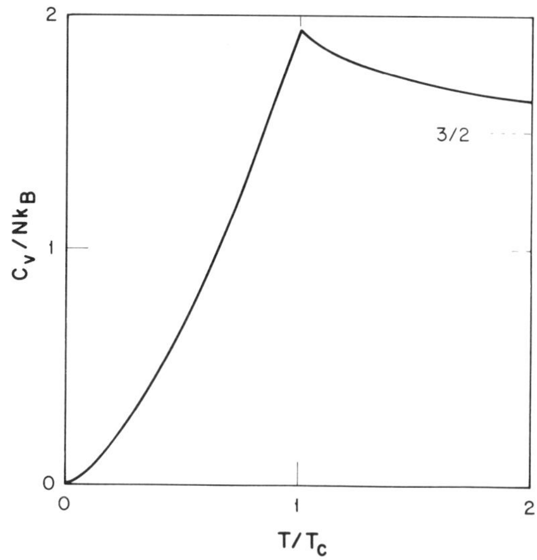

Fig. 4.4. Specific heat constant volume of an ideal Bose gas.

$5Nk_B/2$, the classical limit. However, at the transition the value of C_p shows a discontinuity consistent in form to the behavior of a first-order phase transition.[3]

4.1.4. Vapor Pressure

The discontinuity in specific heat is analogous to that of a first-order phase transition. Another question to ask might be: What is the behavior of the pressure over the same range of temperature? If there is a first-order phase transition, two phases should coexist although the phases would be of different character to classical liquid–gas coexistence. In the case of an ideal Bose gas the two phases would be the vapor state and the condensed vapor state.

To develop an expression for the equilibrium pressure for a Bose gas, it is best to use the fact that the pressure–volume product is simply related to the internal energy for all ideal gases through the expression

$$pv = \frac{2}{3}\frac{E}{N} \tag{4.23}$$

It simply follows that the equilibrium pressure can be written

$$\frac{p}{k_B T} = \frac{1}{\lambda_T^3}\, g_{5/2}(1) \qquad \text{for} \quad v \leqslant v_c \tag{4.24a}$$

$$= \frac{1}{\lambda_T^3}\, g_{5/2}(z) \qquad \text{for} \quad v \geqslant v_c \tag{4.24b}$$

where v_c is the critical specific volume. Note that according to (4.24a) the pressure of an ideal Bose gas is zero at absolute zero. A more subtle fact to consider is that the pressure for $v < v_c$ (4.24a) is independent of the actual volume. This fact can be seen in Fig. 4.5, which is a p-v diagram for an ideal Bose gas. The region below T_c has a horizontal vapor pressure curve indicative of a two-phase region. Two-phase coexistence normally implies the transition is first order; however, a useful way of checking this fact is to compute the latent heat of transformation.

4.1.5. Latent Heat

If the phase transition associated with condensation in the ideal Bose gas is first order, then it should have an associated latent heat. The latent heat λ of transformation is given in terms of the Clausius–Clapeyron equation,

$$\left(\frac{dp}{dT}\right)_{T_c} = \frac{\lambda}{T\,\Delta v} \tag{4.25}$$

where the vapor pressure derivative is taken at the transition temperature and Δv represents the specific volume difference between the two states. Since the condensed phase can be assumed to occupy zero volume, the volume change is the specific volume at the critical temperature,

$$\Delta v = v_c = \frac{\lambda_T^3}{g_{3/2}(1)} \tag{4.26}$$

Taking the derivative of the vapor pressure as given by (4.24a), the expression for the Clausius–Clapeyron equation becomes

$$\left(\frac{dp}{dT}\right)_{T_c} = \frac{1}{T_c v_c}\left[\frac{5}{2} k_B T \frac{g_{5/2}(1)}{g_{3/2}(1)}\right] \tag{4.27}$$

where the quantity in brackets can be taken as the latent heat of transfor-

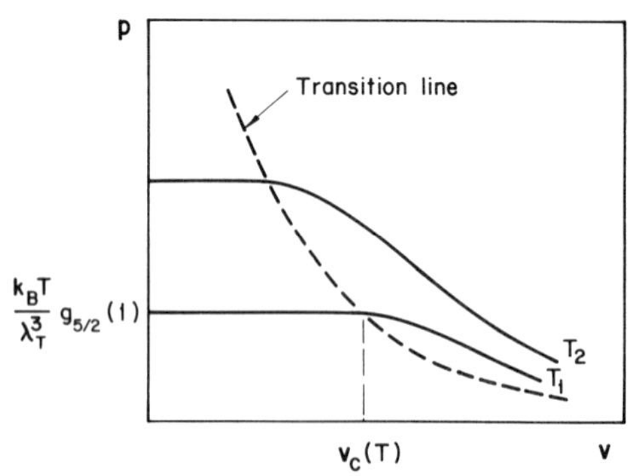

Fig. 4.5. Vapor pressure isotherms of an ideal Bose gas.

mation, $\lambda = 1.284k_BT$, on a per molecule basis. On an absolute scale, this energy is quite large, being comparable to the internal energy of an ideal classical monatomic gas at the same temperature.

The theory of an ideal Bose gas is instructive to consider because it shows how quantum mechanics affects the behavior of an ideal gas. In nature, liquid ^{4}He is the only Bose system known to exist at low temperatures and its behavior is not that close to an ideal gas. Liquid ^{4}He is known to undergo a phase transition to a condensed state, referred to as He II, at a temperature $T_\lambda = 2.172$ K under saturated vapor pressure. Substituting numerical values for ^{4}He in the equation for T_c, (4.17), we can predict a transition to occur at 3.14 K. Although T_c and T_λ are fairly close in value, the correspondence should not be taken as indicating that T_λ marks the beginning of Bose–Einstein condensation in liquid ^{4}He. This is because there are several dissimilarities between the behavior of an ideal Bose gas and liquid ^{4}He. First, the transition in liquid ^{4}He is known to be second order. It is marked by a discontinuity in slope of the entropy. The specific heat has a logarithmic infinity in value at T_λ. Consequently, the Bose–Einstein interpretation of the λ-transition appears to be an oversimplification in that the order of the transitions is not the same. The other noted difference occurs in the low-temperature properties. Below T_c the Bose gas has a specific heat that goes as $T^{3/2}$. This is markedly different from the behavior of liquid He II, which has a specific heat that goes as T^3 or higher.

Although there are several noted differences between the behavior of an ideal Bose gas and that of liquid ^{4}He, there is little doubt that the transition between He I and He II is brought on by Bose–Einstein quantum statistics. He II has a number of unique properties that make it very different from conventional fluids. As will be seen in subsequent sections, many of these properties can be interpreted in terms of physical models that rely heavily on the existence of a condensed phase. The physics of this condensed phase is clearly more complex than that of the condensed Bose gas. This is due, at least in part, to the fact that the excitation spectrum and interactions of the He II system are more complex than assumed in the ideal Bose gas.

4.2. LIQUID He II PROPERTIES

The behavior of liquid He II is considerably different from that of an ideal Bose gas. This fact is at least in part due to its being in the condensed liquid state and therefore having non-negligible interparticle interactions. When considering theoretical descriptions of He II, it is necessary to extend the models beyond that of the ideal gas. Two major features to the theory must be included: (1) a description of the excitation spectrum and (2) a fluid mechanics model that can be used to explain experimental results. The first component of the theory, although having relatively little engineering application, is most effective at interpreting such physical properties as specific heat and entropy. The latter model emphasizes understanding of the transport properties such as viscosity and thermal conductivity. For He II these properties are quite distinct compared to He I or other ordinary liquids.

To provide a better basis for the theoretical description, it is beneficial to begin with survey of the properties of He II. In performing this survey, the emphasis is placed on those properties that are particularly important for the applications. For a further description of other He II properties, the reader should consult one of several references on the physics of liquid helium.[4–6]

As has already been discussed in the introduction to the previous chapter, liquid helium can exist in either of two phases. Normal helium, or He I, exists above T_λ and has properties similar to other monatomic liquids. Superfluid helium as it is commonly called, or He II, is present below T_λ and has a number of outstanding features particularly associated with its transport properties. The λ-line separates these two liquid states as shown in Fig. 3.1, the phase diagram of helium. The λ-line stretches between the liquid–vapor phase transition at low pressure and the solid-liquid phase boundary at high pressure ($p \approx 3$ MPa). The actual values of the pressure, temperature, and density of liquid helium at the λ-transition ae listed in Table 4.1. The transition between these two states of the liquid has associated with it a change in slope of the entropy. Thus, the λ-trans-

Table 4.1. λ-Transition as a Function of Temperature, Pressure, and Density[a]

T (K)	p (bar)[b]	ρ (g/cm^3)
2.172	0.0497	0.1462
2.163	1.00	0.1483
2.15	2.33	0.1506
2.10	6.84	0.1576
2.05	10.91	0.1627
2.00	14.68	0.1669
1.95	18.22	0.1705
1.90	21.55	0.1736
1.85	24.70	0.1763
1.80	27.67	0.1788
1.763	29.74	0.1804

[a] From McCarthy.[7]
[b] 1 bar = 100 kPa = 750 torr.

ition is a second-order phase transition as defined by Ehrenfest and has no latent heat of transformation. The physical significance of a second-order phase transition here is that the two phases cannot coexist in equilibrium. This fact has limited practical significance because near the λ-transition thermal relaxation times are long and small deviations from equilibrium are achieved easily in actual experimental systems.

It is the goal of the present section to survey the properties of He II which are of greatest technical significance. These properties can be categorized into two principal types: (1) state properties such as entropy and specific hear and (2) transport properties. As will become evident, this is a natural separation in light of the theoretical description of He II.

4.2.1. State Properties

The state properties that are particularly useful for practical applications are those of specific heat, entropy, and density. The former two are also quite important for further understanding of the He II excitation spectrum. Consider the specific heat, shown graphically in Fig. 4.6 on a log–log plot. In presenting the specific heat in this form, several interesting features can be seen in addition to the λ-transition. At very low temperatures, $T \lesssim 0.6$ K, the specific heat appears to obey a cubic dependence,

$$C = (0.0204 \pm 0.0004)\, T^3 \text{ J/g} \cdot \text{K} \tag{4.28}$$

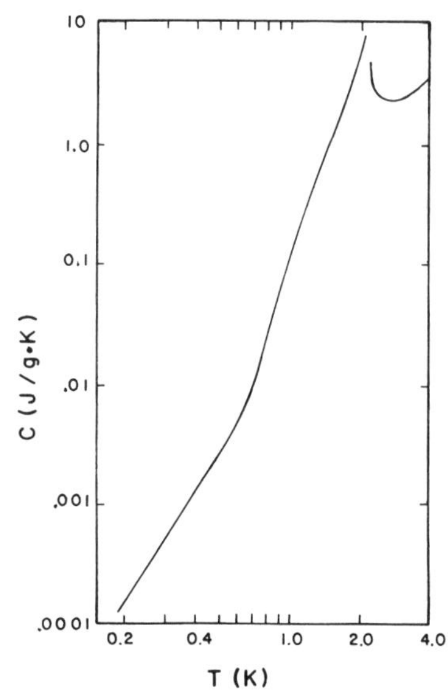

Fig. 4.6. Specific heat of liquid helium under saturated vapor pressure.

where C refers to the specific heat at saturated vapor pressure. At low temperatures the differences between C, C_p, and C_v are small. The physical justification for this dependence is presented in the next section; however, at this point it is worth noting that the low-temperature specific heat of a phonon gas is also proportional to T^3. This effect is important supportive evidence for the Landau theory of He II, which is based on an excitation spectrum dominated by longitudinal phonons at $T \leqslant 0.6$ K.

At higher temperatures, the apparent power law to the specific heat increases by about a factor of 2. Historically, the region has been divided into two where the temperature dependence goes as

$$C = 0.108 T^{6.7} \text{ J/g} \cdot \text{K} \quad \text{for} \quad 0.6 < T < 1.1 \text{ K} \tag{4.29a}$$

$$= 0.117 T^{5.6} \text{ J/g} \cdot \text{K} \quad \text{for} \quad 1.1 \leqslant T \leqslant 2.17 \text{ K} \tag{4.29b}$$

In the Landau theory a new excitation termed the "roton" is introduced to explain this experimental evidence. Clearly, the above behavior is unique to liquid helium.

The entropy shown in Fig. 4.7 exhibits behavior similar to that of the specific heat. At the λ-transition, the discontinuity of slope can be seen quite clearly, indicating a second-order phase transition. Based on the general theory of phase transitions introduced by Ehrenfest, a first-order phase transition has associated with it both a change in entropy and molar

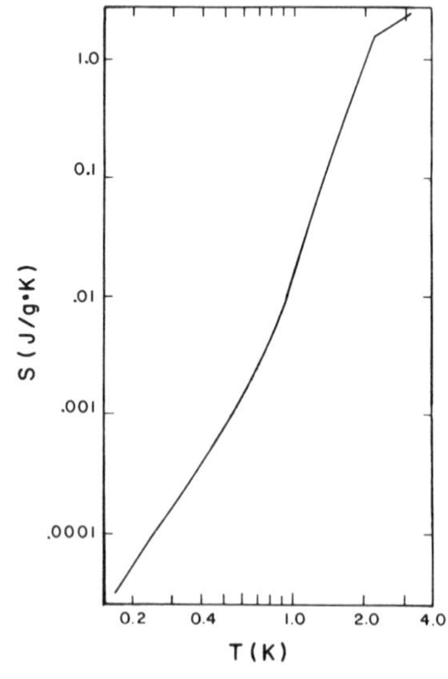

Fig. 4.7. Entropy of liquid helium under saturated vapor pressure.

volume. These quantities are related through the Clausius–Clapeyron equation,

$$\left(\frac{dp}{dT}\right)_{T_c} = \frac{\Delta s}{\Delta v} \tag{4.30}$$

where the changes in s and v occur at the transition. The above expression follows directly from the continuity of the Gibbs potential at the transition. A second-order phase transition must have both continuity of Gibbs potential and entropy. Therefore, the analogous expression to (4.30) becomes

$$\left(\frac{dp}{dT}\right)_{T_c} = \frac{\Delta C_p}{Tv\,\Delta\beta} = \frac{\Delta\beta}{\Delta\kappa} \tag{4.31}$$

where β is the bulk expansivity and κ the bulk isothermal compressibility. It follows that for a second-order phase transition C_p, β, and κ are discontinuous. As discussed in Chapter 3, these effects are observed in the properties of liquid helium near the λ-transition.

Finally, the density of He II as shown in Fig. 4.8 displays some interesting characteristics. At saturated vapor pressure, the density goes through a broad maximum with a discontinuity in slope at 0.006 K above the λ-transition. Below that temperature, He II has a negative thermal expansion coefficient. The density of He II under saturated vapor pressure

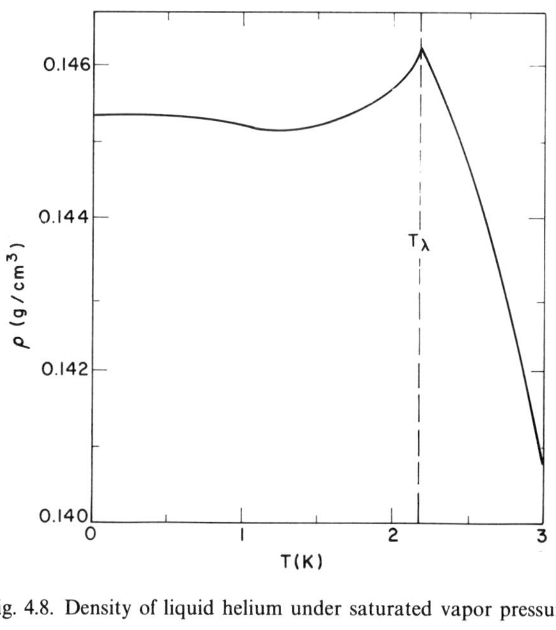

Fig. 4.8. Density of liquid helium under saturated vapor pressure.

approaches a constant value of 0.1453 g/cm^3 which is only 0.55% less dense than at its maximum but is 16% more dense than He I at its normal boiling point, 4.2 K, 1 bar where $\rho = 0.125$ g/cm^3.

The existence of a negative thermal expansion coefficient leaves open some interesting potential applications. One possibility that has been investigated to some degree is that of performing isentropic compression of the fluid to produce cooling. Consider, as an example, the effectiveness of isentropic compressive cooling of He II initially at 2.100 K and saturated vapor pressure. Isentropic compression is achieved by applying pressure to a piston while the system is isolated from the environment. For numerical calculations take the final pressure to be 2.5 bars. The statement of isentropic compression equates the specific entropies:

$$s(T_f, 2.5 \text{ bars}) = s(2.1 \text{ K}, \text{SVP}) \tag{4.32}$$

As a result of this compression, the final temperature should be 2.084 K, which represents a 0.016 K reduction in temperature, a small effect.

4.2.2. Transport Properties

The behavior of He II when subjected to a mass or heat flow clearly demonstrates the unique character of this fluid. The observed effects cannot be treated in terms of classical fluid mechanics. In considering He II under mass flow conditions, the most obvious property to evaluate is the form of the viscosity. For He II this property depends on the method employed to measure it. The apparent viscosity of He II can be measured by essentially two different methods: (1) flow through channels and (2) rotating viscometers. At the outset, it should be noted that both methods yield essentially the same results for ordinary liquids including He I.

To measure the viscosity of He II flowing through a channel, a pressure gradient must be established and a mass flow measured. Since the viscosity is extremely small, very fine channels must be used. Experimental evidence shows that almost independent of channel dimension the viscosity becomes immeasurably small just below the λ-transition. This result is contrary to that observed in rotating viscometers. Here the measurement consists of a change in moment of inertia of a disk rotated in the fluid. For He II, the viscosity measured by a rotating viscometer obeys a more conventional form, actually increasing with decreasing temperature. It is an important test of the theory of He II to describe these different mass transport behaviors.

The heat transport character of He II is also unique. In general, the apparent thermal conductivity of He II is extremely large, being at least several orders of magnitude greater than other liquids or even high-conductivity metals. Furthermore, the heat conductivity cannot be described in terms of a bulk thermal conductivity. This behavior can be seen most clearly by reference to a specific experimental measurement. For a channel containing He II, there exist two heat flow regimes, which are shown schematically in Fig. 4.9. For low heat fluxes up to q_c, the helium obeys a heat transport relationship that has a temperature gradient proportional to heat flux. However, unlike thermal conductivity, the proportionality constant in this regime has a very strong diameter dependence, with the helium getting less conductive as the diameter decreases. Further in this heat flow regime a pressure difference is set up proportional to the temperature difference,

$$\Delta p = \rho s \, \Delta T \tag{4.33}$$

where the proportionality constant is the product of density ρ and specific entropy s. This relationship between pressure and temperature gradients gives rise to the fountain effect which causes He II to flow through very small channels driven by a temperature difference.

Above the critical heat flux q_c, He II enters another heat transport regime where the temperature gradient is now proportional to a higher power of the heat flux and independent of channel dimensions. For reasons that will become apparent later in this chapter, this regime is generally referred to as turbulent heat transport. It should be noted that the transition point between the two heat flow regimes in He II is diameter dependent.

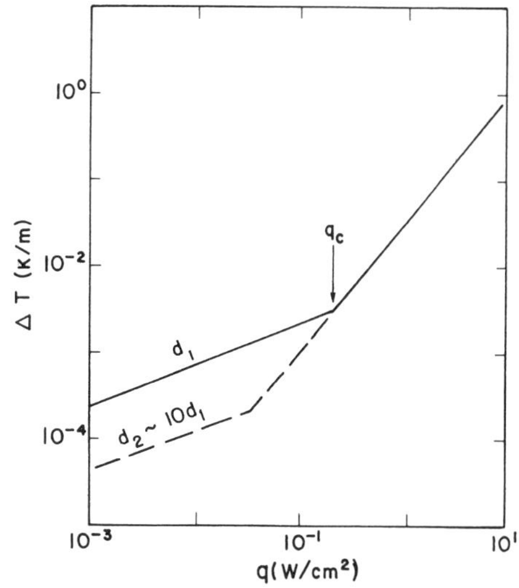

Fig. 4.9. Temperature gradient in a He II filled channel as a function of heat flux density q. The critical heat flux q_c represents a transition between laminar and turbulent flow. Note $d_1 \sim 1\,\text{mm}$ if $T \simeq 1.9\,K$.

4.2.3. Fountain Effect

In addition to the above described heat flow behavior, there are other unique features to the transport properties of He II. One, commonly referred to as the fountain effect, results from the noted relationship between the pressure and temperature difference in He II, (4.33). This effect can have quite spectacular results. Consider, for example, the idealized experiment shown in Fig. 4.10. A vessel containing He II is placed in a bath also of He II initially at the same temperature. A porous material forms a semipermeable separation between the two vessels and the inner vessel is instrumented with a heater. In equilibrium, all temperatures and pressures are equal. However, if heat is applied to the inner vessel, its temperature rises. As a consequence of the unique properties of He II described by (4.33), the pressure rises in the center vessel and helium will squirt out the top channel. Thus, the name "fountain effect" is justified. Although this effect is quite fantastic, it is based on sound physical principles in the theory of He II.

Related to the fountain effect is the existence of what is known as helium film flow. It is a known physical phenomenon that liquids at saturation will form multilayer films adsorbed on surfaces in the vicinity of the bulk liquid. What is unique about He II is that this saturated film, which is typically a few micrometers thick, is mobile. Mobility of the He II film permits the liquid to flow under the influence of the fountain pressure. Again consider an idealized experiment. Assume a dewar of He II contains a separate, smaller vessel with an open top above the bath surface; see Fig. 4.11. As the inner vessel is lowered into the liquid a pressure difference is established due to hydrostatic head, $\Delta p = \rho g \, \Delta h$. But all surfaces are coated with helium film, so that the fluid will flow into the inner vessel. The rate at which this event occurs depends on the details of the experimental

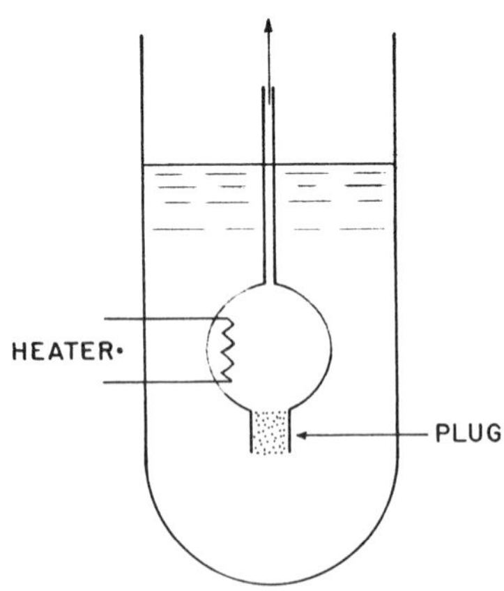

Fig. 4.10. Experiment to demonstrate the fountain effect.

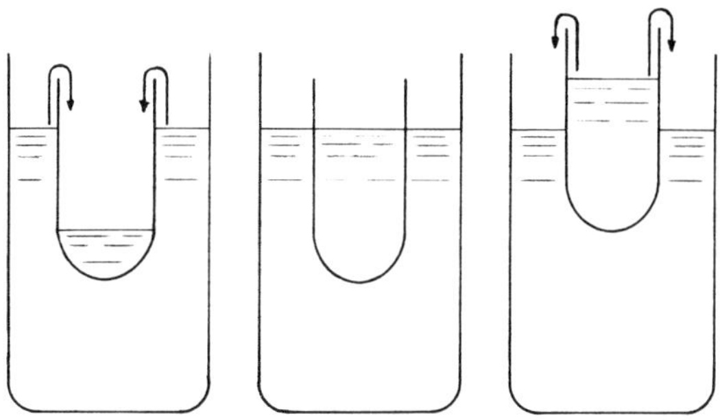

Fig. 4.11. Schematic of a film flow experiment.

conditions; however, it is generally fast enough to see the inner level increase. If the inner vessel is lifted out of the helium bath, the hydrostatic pressure difference works in the opposite direction, causing the liquid to flow out.

The existence of helium film flow was first observed by Kamerlingh Onnes in 1922; however, he interpreted it in terms of an evaporation and condensation mechanism. In 1937, Rollin and Simon were able to interpret the film flow phenomenon based on the theory of He II. As a result, this film flow process is sometimes referred to as the Rollin film. Considerable research has been conducted to expand the understanding of helium film flow. Factors that affect the thickness of the film include height above the helium bath, substrate material, and temperature. Empirical relationships have been established which relate the film thickness d with the height H above the helium bath:

$$d = \frac{K}{H^n} \quad \text{(cm)} \tag{4.34}$$

where $K \simeq 3 \times 10^{-6}\,\text{cm}^{n+1}$ and n is between 0.3 and 0.45. For numerical values, note that the film at about 1 cm above the liquid is approximately 30 nm (300 Å) thick, which corresponds to around 80 atomic layers. Mobility of films of He II has been demonstrated down to thickness of a few atomic layers (see Section 9.3).

The practical significance of the Rollin film is that in addition to mass flow through the film there can also be heat flow. The existence of heat flow through the mobile He II film can affect strongly the system performance. One particular example is the pumping work required to maintain He II at constant temperature. If the helium is contained in a vessel with an opening for pumping at the top, the film will climb the walls of the container. Heat will be conducted between the point on the container which is at T_λ and the bath. If the opening to the container is large, this heat flow can cause a sizable heat inflow to the bath. The best method to reduce this heat inflow is to restrict the free surface of helium on which the pump is working. This will reduce proportionally the helium film flow.

4.3. EXCITATIONS IN He II

Until the 1930's, the unique properties of He II had little theoretical basis. London was the first to point out the close correspondence between the behavior of a Bose gas and He II. This observation resolved only part of the problem because the gas models do not adequately describe the transport properties. Furthermore, the state properties of He II are quite different from those predicted by the Bose gas model, both in terms of the order of the transition and the low-temperature specific heat.

The first attempt to develop a more complete theory of He II was introduced by Tisza in 1938. His work, which is now called the two-fluid model, suggested that liquid He II consists of two interpenetrating fluids. A similar model has also been used by Gorter and Cashmir to describe the behavior of superconductors. Although the two-fluid model was chronologically the first theory to advance significantly the understanding of He II, it does not address satisfactory the state properties. In 1941, Landau introduced a theory of excitations for He II which successfully explained the behavior of such properties as the specific heat and entropy.

Landau approached the problem of He II by introducing an excitation spectrum that might consist of a number of different modes. The spectrum is semiempirical in that it is adjusted to predict experimental data in a consistent manner. Two types of excitation were introduced as part of the model, each having different regions of significance. First, like all liquids, He II is envisioned to possess longitudinal phonon excitations. Transverse phonons can exist only in the solid state. These longitudinal phonons are characterized by a velocity c_l, such that the energy spectrum can be written explicitly,

$$\varepsilon = c_l P \tag{4.35}$$

where P is the momentum of the phonon given explicitly by $P = hk$ with k being the wave number.

The other excitation mechanism unique to He II is referred to as the roton. The name "roton" originally was attached to this mechanism because of the feeling that it may represent some quantized circulation. This possibility is of no practical significance. It is most important to note that the roton excitation brings about a higher-level excitation in the He II spectrum. It is characterized by a minimum in the spectrum and has the general form

$$\varepsilon = \Delta + \frac{(P - P_0)^2}{2\mu} \tag{4.36}$$

where Δ is the minimum energy to excite one roton and P_0 and μ are empirically determined quantities.

Combining the two excitation mechanisms into one continuous spectrum, the result is the dispersion curve shown in Fig. 4.12. The regions described by the two types of excitation are also shown in the curve. Based on experimental data, the parameters associated with the roton minimum are

$$\frac{\Delta}{k_B} = 8.65 \pm 0.04 \text{ K}$$

$$\frac{P_0}{h} = (1.91 \pm 0.01) \times 10^8 \text{ cm}^{-1}$$

$$\mu = 0.16\, m_4$$

where m_4 is the atomic mass of a helium atom.

The test of the theory is in its ability to predict experimental evidence. Considering the phonon part of the spectrum, it is relatively easy to

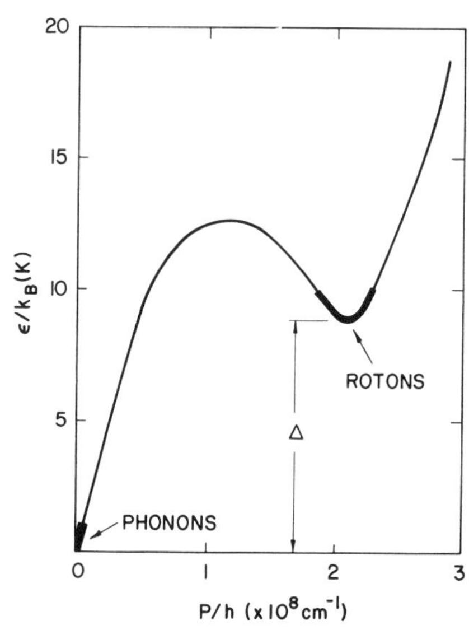

Fig. 4.12. Dispersion curve indicating the elementary excitations of the Landau theory.

calculate the specific heat. Using the Debye model for longitudinal phonons, we can calculate the specific heat at low temperatures as

$$C_v = \frac{2\pi^2 k_B^4}{15\rho h^3 c_l^3} T^3$$
$$= 0.0207 T^3 \text{ J/g} \cdot \text{K} \tag{4.37}$$

This value is an extremely good prediction of experimental results given by the empirical correlation in (4.28).

A further check for the existence of phonon excitations in He II is through the behavior of the thermal conductivity at low temperatures. For $T \lesssim 0.6$ K, the normal fluid density is nearly zero so the heat is not transported by the internal convection mechanism. Rather, the thermal conductivity goes as T^n where n is around 3. Similar behavior occurs in a dielectric solid at low temperatures where phonon transport is the main heat conduction mechanism.

The roton excitations in He II dominate the physical behavior for high temperatures, $T \gtrsim 1$ K. As with phonons, rotons are collective excitations in the liquid. Since the roton region of the dispersion curve is at high temperatures, the physical processes obey Boltzmann statistics. Applying Landau's dispersion relationship to the He II, we can calculate the expressions for the thermodynamic properties. As expected, these expressions are exponential in the roton energy gap Δ. In particular, the free energy,

$$F_r = -\frac{2\mu^{1/2}(k_B T)^{3/2} p_0^2}{(2\pi)^{3/2}\rho\hbar^3} e^{-\Delta/k_B T} \tag{4.38}$$

the entropy,

$$S_r = \frac{2(k_B\mu)^{1/2} p_0^2 \Delta}{(2\pi)^{3/2}\rho\hbar^3 T^{1/2}} \left(1 + \frac{3k_B T}{2\Delta}\right) e^{-\Delta/k_B T} \tag{4.39}$$

and the specific heat contribution,

$$C_r = \frac{2\mu^{1/2} p_0^2 \Delta^2}{(2\pi)^{3/2}\rho\hbar^3 k_B^{1/2} T^{3/2}} \left[1 + \frac{k_B T}{\Delta} + \frac{3}{4}\left(\frac{k_B T}{\Delta}\right)^2\right] e^{-\Delta/k_B T} \tag{4.40}$$

are functions of $k_B T/\Delta$. It is important to note that all expressions have an exponential cutoff at low temperatures, indicating a relatively small roton contribution for $T \ll 1$ K.

It is somewhat difficult to compare the theoretical expression for the roton contribution to the specific heat to the empirical expression for the specific heat, which gives $C \approx T^{5.6}$. Since the specific heat is dominated by rotons only between 1.0 and 2.1 K, the distinction between the power law and an exponential form is not large. Nonetheless, it should be noted that the variation of specific heat near T_λ is one method to measure the roton minimum, Δ, indicating a fairly close correspondence with (4.40). The empirical expression (4.29) is used mostly for calculations of an engineering and applied nature.

4.4. TWO-FLUID MODEL

Although the theory of excitations in He II advanced by Landau and others has considerable physical basis, it lacks a usefulness for attempting to understand some of the unique characteristics of He II. Of greatest note are the transport properties of He II, which cannot be interpreted easily in terms of this theory. Fortunately, in 1938 Tisza developed another phenomenological theory of He II as a simple fluid of two components. Tisza's theory, which was actually developed a few years before Landau's theory of excitation, takes the state functions of entropy and specific heat as empirically determined quantities. The two-fluid model is a description of the fluid mechanics of He II, which is particularly successful at describing heat and mass transport.

The two-fluid model envisions the He II to be comprised of two interpenetrating fluids: normal fluid, which contains the excitations in the liquid, and the superfluid. At this point it is worth making a comment about nomenclature. Often in generic terms, He II is referred to as "superfluid" because this was the word used to describe its behavior before the existence of the two-fluid model. Subsequent to the introduction of the two-fluid model, it has been more correct to refer to liquid helium below T_λ as He II and the component in the two-fluid model as superfluid. This convention, which assumes some knowledge of the physics of He II, will be used throughout this book.

The normal fluid component to the He II is assumed to behave as an ordinary liquid. It is described by having a density ρ_n, viscosity η_n, and specific entropy s_n. On the contrary, the superfluid component has density ρ_s, no viscosity ($\eta_s = 0$), and no entropy ($s_s = 0$). The properties of the fluid must be comprised of a linear combination of the two components. The density is a sum of the two component densities,

$$\rho = \rho_s + \rho_n \tag{4.41}$$

Since the superfluid has no entropy, it follows that the He II entropy can be written in terms of the normal fluid component alone:

$$\rho s = \rho_n s_n \tag{4.42}$$

As was discussed in the previous section, the entropy of He II is strongly temperature dependent, going approximately as $T^{5.6}$ between 1 K and T_λ. For the two-fluid model, it is assumed that $s_n = s_\lambda$, the entropy at the λ-point, and that all the temperature dependence in (4.42) enters through the variation of the normal fluid density. It is therefore possible to write

$$\begin{aligned} \frac{\rho_n}{\rho} &= \left(\frac{T}{T_\lambda}\right)^{5.6} \quad &\text{for} \quad T \leqslant T_\lambda \\ &= 1 \quad &\text{for} \quad T \geqslant T_\lambda \end{aligned} \tag{4.43}$$

as the temperature dependence of the normal fluid density. Because of this strong temperature dependence, the He II is about 99% superfluid component at 1 K. The ratios of normal and superfluid densities to that of the bulk liquid are shown in Fig. 4.13. Recall that for He II the density has a rather weak temperature dependence so these curves closely correspond to the total density profile for each component.

Concerning the viscosity, the model assumes that $\eta_s = 0$ and the dissipative interaction is due only to the normal fluid. This assumption corresponds to the physical fact that the superfluid experiences no resistance to flow and therefore no turbulence. The superfluid can flow through a channel without viscous drag along the boundaries. The velocity of the fluid will be the same at the center of the channel as near the sides. The equivalent mathematical expression for this statement is to say that the superfluid velocity can have no curl:

$$\nabla \times \mathbf{v}_s = 0 \tag{4.44}$$

This description of the superfluid viscous interaction is only partially correct. Above a critical velocity $\mathbf{v}_{sc}$, the superfluid does experience an interactive mechanism but its velocity profile is still consistent with (4.44).

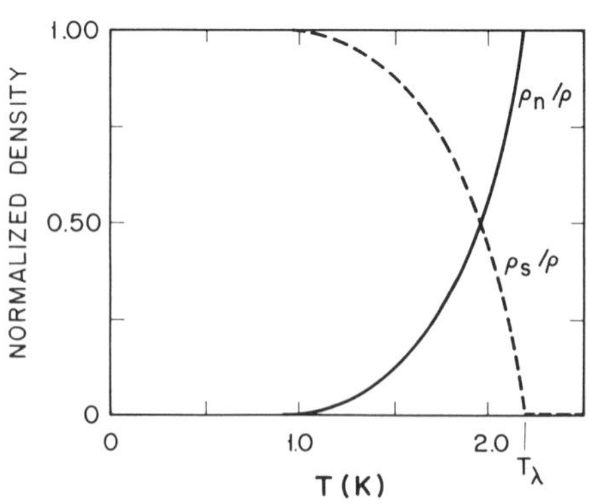

Fig. 4.13. Ratio of normal and superfluid densities in He II.

4.4.1. Equations of Motion

As described by the two-fluid model, He II can be thought to consist of two separate interpenetrating fluids. This picture should not be taken too literally. It is only a model and the existence of superfluid and normal fluid components is a hypothesis. Nonetheless, the existence of this model allows equations of motion to be written that describe the behavior of the fluid.

The momentum density $\mathbf{j}$ of He II can be written as a sum of each momentum density component,

$$\mathbf{j} = \rho\mathbf{v} = \rho_s\mathbf{v}_s + \rho_n\mathbf{v}_n \tag{4.45}$$

where $\mathbf{v}_s$ and $\mathbf{v}_n$ are the superfluid and normal fluid velocities. An interesting special case of (4.45) is when there is no net mass flow and $\mathbf{j} = 0$. Thus, the helium is static although the superfluid and normal fluid components can flow in opposite direction consistent with

$$0 = \rho_s\mathbf{v}_s + \rho_n\mathbf{v}_n \tag{4.46}$$

This expression leads to the notion of internal convection where the two fluids can flow and carry entropy in He II without experiencing macroscopic mass flow.

In addition to the expressions for density (4.41) and momentum (4.45), the two-fluid model relies heavily on conventional fluid mechanics which is used to describe the behavior of each component. The continuity equation describes the rate of change of the density in a given volume of space, that is,

$$\nabla\cdot\mathbf{j} = -\frac{\partial\rho}{\partial t} \tag{4.47}$$

where each component of He II must obey this relationship. Euler's equation describes the acceleration of an ideal fluid. An ideal fluid supports no shear stress even in motion and has negligible viscosity. As a consequence, the only force acting on the fluid is due to pressure gradients:

$$\frac{\partial\mathbf{v}}{\partial t} + \mathbf{v}\cdot\nabla\mathbf{v} = -\frac{1}{\rho}\nabla p \tag{4.48}$$

The above expression is analogous to Newton's second law applied to fluids; stated simply, it defines the rate of acceleration of a fluid subjected to a force per unit volume ∇p.

For the development of the two-fluid model it is instructive to reduce

(4.48) to a linear expression. Since the second term on the left-hand side goes as $\mathbf{v}^2$ for small velocities, it is neglected; this leads to

$$\frac{\partial \mathbf{j}}{\partial t} \simeq -\nabla p \tag{4.49}$$

The suitability of Euler's equation to describe the behavior of the superfluid component should appear obvious. By definition the superfluid does not support shear flow or have a viscous interaction. However, for the normal fluid there is an important viscous term that enters a number of physical processes including heat transport. In the description of these processes, more complex fluid mechanics are required in the form of the Navier–Stokes equation applied to the normal fluid.

Finally, it is necessary to develop an expression for the behavior of the entropy. Since the fluids described by Euler's equation experience no dissipative interaction, all processes are reversible. Reversibility of fluid flow processes implies that entropy must be conserved. It is therefore possible to write an expression for entropy conservation which is analogous to mass conservation in the continuity equation. Consider a volume containing helium with density ρ and specific entropy s; see Fig. 4.14). Here the entropy per unit volume is $\rho s = \rho_n s_\lambda$ since the normal fluid is the only component with nonzero entropy. Furthermore, the rate of flow of entropy into the volume is determined by the normal fluid velocity $\mathbf{v}_n$. This process can be treated in terms of an expression that relates the entropy content to the entropy flow:

$$\frac{\partial}{\partial t}(\rho s) = -\nabla \cdot (\rho s \mathbf{v}_n) \tag{4.50}$$

This expression is analogous to the continuity equation for mass conservation.

The first step in development of the equations of motion for He II is to consider the superfluid component. To do this, it is helpful to imagine an idealized experiment. Assume there exists a volume containing a number of

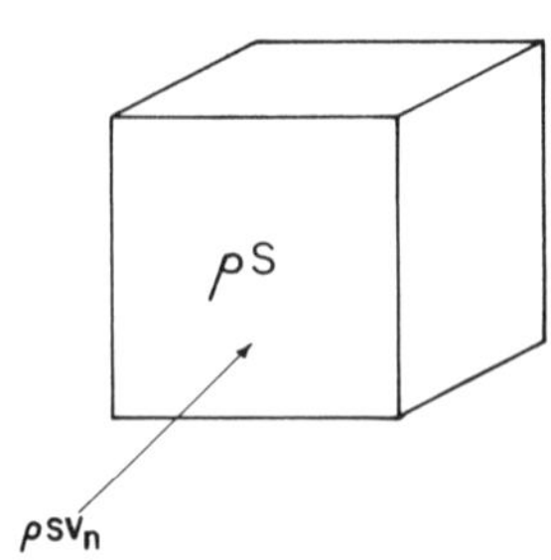

Fig. 4.14. Schematic volume containing He II.

helium atoms with a total mass m. A conceptual picture of this system is shown in Fig. 4.15. The volume is connected to another volume containing helium but there is a special valve between these two which only allows the superfluid component to flow. In practice, such a valve can be made by packing fine powder in a tube which clamps the normal fluid by viscous interaction but allows the superfluid to flow.

The thermodynamic function that is useful in describing the change in the internal energy of the system associated with a change in mass dm is the Gibbs potential per unit mass, g. Assuming the system is open only to the addition of superfluid, we can write the differential form of the first and second laws of thermodynamics as

$$de = T\,ds - p\,dv + g\,dm \tag{4.51}$$

Two initial assumptions can be used to simplify the format of (4.51). First, for a constant volume process, $dv = 0$, and the second term is eliminated. More importantly, the entropy is constant, $ds = 0$, since the fluid that is being added to the box is only superfluid with all particles in the ground state. Therefore, the change in internal energy is proportional to the change in mass, that is,

$$de = g\,dm \tag{4.52}$$

It is sometimes more helpful to think of this process in terms of a thermodynamic driving force that is the gradient of the potential. The acceleration of the superfluid is equal to the negative of the thermodynamic driving force:

$$\frac{d\mathbf{v}_s}{dt} = -\nabla g \tag{4.53}$$

This expression can be written in terms of common thermodynamic quantities by using the differential form of the specific Gibbs potential,

$$dg = -s\,dT + v\,dp \tag{4.54}$$

and since this is a total differential, (4.53) can be inserted to provide an equation of motion for the superfluid:

$$\frac{\partial \mathbf{v}_s}{\partial t} = s\,\nabla T - \frac{1}{\rho}\nabla p \tag{4.55}$$

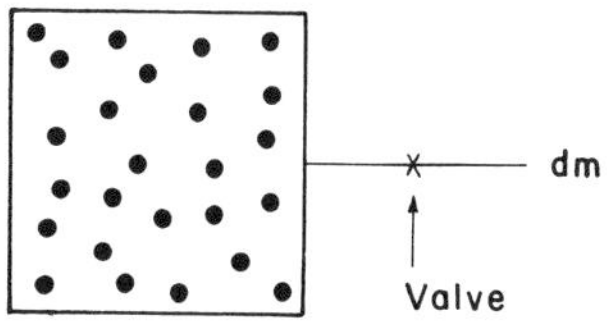

Fig. 4.15. Isentropic process involving He II.

Note that between (4.53) and (4.55) the velocity derivative has been changed from a total to partial derivative. This approximation, which is good to terms of order v^2, is useful in placing the equation in its well-known form. The above expression, (4.55), is of fundamental importance to understanding the behavior of He II. Although its existence seems to be based on an idealized example, it is extremely effective at predicting numerous physical aspects of He II including the fountain effect and sound propagation which are discussed next.

4.4.2. Thermomechanical Effect

The thermomechanical effect, or fountain effect as it is more commonly known, is predictable from the equation of motion for the superfluid component in the two-fluid model. As has already been discussed in the properties section of this chapter, there exists a relationship between temperature difference and pressure difference in He II. Furthermore, the proportionality factor is the volumetric entropy ρs. To see how this relationship originates, consider the experimental setup shown in Fig. 4.16. Two containers each filled with He II are connected together through a superfluid filter or fine capillary having a diameter of the order of 1 μm. The capillary allows superfluid to flow from one container to the other but the normal fluid is clamped by the viscous interaction to the walls of the channel.

In the steady state, the two containers have equal Gibbs potentials, and, furthermore, the fluid will not be accelerating; that is $\partial v_s/\partial t = 0$. It follows that by setting the left-hand side of (4.55) to zero, the fountain pressure relationship is derived:

$$\frac{dp}{dT} = \rho s \tag{4.56}$$

If, as in Fig. 4.16, the temperature on either side of the capillary is not

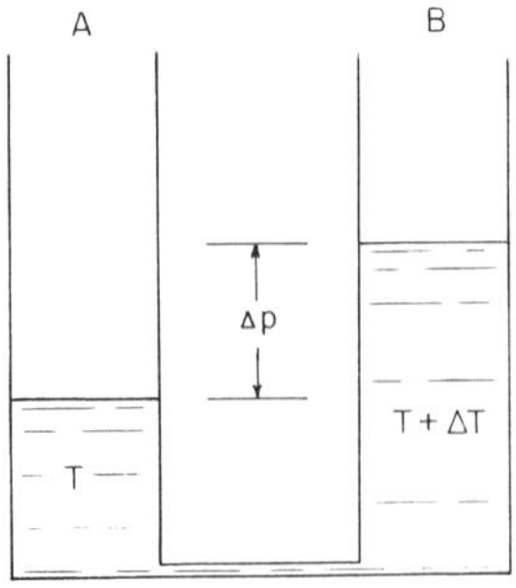

Fig. 4.16. Thermomechanical effect.

equal, there will be a flow of helium under the influence of the induced pressure difference. This process will continue until the hydrostatic head ($\rho g\,\Delta h$) equals the fountain pressure ($\rho s\,\Delta T$). It is worth noting the order of magnitude of this effect. At 1.8 K, $dp/dT = 0.6$ torr/mK or about 5% of the vapor pressure per millikelvin. Thus, a 1 mK temperature difference between the two sides in Fig. 4.16 will cause a 0.6 torr pressure difference which corresponds to about 6 cm of helium liquid.

A corollary to the thermomechanical effect is something called the mechanocaloric effect, which relates the heat flow brought on by physically forcing the helium from one side to the other. Consider the system as described above in equilibrium where $T_A = T_B$ and the two heights are the same. Now imagine that a movable, insulating piston on the left-hand side pushes the helium down and through the capillary; see Fig. 4.17. Since only the superfluid can pass through the capillary, the effect of the flow will be to reduce the entropy per unit volume on the right, *B*, and increase it on the left, *A*. But this is contrary to isentropic conditions since the higher level should have the greater entropy per volume because the temperature is higher. To bring this system back to equilibrium an amount of heat Q^* must be absorbed from the isothermal bath. Since the process is isentropic, the heat must originate from some point in the system. It turns out that Q^* is liberated by side *A* into the isothermal bath. As a result, the heat is transmitted through the isothermal bath while the mass flow is through the capillary! The quantity of heat that is transferred between baths is given in terms of the entropy,

$$Q^* = \rho s T \tag{4.57}$$

per volume of helium flowing between these two vessels.

As a final comment, the above thought experiment is definitely idealized. It requires two frictionless, massless, insulating pistons. This is because if the helium is in contact with its vapor, heat will also be transferred through the vapor phase by evaporation and condensation

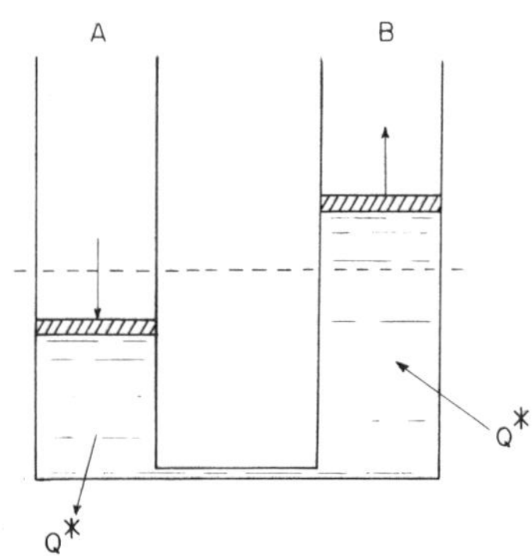

Fig. 4.17. Mechanocaloric effect.

processes. The lower-entropy fluid on the right will absorb heat from the vapor by condensation while the higher-entropy fluid on the left will reduce its temperature by evaporating some of the liquid into the vapor phase.

4.4.3. Sound Propagation

One of the most interesting characteristics of He II is ability to transmit more than one type of sound. In addition to ordinary or first sound, which is a density variation brought on by local pressure gradients, there is second sound which is propagation of thermal waves as a result of fluctuations in the local entropy. This latter transport mechanism is predicted by the two-fluid model.

To begin, consider briefly the origin of ordinary first sound. Sinusoidal wave propagation can be obtained analytically by combining the continuity equation (4.47) with Euler's equation for an ideal fluid with linear response (4.48). The result is a wave equation that associates pressure and density:

$$\nabla^2 p = \frac{\partial^2 \rho}{\partial t^2} \tag{4.58}$$

This differential equation has the standard solution of a sinusoidally varying function. The wave packet has a characteristic velocity c_1, the velocity of sound, given by

$$c_1^2 = \left(\frac{\partial p}{\partial \rho}\right)_s = \frac{\gamma}{\rho\kappa} \tag{4.59}$$

where the partial differentiation is taken at constant entropy and γ is the ratio of the specific heats, C_p/C_v. The velocity of first sound in liquid helium is of the order 200 m/s.

Second sound is the result of entropy fluctuation rather than density fluctuations as in ordinary sound. Beginning with the entropy conservation equation (4.50) and the equation of motion for the superfluid component (4.55) the pressure gradient may be written

$$-\nabla p = \rho\left(\frac{\partial \mathbf{v}_s}{\partial t} - s\,\nabla T\right) \tag{4.60}$$

The time rate of change of the momentum is equal to the negative of the pressure gradient. Taking the derivative of (4.45) and dropping all except linear terms in velocity, we obtain

$$\frac{\partial \mathbf{j}}{\partial t} = -\nabla p \simeq \rho_s \frac{\partial \mathbf{v}_s}{\partial t} + \rho_n \frac{\partial \mathbf{v}_n}{\partial t} \tag{4.61}$$

which combined with (4.60) results in an expression for the difference in the acceleration of the two fluid components:

$$\frac{\partial \mathbf{v}_n}{\partial t} - \frac{\partial \mathbf{v}_s}{\partial t} = -\frac{\rho}{\rho_n} s\,\nabla T \tag{4.62}$$

The above expression can be used to illustrate some important physical facts. In nondissipative He II, a temperature gradient produces a relative acceleration of the two fluid components. Because the relative densities of the two fluid components are related to the entropy, a temperature gradient implies a variation in ρ_s/ρ_n.

To make (4.62) approach more closely a wave equation, it is necessary to take the divergence of each side and then replace the superfluid velocity with the continuity equation. The results are simpler if it is assumed that the time rate of change of the total density is small, that is, $\partial\rho/\partial t \approx 0$. A relationship is obtained for the normal fluid velocity in terms of the second derivative of the temperature,

$$\frac{\rho_n}{\rho_s}\frac{\partial}{\partial t}(\rho\nabla\cdot\mathbf{v}_n) = -\rho s\,\nabla^2 T \tag{4.63}$$

The left-hand side may be written in terms of the fluid entropy by the conservation of entropy equation, (4.50). This result is reduced to a fairly simple form if the entropy does not vary significantly with time and position; that is,

$$\frac{\partial^2 s}{\partial t^2} = \frac{s^2\rho_s}{\rho_n}\nabla^2 T \tag{4.64}$$

which is referred to as the linearized second sound equation. It predicts a sinusoidally varying function propagating with velocity c_2 given by

$$c_2^2 = \frac{s^2\rho_s}{\rho_n}\left(\frac{\partial T}{\partial s}\right)_\rho \simeq \frac{\rho_s}{\rho_n}\frac{Ts^2}{C_v} \tag{4.65}$$

The expression for the velocity of second sound turns out to be quite accurate and is even used as a method of measuring the entropy of He II. Typical values for c_2 are around 20 m/s between 1 and 2 K.

It is worth noting at this juncture the methods available for measurement of second sound. As might be expected, these methods have analogues in the measurement of first sound, except that a travelling temperature wave is detected with a thermometer rather than a travelling pressure wave detected with a pressure transducer. There are principally two methods of measuring the velocity of second sound. The most obvious is the time-of-flight method shown schematically in Fig. 4.18a. A heater is located some distance away from the source providing a pulse that travels the length of the channel. The pulse is detected at some time later, $t = L/c_2$. This method, although effective, is cumbersome since the channel must be fairly long to have good resolution. Consequently, the time-of-flight method is not the most convenient for measuring c_2.

Second sound may also be measured by a resonance method. In this case, a channel is constructed with a closed end, as in Fig. 4.18b. Within the channel there is a heater that can be driven at variable frequency. Also installed in the channel is a movable thermometer located between the heater and closed end of the tube. The principle, which was originally employed by Peshkov[8] to determine c_2, is to drive the heater at a resonant frequency of the channel, $\nu = nc_2/2L$, where n is an integer. The temperature sensor is then moved along the tube, detecting the resultant standing wave. Besides simplicity, there are other advantages of this resonant method of detecting second sound. Since the geometry is less restricted, it is possible to make quite compact second sound detectors, which is desirable when measuring the attenuation of second sound in turbulent He II.

Second sound has been measured extensively everywhere from T_λ to near absolute zero and as a function of pressure.[8] Values for c_2 are compared to c_1 in Fig. 4.19. Note that c_2 is approximately constant over the range 1–2 K. At lower temperatures the second sound velocity increases rapidly as $T \to 0$ K. Theory predicts that the limiting low-temperature value of second sound is $c_1/\sqrt{3} \simeq 137$ m/s.

In summary, the important factors controlling first sound and second sound are listed in Table 4.2. For He II and the two-fluid model, first sound requires that the two component velocities be in the same direction, while second sound is associated with a counterflow having $\mathbf{v}_s$ in the opposite sense to $\mathbf{v}_n$. The existence of second sound in He II can be used as a diagnostic tool since it is attenuated by forces of interaction between the

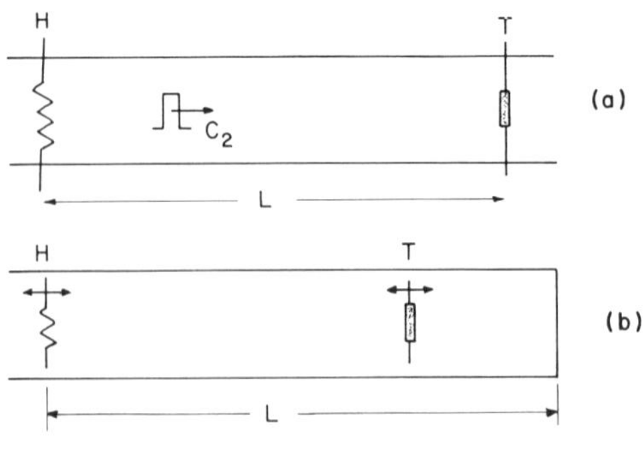

Fig. 4.18. Measurements of the velocity of second sound: (a) time-of-flight method and (b) resonance method.

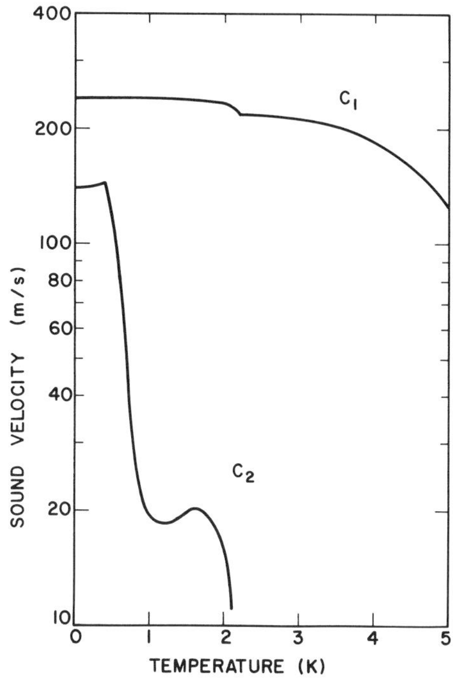

Fig. 4.19. Velocities of sound in liquid helium.

two fluid components. Thus, the variation of the second sound amplitude can give information about the local state of the helium.

4.4.4. Viscous Flow

Until now the two-fluid model has been applied to He II in a slightly idealized manner. In the prediction of second sound the viscosity of each

Table 4.2. Comparison of Sound Propagation in He II

	First sound	Second sound
Driving force	δp	δT
Propagator	$\delta\rho$	δs
Density (ρ)	Wavelike	~constant ($\rho_n \mathbf{v}_n \approx -\rho_s \mathbf{v}_s$)
Temperature (T)	~constant ($\mathbf{v}_s \approx \mathbf{v}_n$)	Wavelike
Velocity	$c_1 = \left(\dfrac{\gamma}{\rho\kappa}\right)^{1/2} \approx 200$ m/s	$c_2 = \left(\dfrac{\rho_s}{\rho_n}\dfrac{Ts^2}{C_v}\right)^{1/2} \approx 20$ m/s
Relationship	$\nabla^2 p = \dfrac{\partial^2 \rho}{\partial t^2}$	$\nabla^2 T = \dfrac{\rho_n}{s^2 \rho_s}\dfrac{\partial^2 s}{\partial t^2}$

component is assumed to be negligible, allowing the use of a linearized form of Euler's equation. On the other hand, in analyzing the thermomolecular fountain effect, the normal component has been given sufficient viscosity that it cannot flow through a capillary with $d \approx 1\ \mu$m. In actual fact, neither of these assumptions are correct since the normal fluid has a measurable viscosity which is of the order of 10–15 μP in the regime of interest. The present section examines how the introduction of the normal viscosity η_n into the two-fluid model aids in the prediction of several important physical phenomena.

To begin, a viscous term is introduced into the linearized Euler's equation (4.49), resulting in a somewhat more complex expression for the rate of change of the momentum,

$$\frac{\partial \mathbf{j}}{\partial t} = -\nabla p + \eta \nabla^2 \mathbf{v} \tag{4.66}$$

where the viscosity η enters along with the second-order derivative of the velocity. Note that the superfluid viscosity is zero in the two-fluid model and that the momentum can be written as a sum of the two components, $\mathbf{j} = \rho_s \mathbf{v}_s + \rho_n \mathbf{v}_n$. Combining these facts and keeping only terms which are linear we can rewrite (4.66) as

$$\rho_s \frac{\partial \mathbf{v}_s}{\partial t} + \rho_n \frac{\partial \mathbf{v}_n}{\partial t} = -\nabla p + \eta_n \nabla^2 \mathbf{v}_n \tag{4.67}$$

where the left-hand side is simplified by the assumption that ρ_s and ρ_n are weak functions of time. Recall that a relationship for the time derivative of the superfluid component has already been derived (4.55). Combining that relationship with (4.67) gives the two linearized equations that describe the flow of He II including the normal fluid viscosity,

$$\rho_s \frac{\partial \mathbf{v}_s}{\partial t} = \rho_s s\, \nabla T - \frac{\rho_s}{\rho} \nabla p \tag{4.55}$$

and

$$\rho_n \frac{\partial \mathbf{v}_n}{\partial t} = -\rho_s s\, \nabla T - \frac{\rho_n}{\rho} \nabla p + \eta_n \nabla^2 \mathbf{v}_n \tag{4.68}$$

The above expressions generally are referred to as the Navier–Stokes equations for the two fluid components. It is of interest here to determine how well these equations predict the transport properties of He II.

There are two principal methods for measuring the viscosity of a fluid: Poiseville type flow through a channel and the damping of an oscillating disk. For the case of ordinary liquids, these two methods yield essentially identical results. However, in the case of He II, very different results are obtained which can be understood only in terms of the two-fluid model.

First consider the Poiseville flow. If two vessels containing He II at different temperatures are connected together through a capillary of small diameter, the apparent viscosity of the helium is extremely small if not zero. Some measuraments have indicated that $\eta \lesssim 10^{-11}$ P, which is about six orders of magnitude less than the viscosity of normal helium. A further surprising feature of these measurements is the relationship between mass flow or fluid velocity and pressure difference. To first order, the mass flow is independent of Δp. These results can be interpreted in terms of the two-fluid model. In the channel, the normal fluid interacts strongly with the boundaries and is therefore locked in place. The superfluid component has zero viscosity and can flow through the channel without losses. Thus, the measured viscosity should be extremely small. This result is consistent of course with the fountain effect, requiring that the pressure difference be proportional to temperature difference. If the above description were complete, the superfluid component would be expected to accelerate indefinitely. However, this does not occur and the superfluid appears to transport itself at constant velocity nearly independent of the pressure difference. This velocity is termed the critical velocity $\mathbf{v}_c$ and also determines the behavior of the heat transport. Briefly, above the critical velocity the superfluid becomes turbulent which brings about new interactive mechanisms.

The alternative method of measuring the viscosity of He II is to oscillate a disk in the liquid and measure the damping; see Fig. 4.20. Since the damping of the oscillation is a function of moment of inertia and drag with the fluid, the oscillatory mass is related closely to the viscosity of the liquid. The results of this measurement are substantially different from the Poiseville experiment. In fact, what is observed is more traditional behavior for the viscosity; see Fig. 4.21. Below the λ-point the measured viscosity decreases, forming a broad minimum at a value of 14 μP between 2.0 and 1.5 K. Below this temperature the viscosity increases almost exponentially with decreasing temperature. It is of primary interest to interpret these

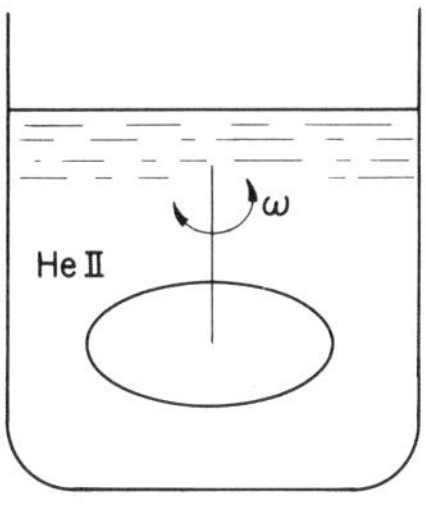

Fig. 4.20. Oscillating disk for viscosity measurement.

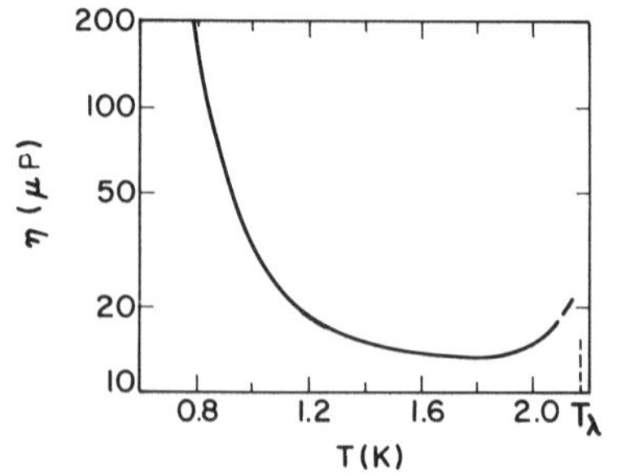

Fig. 4.21. Viscosity of He II as measured by oscillating disk.

results in terms of the two-fluid model. Unlike Poiseville flow, the oscillating disk technique forces the normal fluid to flow against its own viscous drag forces. Since the normal fluid interacts strongly with the boundaries, the oscillating disk directly measures its viscosity. Therefore, the results plotted in Fig. 4.21 actually represent the behavior of the normal fluid viscosity η_n.

One of the landmark experimental investigations of He II was performed by Andronikashvili on the measurement of rotating fluids.[10] This work is of great significance for it clearly supports the concept of the two-fluid model and agrees very well with other measurements of the He II properties. The basic idea of the experiment is shown schematically in Fig. 4.22. It employs a rotating viscometer; however, unlike the measurement described previously, this device is fitted with a number of plates with narrow spacings. The spacings were chosen so that they were less than the viscous penetration length, $d < (2\eta_n/\rho\omega)^{1/2}$. This condition requires all the fluid to rotate provided it has a viscous interaction with the plates. Since the two-fluid model demands that the fluid in motion be proportional to ρ_n, the Andronikashvili experiment is a direct measurement of the normal fluid density. The results of this experiment are shown in Fig. 4.23, which is a plot of ρ_n/ρ versus temperature. Also indicated on the graph are the values for ρ_n/ρ as measured by second sound techniques (4.65). The agreement between these two experimental findings is remarkable and adds considerable support to the physical basis of the two-fluid model.

4.4.5. Heat Transport

The behavior of He II under heat flow conditions can be interpreted similarly in terms of the two-fluid model. Again consider two reservoirs containing He II and connected by a channel of diameter d. At this point it is important to require that d be small, of the order 10 μm; however, the actual physical limitations are defined more clearly in the next section. If

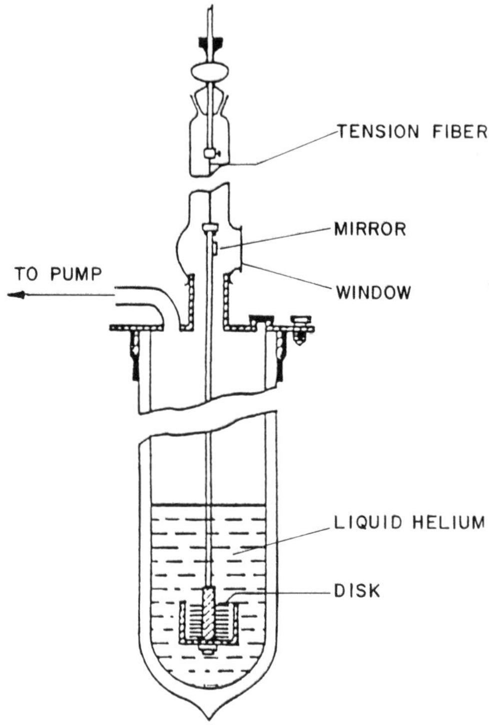

Fig. 4.22. Schematic diagram of Andronikashvili's experiment.

heat Q is applied to one side of the system, the temperature rises and heat is conducted through the insulated channel. As long as the helium obeys ideal two-fluid hydrodynamics, the relationship between the temperature gradient and pressure gradient is established from the thermomechanical effect, $\nabla p = \rho s \, \nabla T$.

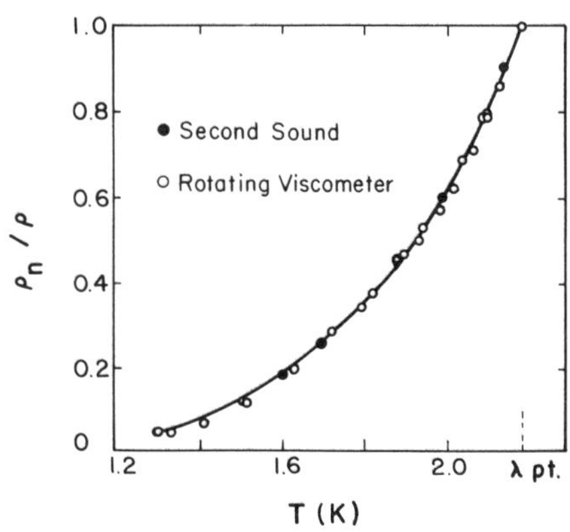

Fig. 4.23. ρ_n/ρ as a function of temperature compared with experimental results of Andronikashvili: ●, from second sound measurements and ○, from rotating viscometer.

To solve the heat flow problem in the steady state, the two fluid equations are combined, noting that $\partial \mathbf{v}_s/\partial t = \partial \mathbf{v}_n/\partial t = 0$, to yield

$$\nabla p = \eta_n \nabla^2 \mathbf{v}_n \tag{4.69}$$

which is known as the Poiseville equation. It is interesting to note that (4.69) inicates a nonzero normal fluid velocity no matter how small the channel diameter. Thus, the assumptions that lead to the fountain effect are oversimplified. Given a channel of constant cross section, (4.69) can be solved to relate the normal fluid velocity to the pressure and temperature gradients:

$$\nabla p = \rho s \, \nabla T = \frac{\beta \eta_n \mathbf{v}_n}{d^2} \tag{4.70}$$

where β is a numerical constant determined by the geometrical conditions. Note that $\beta = 12$ for parallel plates or large aspect ratio rectangular cross sections and $\beta = 32$ for circular tubes.

It is important to emphasize that the heat flow can be carried only by the normal fluid component under internal convection. Thus, for the case where the net mass flow is zero, $\mathbf{v} = 0$, a relationship exists between q, the heat flux density, and $\mathbf{v}_n$:

$$q = \rho s T \mathbf{v}_n \tag{4.71}$$

Subsequently, (4.71) will be shown to have a more general form when considering nonzero net mass flow. Eliminating $\mathbf{v}_n$ from the Poiseville equation, (4.70), we obtain a heat conductivity equation that should describe the behavior of the ideal He II:

$$q = \frac{d^2 (\rho s)^2 T}{\beta \eta_n} \nabla T \tag{4.72}$$

Note that this expression has a form similar to pure conduction although the function of proportionality, which is the effective thermal conductivity k_{eff}, varies as the square of the diameter. Consider as an example heat flow in a 10 μm diameter circular tube containing He II at 1.9 K. By inserting the appropriate physical properties for He II, we can calculate an effective thermal conductivity of 4.9×10^4 W/m · K. This quantity increases strongly with temperature dominated by the dependence of $(\rho s)^2 T \approx T^{12}$. Thus, although k_{eff} is around 100 times greater than that of high-purity copper at 1.9 K it is only about 25 W/m · K at 1 K.

If (4.72) were the only expression limiting the heat transport in He II, it would be possible to increase the effective thermal conductivity indefinitely by simply increasing the diameter, since $k_{eff} \approx d^2$. However, there is a physical limitation resulting from the existence of a critical velocity in the superfluid component. This critical velocity is the same one that limits the rate of helium flow through a channel under Poiseville flow. Above this critical velocity a second heat transport controlling mechanism begins to be significant. Mostly, these effects are determined by the existence of turbulence in both the normal and superfluid components in the He II, which is the subject of the next section.

4.5. VORTICES AND TURBULENCE IN He II

As outlined in the previous section, He II has a number of unique physical properties, the most notable being the transport properties. However, the picture portrayed is incomplete because it contains no limit for the special heat and mass flow expressions. Such behavior is not observed experimentally. Rather there exists a critical velocity $\mathbf{v}_c$ above which the He II begins to exhibit viscous flow and nonideal heat transport. The subject of the present section is to convey the significance of this critical velocity and how it affects the unique properties of He II. In particular, there is a need to introduce a new loss mechanism that limits heat transport in large systems.

4.5.1. Rotating Helium

In considering the flow of He II in the two-fluid model, an assumption was made that the superfluid component has zero viscosity, that is $\eta_s = 0$. This assumption leads to the requirement that superfluid velocity cannot have a gradient perpendicular to the flow direction, that is $\nabla \times \mathbf{v}_s = 0$. An interesting contradiction occurs when this condition is applied to He II in rotation.

Consider a vessel containing He II rotating with an angular frequency ω. For the case in discussion, the temperature of the fluid will be taken to be below 1 K so there is hardly any normal fluid present and $\rho_s \simeq \rho$. Also to simplify the description of the fluid, it will be assumed to be incompressible allowing $\rho \simeq$ constant and $\partial\rho/\partial t = 0$. It follows from the continuity equation that $\nabla \cdot \mathbf{j} = 0$, and since the normal fluid density is essentially zero,

$$\nabla \cdot \mathbf{v}_s = 0 \tag{4.73}$$

or the divergence of the superfluid velocity must vanish also. Combining

(4.73) with $\nabla \times \mathbf{v}_s = 0$ requires that the superfluid obey Laplace's equation, with the curl-free relationship for the superfluid velocity,

$$\nabla^2 \mathbf{v}_s = 0 \tag{4.74}$$

which for a simple rotating vessel like in Fig. 4.24 has only one solution—that the He II must be at rest, that is, $\mathbf{v}_s = 0$.

It is a straightforward argument to show that $\mathbf{v}_s = 0$ for any fluid that obeys Laplace's equation. Introducing the concept of circulation, we obtain

$$C = \int_c \mathbf{v}_s \cdot d\mathbf{l} \tag{4.75}$$

where the integral is taken around any simple path in the fluid. The circulation is a measure of the rotation of the fluid. Stokes' law can be applied to (4.75), yielding

$$\int_c \mathbf{v}_s \cdot d\mathbf{l} = \int_s \nabla \times \mathbf{v}_s \cdot d\mathbf{s} \tag{4.76}$$

where $d\mathbf{s}$ is the integral over the surface enclosed by c. But for the superfluid, $\nabla \times \mathbf{v}_s = 0$, which can be satisfied only by

$$\int_c \mathbf{v}_s \cdot d\mathbf{l} = 0 \tag{4.77}$$

a condition that requires $\mathbf{v}_s = 0$.

The above conclusion is in contradiction to experiment. In 1950, Osborne[11] reported an experiment that measured the rotation of a single bucket of He II as shown in Fig 4.24. Under all conditions, he found the helium to rotate with a parabolic meniscus. Furthermore, if all the helium

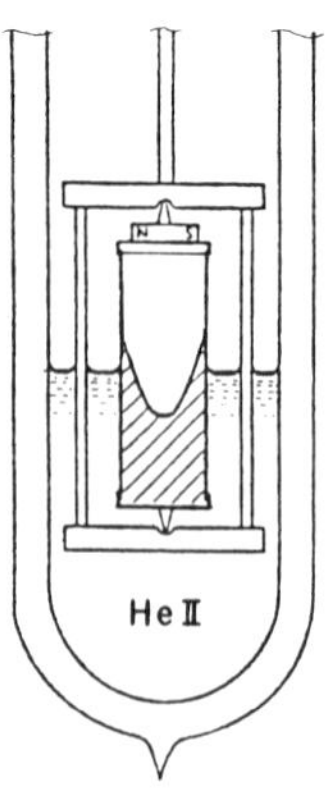

Fig. 4.24. Simple rotating bucket as used by Osborne.[11]

rotated, the vertical height z of the meniscus should be given by the simple relationship

$$z = \frac{\omega^2}{2g} r^2 \tag{4.78}$$

where g is the acceleration of gravity and r is the radial coordinate. If only the normal fluid rotates there would be a ratio ρ_n/ρ multiplying the right-hand side of (4.78). Osborne's experiments were in general agreement with (4.78), which leads to the conclusion that He II behaves as a normal liquid when rotating in a simple bucket.

The explanation for the apparent contradiction between experiment and theory of He II lies in the method by which turbulence is introduced into liquid He II. The qualitative picture of He II above its critical velocity imagines the existence of an array of vortex lines brought on by the rotation of the fluid. Each vortex line contains a unit of angular momentum and can be used to explain numerous physical phenomena associated with He II. London first postulated the existence of vortex lines, but their actual observation did not occur until nearly 30 years later.

The vortex line array exists as follows. Rotating He II contains vortex cores shown in Fig. 4.25. The array is similar to the fluxoid array in a Type II superconductor. Each vortex core contains a unit of circulation. The behavior of this array is governed by quantum mechanics to the extent that each unit of circulation is determined by the quantization rule,

$$\int \mathbf{p} \cdot d\mathbf{q} = nh \tag{4.79}$$

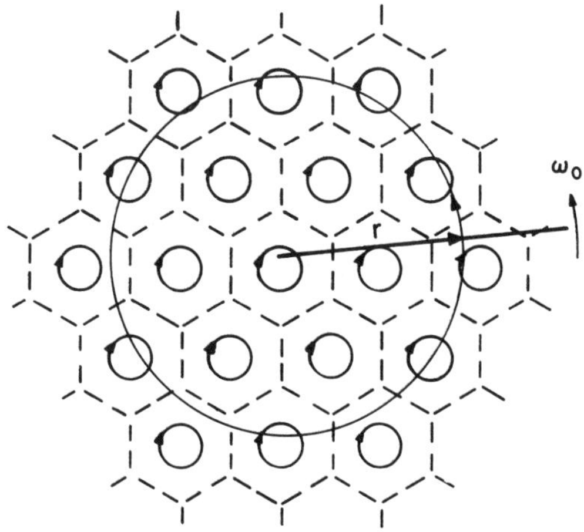

Fig. 4.25. Idealized vortex array.

where $\mathbf{p}=m\mathbf{v}$ is the momentum in a linear system and $d\mathbf{q}=d\mathbf{l}$. It is apparent that the circulation C is quantized based on (4.79) so that

$$\int \mathbf{v}_s \cdot d\mathbf{l} = n\frac{h}{m} \tag{4.80}$$

where h/m is a unit of circulation. For a regular array, each vortex core contains one unit of circulation equal to h/m.

The existence of a regular array of vortex lines is helpful in resolving the apparent contradiction between experiment and the two-fluid model. In particular, if the array is as shown in Fig. 4.25, the circulation cancels locally because each vortex borders another and the line separating the two has zero velocity. Therefore, taking an arbitrary path containing numerous vortices, we find that the net circulation within this path is zero. However, at the boundary of the container, the circulation does not cancel, leaving a net value around this path consistent with the entire fluid in rotation. Thus, the contradiction can be resolved by assuming the existence of a vortex array.

The density of vortex lines depends on the angular frequency. To compute this density n_0, begin by determining the total circulation of a path of radius r in the rotating He II. For the case with the rotation at constant angular frequency ω_0, integration of (4.75) yields

$$C = 2\pi r^2 \omega_0 \tag{4.81}$$

Because this path contains a total number of vortex lines $N = n_0(\pi r^2)$, the total circulation is $N(h/m)$ or

$$C = n_0 \frac{h}{m} \pi r^2 \tag{4.82}$$

Equating (4.81) and (4.82), we can calculate density of vortex lines that depends only on the angular frequency:

$$n_0 = \frac{2\omega_0 m}{h} = 2000\ \omega_0\ \text{cm}^{-2} \tag{4.83}$$

where ω_0 is in rad/s. For a reasonable angular frequency of perhaps 1 rad/s, (4.83) indicates a sizable vortex line density. It would therefore appear possible to observe vortex lines in rotating helium if they could be made visible.

Visual observation of vortex lines in rotating helium was first achieved by Williams and Packard in 1974.[12] In their experiment, the vortex lines were made visible by trapping electrons in the vortex core. It is an established characteristic to the behavior of electrons in liquid helium that they will form small voids or bubbles of the order of tens of angstroms in diameter. These electron bubbles are attracted to the vortex core as a result of the Bernoulli potential, caused by the electron bubbles' desire to reduce their surface energy. In the vortex line visualization experiment, electrons emanating from a β-decay of a radioactive source were trapped on the vortices in a rotating bucket of liquid He II. The electrons subsequently were accelerated through an electrostatic potential, striking a phosphorus screen. Thus, the existence of a vortex line was recorded as a flash of light on the phosphorus screen.

The observations accrued from this and subsequent experiments on visualization of vortex lines appear consistent with the description of the turbulent state. The number of vortex lines was proportional to angular frequency ω_0, as predicted by (4.83). Furthermore, when many lines were present in the experiment, the spacing between lines was approximately equal to $(h/2\omega_0 m)^{1/2}$, also consistent with theory. One aspect to these measurements which was not predictable was the mobility of the vortex lines. A vortex appeared to be created and destroyed at boundaries. Also, the vortices moved throughout the fluid at a rate which made it necessary to add some ^{3}He to produce viscous damping. Shown in Fig. 4.26 are photographic reproductions of the vortex line array created in these experiments.

4.5.2. Critical Velocities

At this point a contradiction becomes evident in the description of experiments which involves rotating He II. Recall the Andronikashvili experiment which consisted of rotating closely spaced disks in liquid helium. The principal observation of that work was that only the normal fluid component rotated with the device. This result appears to be in contradiction with the rotating bucket experiments where the entire helium fluid is seen to be in motion. A distinction between these two observations can be made in terms of the magnitude of rotation velocity. To establish the entire bucket in rotation it is necessary to exceed a critical velocity for the interaction between the normal and superfluid components. The critical velocity is also of prime importance in the development of a complete picture of the heat transport of He II.

The critical velocities of the two components in He II have different origins. For the superfluid, the critical velocity $\mathbf{v}_{sc}$ is that which is necessary to create turbulence. This turbulence takes the form of a vortex line tangle. The normal fluid component can be thought of as having a critical velocity $\mathbf{v}_{nc}$ associated with the transition from ordinary laminar to turbulent flow. The normal fluid critical velocity can be considered in terms of classical

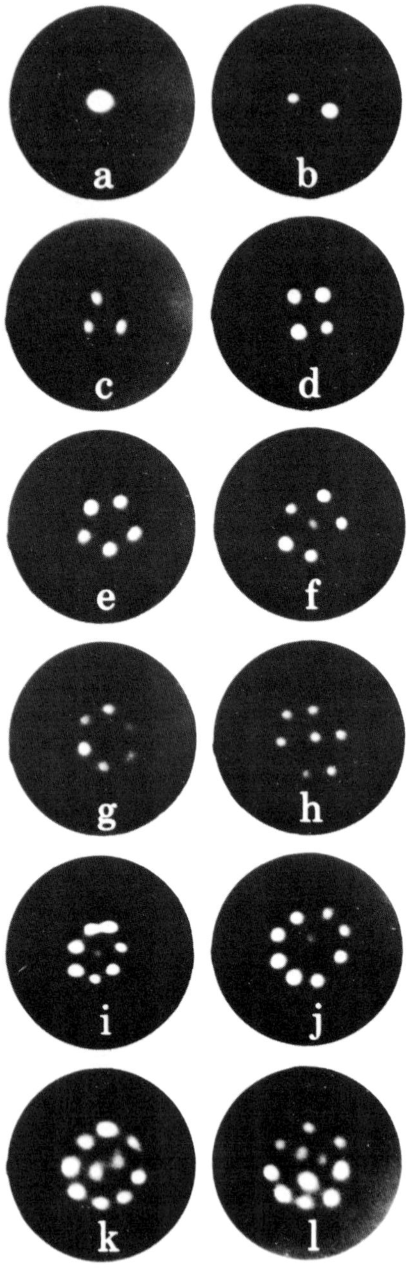

Fig. 4.26. Photographic reproduction of vortex line array in rotating He II (from Yarmchuk and Packard[12]): (a) through (l) indicate increasing angular frequency.

hydrodynamic principles. Finally, with the two-fluid model there exists the possibility of a critical velocity associated with the relative motion of the two components, $|\mathbf{v}_n - \mathbf{v}_s|_c$. It should be emphasized that all three of these critical velocities may impact the behavior of mass and heat flow in He II.

First, consider the superfluid critical velocity $\mathbf{v}_{sc}$. Recall that the superfluid velocity is more or less independent of pressure difference for flow within a channel. The equation of motion for the superfluid component predicts an acceleration of this component under the influence of a force ∇p. Clearly, the acceleration cannot go no indefinitely; the superfluid has some limiting or critical velocity above which frictional forces may contribute. It is possible to understand the concept of a superfluid critical velocity by invoking semiclassical arguments. Imagine that the He II is confined to flow in a channel of dimension d. At very small velocities, the flow will be laminar. However, with increasing $\mathbf{v}_s$, there comes a point where the ideal laminar flow breaks down and turbulence is created. Thus, the critical velocity is that which is necessary to create turbulence or vortices in the channel.

In He II the behavior of vortices is controlled to a large extent by quantum mechanics. In particular, a vortex contains one quantum of circulation. One might then ask, what is the characteristic size of a vortex? Employing the Heisenberg uncertainty principle, $\Delta p\, \Delta x \approx h$, the position of a helium molecule in rotation around the vortex must be defined to within some fraction of the radius. Furthermore, the vortex must fit within the channel which has a characteristic dimension d. For the sake of discussion, we choose $\Delta x \approx d/10$ for the vortex to exist. This numerical assumption leads to an equation for $\mathbf{v}_{sc}$:

$$\mathbf{v}_{sc} \approx \frac{10h}{md} \approx \frac{10^{-2}}{d}\ \text{cm/s} \tag{4.84}$$

where m the mass of the helium molecule. Clearly, the above picture is oversimplified. A more detailed theoretical development due to Feynmann[13] results in a similar expression for $\mathbf{v}_{sc}$:

$$\mathbf{v}_{sc} = \frac{h}{\pi md}\left(\ln\frac{4d}{a_0} - K'\right) \tag{4.85}$$

where a_0 is the size of a vortex core and K' is a numerical constant. It is interesting to note that although the Feynmann equation (4.85) has a stronger theoretical basis than (4.84) it still is dominated by an inverse diameter dependence.

There have been considered theoretical and experimental efforts to solve the problem of superfluid critical velocity. Unfortunately, these efforts have not achieved a great deal of success. Plotted in Fig. 4.27 are different theoretical approaches to $\mathbf{v}_{sc}$ as well as an empirical fit to existing data in the literature. The experimental data appears to follow a $d^{1/4}$ law,

$$\mathbf{v}_{sc} \simeq d^{-1/4} \quad \text{(cgs)} \tag{4.86}$$

based on results in small-diameter channels. It should be noted that the measurement of critical velocity in the superfluid component is sometimes disturbed by the existence of a similar transition in the normal fluid. Consequently, some discrepancy between theory and experiment may be attributed to experimental variations.

In addition to superfluid turbulence it is possible to create turbulence in the normal fluid component of He II. This effect can be described in terms of more conventional fluid dynamic models. It is generally assumed that the nonzero normal fluid viscosity ($\eta_n \neq 0$) that causes a transition from laminar to turbulent flow in the normal fluid component. In this picture, there exists a critical velocity $\mathbf{v}_{nc}$ that describes the unstable transition from laminar to turbulent flow. In classical fluids this transition is defined in terms of a critical Reynolds number Re_c, which is a dimensionless quantity made up of the ratio between inertial and viscous forces. For most classical fluids, the critical Reynolds number for flow in a tube occurs near

$$\mathrm{Re}_c = \frac{d\rho \mathbf{v}_c}{\eta_n} \approx 1200 \tag{4.87}$$

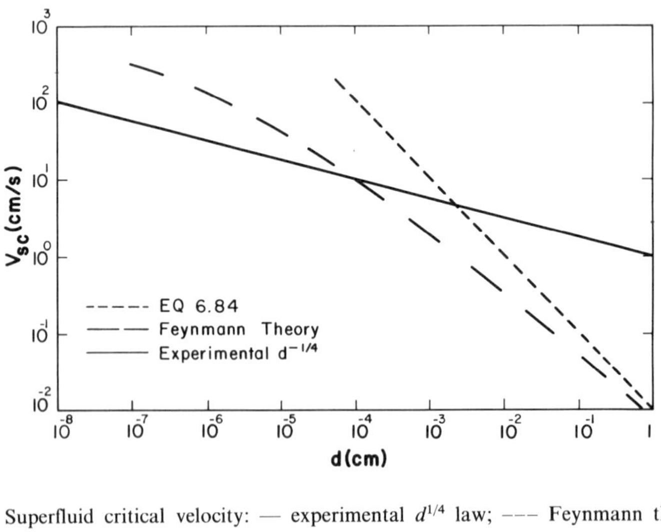

Fig. 4.27. Superfluid critical velocity: — experimental $d^{1/4}$ law; ––– Feynmann theory; --- based on (6.84).

For He II, the critical Reynolds number is determined by the normal fluid velocity and viscosity,

$$\mathrm{Re}(\mathbf{v}_{nc}) = \frac{d\rho \mathbf{v}_{nc}}{\eta_n} \tag{4.88}$$

Note that the expression is determined by the total density of He II and not just the normal fluid component. Equation (4.88) predicts a fairly strong temperature dependence of the normal fluid critical velocity. At low temperatures, $T < 1.5$ K, $\mathbf{v}_{nc}$ is dominated by the normal fluid viscosity η_n which goes as T^{-5}.

Before leaving the subject of critical velocities, it is worth estimating their value for systems of practical significance. To begin, assume the critical velocities of the normal and superfluid components in a channel of diameter 10 mm at 1.8 K are to be estimated. For the sake of discussion, the helium will be considered to be undergoing internal convection only; that is there is no net mass flow. As a consequence, the helium is made to flow by the introduction of a heat flux q. Since the superfluid component can carry no entropy, the normal fluid velocity $\mathbf{v}_n$ is determined entirely by the heat transport expression in internal convection, (4.71). The critical heat flux for the introduction of normal fluid turbulence is given by the combination of (4.71) and (4.88) such that

$$q_{nc} = sT \frac{\eta_n}{d} \mathrm{Re}(\mathbf{v}_{nc}) \tag{4.89}$$

which for 1.8 K, a tube diameter of 10 mm, and a critical Reynolds number given by (4.87) yields $q_{nc} \simeq 15$ mW/cm^2.

The value of the critical heat flux necessary to induce turbulence in the superfluid component is determined by rearrangement of (4.71). Keeping in mind that zero net mass flow implies $\rho_s \mathbf{v}_s = \rho_n \mathbf{v}_n$, we calculate the critical heat flux for the onset of superfluid turbulence as

$$q_{sc} = \frac{\rho_s \rho}{\rho_n} sT\mathbf{v}_{sc} \tag{4.90}$$

For the empirically determined values for $\mathbf{v}_{sc}$, (4.86) gives $q_{sc} \simeq$ 300 mW/cm^2 at 1.8 K, a value at least one order of magnitude higher than q_{nc}.

The practical significance of both critical velocities is in their signaling transitions in the behavior of $\mathbf{v}_s$ and $\mathbf{v}_n$. In particular, the onset of turbulence is expected to affect heat transport and fluid flow. Above these critical velocities, additional interactive terms must be included in the fluid flow equation which describes the temperature and pressure gradients. The two contributions of greatest importance are the turbulent pressure gradient and the mutual friction interaction in the temperature gradient.

The first issue to consider when introducing turbulence into the problem is how its existence affects the pressure gradient measured within the fluid. As already noted, in the laminar regime, the pressure gradient is proportional to the normal fluid velocity which in turn is proportional to the temperature gradient. This relationship ceases to apply above the critical velocity of the normal fluid, $\mathbf{v}_{nc}$, where additional mechanisms modify the ideal behavior of the fluid.

One way to look at the normal fluid pressure gradient is in terms of classical fluid mechanics. By defining a Reynolds number dependent on pressure gradient,

$$\mathrm{Re}(p) = \frac{\rho d^3}{\beta \eta_n^2} \nabla p \tag{4.91}$$

where β is the same numerical quantity introduced in laminar heat transport, the London equation is still obeyed. This format allows the laminar equation to be obtained simply by equating (4.91) with (4.87). Furthermore, in the turbulent regime, Blasius' rule for classical turbulence applies,

$$\mathrm{Re}(p) = 0.00494[\mathrm{Re}(\mathbf{v}_n)]^{1.75} \tag{4.92}$$

which is based on empirical evidence for classical fluids. Combining this with the heat flux relationship for the normal fluid turbulence (4.71), we obtain the pressure gradient,

$$\nabla p = \frac{0.00494 \beta \eta_n^{1/4}}{\rho (sT)^{7/4} d^{5/4}} q^{7/4} \tag{4.93}$$

where β takes on values of 12 for parallel plates and 32 for tubes. This relationship has been tested only for channels of relatively small diameter ($d \lesssim 100\ \mu$m).

4.5.3. Mutual Friction

There is one more interaction that exists in He II and has profound significance to the heat transport character. This interaction is hypothesized to be the result of a viscous drag between the normal fluid component containing the excitations and the turbulent superfluid component with vortices. The existence of this force was first introduced to help explain heat transport data by Gorter and Mellink.[14] It has since been given a physical basis by several experimental and theoretical studies.

The basic idea of the mutual friction interaction is as follows. In turbulent He II there are two fluid components, a normal fluid containing excitations and the superfluid component with associated vortices. For the present argument's sake, these two components will be assumed to be travelling in opposite directions as a result of internal convection. The vortex will be taken to be travelling at $\mathbf{v}_L$, vortex line velocity, which is of the order of $\mathbf{v}_s$. A schematic representation of this situation is shown in Fig. 4.28.

The force per unit length of vortex is dependent on the relative velocities of $\mathbf{v}_L$ and $\mathbf{v}_n$ as well as the densities of each. Theoretical justification for the exact form of this quantity is incomplete. This is due to insufficient understanding of the interactive mechanisms. On dimensional grounds alone, it would be expected that the force per unit length of the vortex would be expected to take the form

$$\mathbf{f} = \frac{\rho_s \rho_n}{\rho^2} \eta_n \, |\mathbf{v}_n - \mathbf{v}_s| \tag{4.94}$$

where η_n is the only possibility relevant viscosity. It is assumed that $\mathbf{f}$ is proportional to the product of ρ_s and ρ_n, so that the interaction is maximum where these two quantities are equal. Other arguments can be made to justify $\mathbf{f}$ going as $|\mathbf{v}_n - \mathbf{v}_s|$, which are based on classical turbulent interactive mechanisms.

To determine the total interaction per unit volume of He II, it is necessary to consider the macroscopic picture of the turbulent state. As envisioned, the He II consists of a tangled mass of vortex lines. Unlike the picture of rotating He II with a regular matrix of vortex lines, heat-induced vortex lines form an isotropic mass which is assumed to be spacially non-varying. This assumption is not true generally but it simplifies the physical explanation. A schematic representation of the vortex array is shown in Fig. 4.29. The vortex array is described by two quantities: L, the vortex line length per unit volume of fluid, and δ, the mean spacing between vortex lines. For a given set of conditions, the vortex line length has one stable

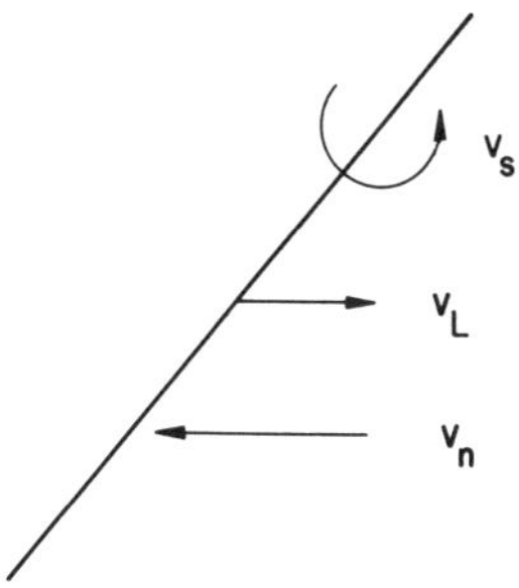

Fig. 4.28. Schematic of the normal fluid and superfluid vortex interaction.

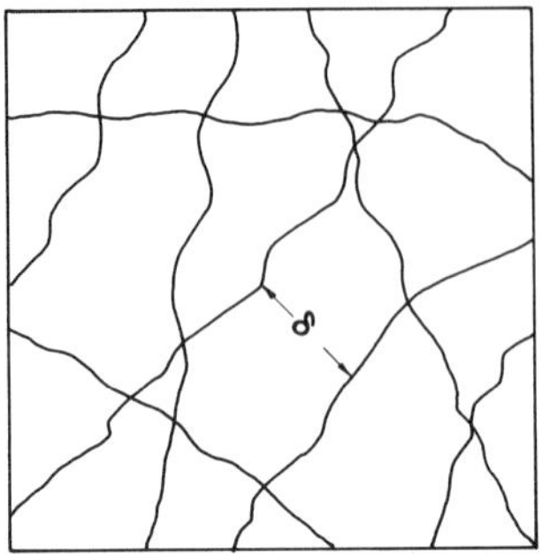

Fig. 4.29. Schematic representation of a vortex array.

value. In the model first proposed by Vinen,[15] the steady-state length is seen as a balance between the rate of generation of vortex lines and the rate of decay. Therefore, in the steady state,

$$\left.\frac{dL}{dt}\right|_{\text{gen}} = \left.\frac{dL}{dt}\right|_{\text{decay}} \tag{4.95}$$

and the line length takes on the value L_0. The total force per unit volume of He II is the product of the line force $\mathbf{f}$ and the line length L_0.

Vinen used dimensional grounds to show the variation of the vortex line length with velocity. For the two quantities defined in (4.95), the fundamental dependencies may be written

$$\left.\frac{dL}{dt}\right|_{\text{gen}} = \frac{\chi_1 B \rho_n}{2\rho} \mathbf{v}_{ns} L^{3/2} \tag{4.96}$$

where $\mathbf{v}_{ns}$ is the relative velocity of the normal and superfluid components and χ_1 and B are parameters describing the interaction between normal fluid and a vortex. The decay of the vortex lines also has a simple functional form,

$$\left.\frac{dL}{dt}\right|_{\text{decay}} = -\frac{\chi_2 h}{2\pi m} L^2 \tag{4.97}$$

where χ_2 is of order unity and a function of temperature. Equating (4.96) and (4.97) in the appropriate form, we can write the steady-state vortex line length, to within numerical constants, as

$$L_0 \approx \left(\frac{\rho_n}{\rho} \mathbf{v}_{ns}\right)^2 \tag{4.98}$$

The velocity dependence of the vortex line length may be understood in a somewhat different manner. Assume the vortex has an average radius of curvature **R**. The drift velocity of a line is then given by

$$\mathbf{v}_L = \beta \mathbf{S}' \times \mathbf{S}'' \tag{4.99}$$

where $\mathbf{S}'$ is the first spatial derivative of the vortex line and $\mathbf{S}''$ is the second. Since the local radius of curvature is the only relevant length in the problem, it follows that $\mathbf{S}' \approx R^{-1}$ and $\mathbf{S}'' \approx$ constant. The line length per unit volume is related to the radius by $L_0^{1/2} \approx R^{-1}$. Therefore, $\mathbf{v}_L \approx L_0^{1/2}$, a relationship similar to that derived above.

Using the above form for the vortex line length and force per unit length, we can write the total mutual friction force per unit volume of He II. The treatment developed originally by Vinen and more recently by Schwarz[16] includes a small velocity offset $\mathbf{v}_o$, such that the final equation has the form

$$\mathbf{F}_{ns} = L_0 \mathbf{f} = A \rho_n \rho_s (\mathbf{v}_n - \mathbf{v}_s - \mathbf{v}_o)^2 (\mathbf{v}_n - \mathbf{v}_s) \tag{4.100}$$

where A is termed the Gorter–Mellink mutual friction coefficient. According to the model presented above, $A \approx \rho_n^2/\rho^3$, but because the physical quantities χ_1 and χ_2 are undetermined, the coefficient must be evaluated empirically.

There have been a wide variety of measurements of the Gorter–Mellink parameter.[17] The most accepted values for A, measured by Vinen,[18] are displayed in Fig. 4.30 along with several other measurements.[19,20] Vinen employed two methods to measure the Gorter–Mellink coefficient, steady-state heat flow techiques and attenuation of second sound. Both techniques give essentially the same value for A, which

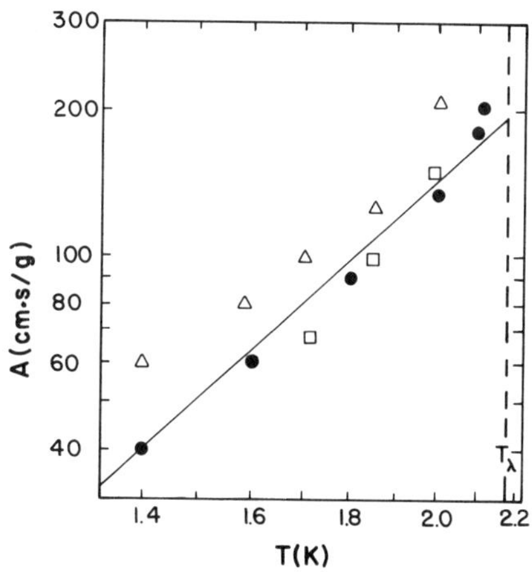

Fig. 4.30. Gorter–Mellink mutual friction parameter. (Data are from Ref. 18–20.)

adds strong support to the Vinen theory of turbulent He II. The theory presented by Vinen predicts a temperature and pressure dependence which should be dominated by the square of the normal fluid density ρ_n. The quantity ρ_n^2/ρ^3 has a strong temperature dependence going roughly as $(T/T_\lambda)^{13}$, while the experimentally determined value for A exhibits a much weaker temperature dependence, that is $A \approx T^3$.

The pressure dependence of the Gorter–Mellink coefficient is less well established. Some work has shown that the variation of the heat conductivity of turbulent He II with pressure can be interpreted only by a pressure dependence of A. Since this quantity varies as ρ_n^2/ρ^3, its pressure variation is not a large effect except near the λ-transition. For example, at $T = 2.0$ K, the ratio between ρ_n^2/ρ^3 at 2.5 bars and saturated vapor pressure is 1.17. This should be compared to the experimental ratio of $A(2.5)/A(0)$ of 1.19 ± 0.05 at 2 K.[19] Such a comparison, although within experimental error, does not established fully the pressure dependence of A. However, for the sake of further discussion of this effect, the theoretical pressure dependence $A \approx \rho_n^2/\rho^3$ will be assumed.

Given the existence of this additional interactive mechanism between the two components in He II, the two-fluid hydrodynamic equations can be modified to include these terms. The result of this modification is equivalent to the Navier–Stokes equations for the two fluids:

$$\rho_n \frac{\partial \mathbf{v}_n}{\partial t} = -\frac{\rho_n}{\rho} \nabla p - \rho_s s \nabla T + \eta_n \nabla^2 \mathbf{v}_n + A \rho_n \rho_s |\mathbf{v}_n - \mathbf{v}_s|^3 \tag{4.101}$$

$$\rho_s \frac{\partial \mathbf{v}_s}{\partial t} = -\frac{\rho_s}{\rho} \nabla p + \rho_s s \nabla T - A \rho_n \rho_s |\mathbf{v}_n - \mathbf{v}_s|^3 \tag{4.102}$$

where the above two expressions are similar to (4.55) and (4.68) except that they add the terms due to mutual friction. The expressions as given above are somewhat simplified over those used by Vinen and others. In particular, (4.101) and (4.102) lack any explicit indication of either critical velocity, $\mathbf{v}_{sc}$ or $\mathbf{v}_{nc}$. Critical velocity can be taken into account by introduction of a small velocity offset $\mathbf{v}_o$, as shown in (4.100). However, since this correction is quite small, it is neglected in the present treatment of turbulent He II.

4.5.4. Steady-State Heat Transport

Consider the problem of steady-state heat transport, $\partial \mathbf{v}_s/\partial t = \partial \mathbf{v}_n/\partial t = 0$, in a channel containing He II. This problem, which has already been discussed for the nonturbulent case, can be solved analytically to give an expression for the temperature gradient in terms of the velocities of each component:

$$\nabla T = \frac{\beta \eta_n |\mathbf{v}_n|}{\rho s d^2} + \frac{A \rho_n}{s} |\mathbf{v}_n - \mathbf{v}_s|^3 \qquad (4.103)$$

Note that (4.103) is a scalar equation established with the understanding that ∇T is negative in the direction of $\mathbf{v}_n$. If there is no net mass flow, that is, internal convection is the only important transport mechanism, then $\rho_s \mathbf{v}_s = -\rho_n \mathbf{v}_n$ and the temperature gradient can be reduced to a simple form that depends only on the heat flux:

$$\nabla T = \frac{\beta \eta_n q}{d^2 (\rho s)^2 T} + \frac{A \rho_n}{\rho_s^3 s^4 T^3} q^3 \qquad (4.104)$$

where use has been made of $q = \rho s T \mathbf{v}_n$. Note that the first term in (4.104) is the same as developed to describe the viscous flow of nonturbulent He II. The second term describes the mutual friction interaction. Because it lacks diameter dependence and goes as the cube of the heat flux, the second term dominates the temperature gradient at high heat fluxes and large diameter channels, that is, $d \gtrsim 1$ mm. For example, at 1.9 K, a 1 mm diameter circular cross-section channel will have the largest contribution to the temperature gradient from the mutual friction for heat fluxes in excess in 10 mW/cm^2. For practical applications, larger heat fluxes are of interest indicating that under many circumstances the first term can be ignored in the description of heat transport in He II.

There is a fundamental assumption that enters into the description of He II heat transport as given by (4.104). The concept of critical velocity has been introduced to separate the regions of turbulent and nonturbulent heat transport. However, simply adding the laminar and turbulent terms to describe heat transport in He II assumes that the transition between these two regions is continuous. This is not what is observed experimentally.

The actual situation concerning the heat flux dependence of the temperature and pressure gradients in turbulent and nonturbulent He II is shown in Fig. 4.31. At very low heat fluxes, much less than that necessary to create turbulence, the laminar expression suitably describes the heat transport and the temperature and pressure gradients are proportional to heat flux. Similarly, at very high heat fluxes, the turbulent expression does an adequate job of explaining experimental data. However, the behavior is much more complex in the transition region. Around q_c, there is often hysteresis in the relationship between ∇T and q. Sometimes the turbulent character is suppressed beyond q_c because of the need to nucleate vortices. Once the turbulent state exists, the vortices do not dissipate easily, per-

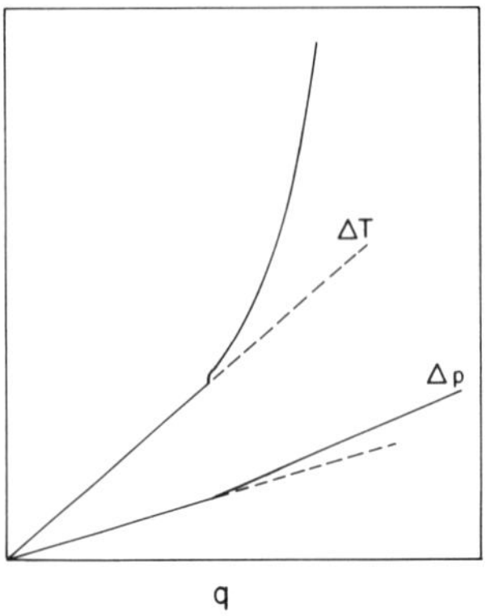

Fig. 4.31. Typical behavior of ∇T and ∇p near the critical velocity in He II.

sisting below q_c. Hysteresis is common in the temperature gradient behavior. Similar character is seen often in the pressure gradient near q_c. At this point it is not clear whether a simple addition of terms as in (4.104) is correct. Far enough away from q_c either the laminar or turbulent expression dominates so that the details of the transition region are of small consequence. This is often the case in practical applications of He II.

The entire heat flow regime of He II can be summarized graphically as in Fig. 4.32 by assuming constant properties. Here both the laminar and turbulent characteristics are shown for varying diameter channels from 1 μm to 100 mm. Also shown are the several critical velocities, Feynmann's

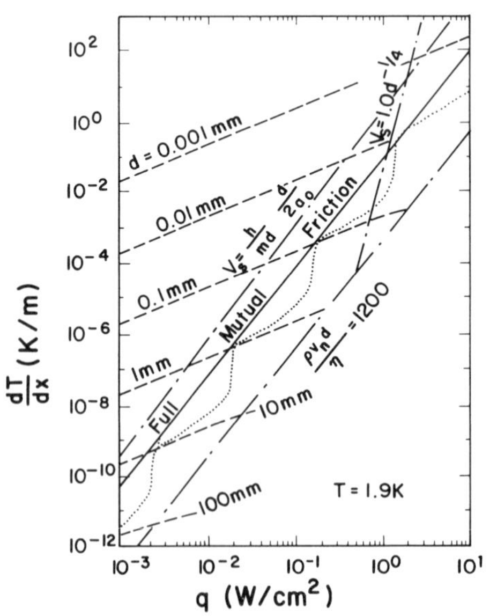

Fig. 4.32. Heat transport regimes of He II at 1.9 K (from Arp[17]).

and the empirical form for $\mathbf{v}_{sc}$, and the constant Reynolds number ($\text{Re}_c = 1200$) form for $\mathbf{v}_{nc}$. Shown by the dotted line is the behavior of the transition region between the two heat transport characters. This dependence is associated with the potential suppression of turbulent heat transport above q_c. In general, the transition regime of heat transport is quite complex and not amenable to simple interpretation.

4.5.5. Forced Convection Heat Transport

Up to now, the heat transport has been assumed to occur in static He II; that is, there has been no net momentum ($\mathbf{j} = 0$). This is an important assumption leading to the derivation of the heat flow equations. It is of interest, however, to consider the effect of net mass flow when the entire fluid is driven by a pressure gradient, resulting in a total fluid velocity $\mathbf{v}$. Pressure-gradient-induced flow in He II is really only possible if the thermomechanical effect is small because the normal fluid velocity can be affected significantly by viscous drag. For small-diameter channels where the thermomechanical effect is considerable, mass flow is controlled essentially by the superfluid velocity. As this situation leads to considerable complication, it will not be considered here. Rather, the other heat transport regime will be considered, which is demonstrated by the mutual friction term. There are numerous examples in the literature where $\mathbf{v}_s$ and $\mathbf{v}_n$ are independently controlled variables.[21]

Neglecting the laminar term in the temperature gradient expression (6.103), we obtain a simplified expression that does not depend on net mass flow:

$$\nabla T = \frac{A\rho_n}{s} |\mathbf{v}_n - \mathbf{v}_s|^3 \qquad (4.105)$$

Now since momentum conservation must be obeyed in the two-fluid model (4.45), a more general relationship for q as a function of velocity is obtained:

$$q = \rho s T(\mathbf{v}_n - \mathbf{v}) \qquad (4.106)$$

which reduces to the previous form when $\mathbf{v} = 0$. This relationship can be derived in general by applying conservation of momentum and energy in a moving reference frame. It can also be shown to be the correct assumption when solving the problem of forced flow with no net heat flux, that is, $q = 0$.

Substitution of (4.106) into (4.105) leads to the somewhat surprising result that the temperature gradient in forced flow He II is independent of velocity; that is,

$$\nabla T = \frac{A\rho_n}{\rho_s^3 s^4 T^3} q^3 \tag{4.107}$$

which is identical to the mutual friction term in (4.104). It should be emphasized that this derivation has not included the potentially important term associated with the net enthalpy transport of the flowing helium, $\rho C\,\Delta T$. Because enthalpy transport is actually more relevant to the practical heat transfer character of forced flow He II, this subject is considered further in the next chapter. It is the purpose of the present discussion to principally identify the invariance of (4.107).

4.5.6. Attenuation of Second Sound

An alternative method exists for the study of the turbulent state in He II which also represents a reassuring check on the theory of He II. The principle is based on the fact that second sound is attenuated in turbulent He II and that the attenuation is a direct measure of the strength of the mutual friction interaction. In the turbulent regime, the solution of the time-varying wave propagation equation (4.64) is modified. It can be shown by using (4.101) and (4.102) that second sound obeys the expression

$$\frac{\partial^2 T}{\partial t^2} = c_2^2 \nabla^2 T + \frac{A\rho q^2}{\rho_s^2 s^2 T^2} \frac{\partial T}{\partial t} \tag{4.108}$$

where the second term on the right-hand side causes the attenuation and is additional to the normal linearized second sound equation. The rate at which this attenuation takes place leads to an exponential process with a coefficient

$$\alpha = \frac{A\rho q^2}{2\rho_s^2 s^2 T^2 c_2} \tag{4.109}$$

The Gorter–Mellink coefficient A enters explicitly in (4.109), allowing its value to be determined directly by measurement of α.

Experimentally, the amplitude of a traveling wave is proportional to $e^{-\alpha x}$. The quantity α is found generally to consist of two terms,

$$\alpha = \alpha_0 + \alpha'(q) \tag{4.110}$$

where the first term is more or less independent of heat currents and the second term is due to the mutual friction interaction. The second term in (4.110) is seen to vary as the square of the heat flux q, such that

$$\alpha' = C(q - q_0)^2 \tag{4.111}$$

where the proportionally factor is a function of temperature and pressure. Measurements by Vinen which show this result are displayed in Fig. 4.33. Quite apparent in these measurements is the onset of attenuation at the critical velocity. Below q_c, there is no heat-flux-dependent attenuation. Also, note the approximate magnitude of $q_0 \approx 5 \mathrm{mW/cm^2}$ by extrapolation of the attenuation coefficient to $\alpha' = 0$. This value is in close agreement with the critical heat flux q_c estimated from Fig. 4.27.

The second sound attenuation method is a powerful probe of the mutual friction interaction. It represents an effectively independent method for determining the mutual friction coefficient A. This point was exploited by Vinen who made a simultaneous measurement of A by both methods on the same channel.[18] The results of these measurements strongly demonstrate that the two methods for determining A are for all practical purposes equivalent. However, the possibility of geometrical factors affecting A is somewhat less well established.

4.5.7. Development of Turbulence

In the preceding development, it has been assumed generally that the He II is either in the nonturbulent state, obeying isentropic laminar flow models, or in the turbulent state with the additional mutual friction interaction. As a conclusion to the physical description of He II, consider the rate at which the turbulent state is created when the fluid is initially in the ideal vortex-free condition. This problem has received extensive experimental investigation.[23,24] The general approach is to measure the time rate of change of the second sound attenuation after a turbulence-creating heat flux is applied, that is, $q > q_c$. The general character of the

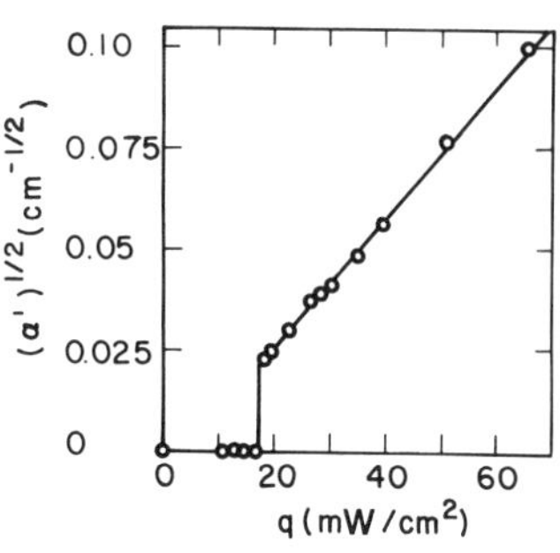

Fig. 4.33. Measurements of the attenuation of second sound in He II (from Vinen[22]).

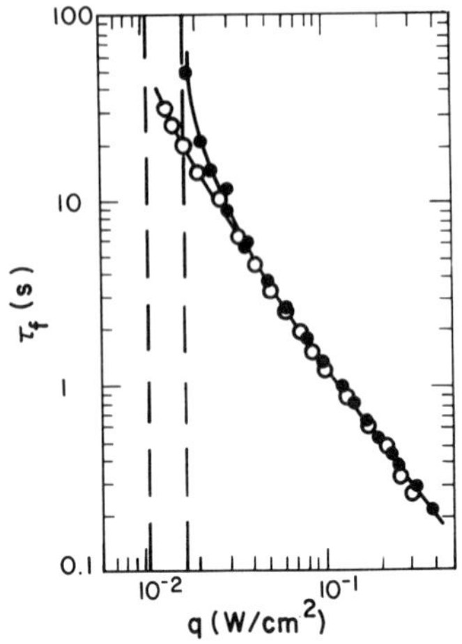

Fig. 4.34. Characteristic time for buildup of turbulence in He II at 1.41 K. The symbols represent different channel configurations (from Vinen[22]).

results indicates a characteristic time for the development of the fully turbulent state. This characteristic time τ has been shown to be both heat flux and temperature dependent. Typical results for an experiment by Vinen are shown in Fig. 4.34. An empirical fit to these data indicate that the time constant has the simple form

$$\tau = aq^{-3/2} \tag{4.112}$$

where a is a temperature- and geometry-dependent quantity. Chase[24] has measured this coefficient for a number of channel configurations ranging from open ducts to complex geometries. Figure 4.35 displays the empirically determined coefficients a. The variation is between a factor of 3 and 5. It should be noted that this quantity is not always reproducible and will be shorter if the helium is not allowed to reach equilibrium without

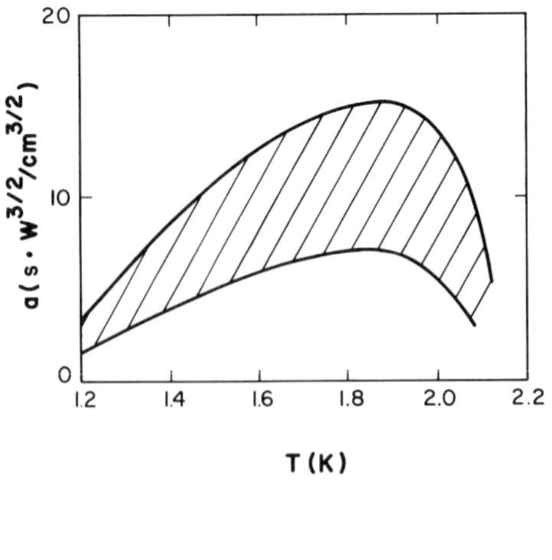

Fig. 4.35. Coefficient of the turbulence buildup expression for different channels (from Chase[23]).

turbulence before a measurement. Full recovery to the nonturbulent state can take on the order of hundreds of seconds.

As a final comment, it should be emphasized that there is by no means universal agreement in the actual value of the Gorter–Mellink coefficient. Although the Vinen values are most accepted, numerous others are published in the literature for channel diameters from 100 μm to tens of millimeters. Furthermore, although the theory does not predict either surface or dimensional dependence of this quantity, the wide variation in A may be in part due to these effects. There is also some controversy over whether the actual power law of the mutual friction term should be cubic or perhaps some other noninteger value. Theory supports the cubic dependence, but many experiments, particularly near the λ-point, have shown the power laws to be higher, perhaps approaching q^4. Some authors[21,25] have argued that this noncubic dependence is an experimental artifact. These issues are considered in more detail in Chapter 5, which specifically treats heat transport in He II.

PROBLEMS

4.1. Make a plot of the temperature and pressure gradients versus heat flux in a 50 μm diameter tube containing He II. Consider both the normal fluid viscous and mutual friction interactions. Indicate the location of and the importance of critical velocities.

4.2. Starting with the equation for the vapor pressure of an ideal Bose gas (6.24a), cast it in the form of the virial expansion as discussed in Chapter 5. Placed in this form, the second term represents the ideal quantum gas correction which appears as part of the second virial coefficient. Compare the value of this correction to the difference between the classical and quantum second virial coefficients at 4 K (see Fig. 5.3).

4.3. Calculate the latent heat per mole of an ideal Bose gas at 2 and 4 K. Compare these values with the latent heat of liquid helium at the same temperatures.

4.4. Using the empirical relationships for the entropy, specific heat, and normal and superfluid densities, derive an expression for the temperature dependence of the second sound velocity c_2. Plot your result and compare it to Fig. 4.19.

4.5. Calculate the angular frequency at which one vortex could exist in a 1 cm diameter tube containing He II and spinning about its longitudinal axis. One method to prevent vortex formation in rotating He II is to subdivide the helium pool into many cells, perhaps with a honeycomb structure or porous medium. Calculate the cell radius necessary to prevent the formation of vortices in a bucket rotating at an angular frequency of 1 rad/s.

4.6. Consider a rectangular cross section chamber containing He II. Second sound can be generated and detected across the 1 cm width of the chamber, while a steady-state heat can be applied perpendicularly down the length of the chamber. Assume that the helium bath temperature is 1.8 K.

  a. Calculate the second sound velocity and the frequency of the lowest harmonic standing wave which could be set up within the chamber;

b. If the standing wave has an amplitude of 5 mK at the generator, what will be the amplitude at the detector if a 20 mW/cm^2 steady-state heat flux is applied perpendicular to the second sound standing wave.

4.7. Calculate the pressure drop over a 1 m long, 100 μm diameter tube containing He II at 1.8 K subjected to a 2 W/cm^2 heat flux. Consider only the Δp resulting from the normal fluid turbulence. Compare the result of this calculation with that obtained for classical fluids,

$$\Delta p = \frac{2f\rho u^2 L}{D}$$

where $f = 0.0791/\mathrm{Re}^{1/4}$ and $u = \mathbf{v}_n$.

REFERENCES

1. L. D. Landau and E. M. Lifshitz, *Statistical Physics*, Addison-Wesley, Reading, MA, 1969.
2. K. Huang, *Statistical Mechanics*, Wiley, New York, 1963.
3. F. London, *Superfluids*, Vol. II, Dover Publications, New York, 1954.
4. K. R. Atkins, *Liquid Helium*, Cambridge University Press, Cambridge, 1959.
5. J. Wilks, *Liquid and Solid Helium*, Clarendon Press, Oxford, 1967.
6. W. E. Keller, *Helium-3 and Helium-4*, Plenum Press, New York, 1969.
7. R. D. McCarthy, *The Thermodynamic Properties of He II from 0 K to the Lambda Transitions*, NBS Technical Note 1029, U.S. Government Printing Office, Washington, DC, 1980.
8. V. Peshkov, Determination of the Velocity of Propagation of the Second Sound in He II, *J. Phys.* (*USSR*) **10**, 389 (1946).
9. J. S. Brooks and R. J. Donnelly, The Calculated Thermodynamic Properties of Superfluid Helium-4, *J. Phys. Chem. Ref. Data* **6**, 51 (1977).
10. E. L. Andronikashvili, *Zh. Esksp. Theor. Fiz.* **16**, 780 (1946); **18**, 424 (1948).
11. D. V. Osborne, The Rotation of Liquid Helium II, *Proc. Phys. Soc.* **A63**, 909 (1950).
12. G. A. Williams and R. E. Packard, Photographs of Quantized Vortex Lines in Rotating He II, *Phys. Rev. Lett.* **33**, 280 (1974); E. A. Yarmchuk and R. E. Packard, Photographic Studies of Quantized Vortex Lines, *J. Low Temp. Phys.* **46**, 479 (1982).
13. R. P. Feynmann, Application of Quantum Mechanics to Liquid Helium, in *Progress in Low Temperature Physics*, C. J. Gorter (Ed.), Vol. I, p. 17, North Holland Publishing, Amsterdam, 1955.
14. C. J. Gorter and J. H. Mellink, On the Irreversible Processes in Liquid Helium II, *Physica* **XV**, 285 (1949).
15. W. F. Vinen, Theory of the Mutual Friction, *Proc. R. Soc.* (*London*) **A242**, 493 (1957).
16. K. W. Schwarz, Turbulence in Superfluid Helium: Steady Homogeneous Counterflow, *Phys. Rev.* **18**, 245 (1978).
17. V. Arp, Heat Transport through Helium II, *Cryogenics* **10**, 96 (1970).
18. W. F. Vinen, Experiments on Steady Heat Currents, *Proc. R. Soc. London* **A240**, 114 (1957).
19. S. W. Van Sciver, Kapitza Conductance of Aluminium and Heat Transport Through Subcooled He II, *Cryogenics* **18**, 521 (1978).
20. D. F. Brewer and D. O. Edwards, Heat Conduction by Liquid Helium II in Capillary Tubes. III. Mutual Friction, *Philos. Mag.* **7**, 721 (1962).
21. J. T. Tough, Superfluid Turbulence, in *Progress in Low Temperature Physics*, D. F. Brewer (Ed.), Vol. VIII, Chap. 3, North Holland Publishing, Amsterdam, 1982.

22. W. F. Vinen, Critical Heat Currents in Wide Channels, *Proc. R. Soc. London* **A243**, 400 (1957).
23. W. F. Vinen, Experiments on Transient Effects, *Proc. R. Soc. London* **A240**, 128 (1957).
24. C. E. Chase, Thermal Conduction in Liquid Helium II. I. Temperature Dependence Effects of Channel Geometry, *Phys. Rev.* **127**, 361 (1962); Thermal Conduction in Liquid Helium II. *Phys. Rev.* **131**, 1898 (1963).
25. P. E. Dinotakis and J. E. Broadwell, Local Temperature Measurements in Supercritical Counterflow in Liquid Helium II, *Phys. Fluids* **16**, 1787 (1973).

Chapter 5

He II Heat Transfer

Chapter 4 emphasized the physical understanding of He II including heat transport in the laminar flow and turbulent mutual friction regimes. These mechanisms are fundamental to the behavior of He II, although they represent rather idealized conditions. Therefore, it is of interest to apply the treatment of He II to practical heat transfer problems. In doing so, the concepts already developed must be extended into regimes that are usable in more practical situations. To be more specific, the emphasis of Chapter 4 has been to understand the interactive mechanisms. Thus, of principal concern are the behavior of the transport properties including mainly the normal fluid viscosity η_n and the mutual friction parameter A. Of interest now is to use these concepts in understanding such phenomena as the maximum heat flux q^*, the maximum energy deposition ΔE^*, and the maximum temperature difference ΔT_m, which can be either within the fluid or across a solid–fluid interface. The goal of the present chapter is to establish a connection between the engineering parameters q^*, ΔE^*, and ΔT_m and the physical properties of the fluid and solid–fluid boundaries. In establishing this connection there are a number of subjects of practical interest which must be addressed. These include steady-state heat transport, transient heat transport, Kapitza boundary conductance, and film boiling. Some of these phenomena are also important in pool boiling He I heat transfer, which is the subject of Chapter 6.

Before delving into these individual subjects, it is worth describing, in a general way, the heat transfer character of liquid He II. This character in fact does not deviate qualitatively from that of He I, albeit the numerical values and physical explanation are considerably different. A typical steady-state heat transfer curve for a metal surface at the end of a duct containing He II is plotted in Fig. 5.1. These results are intended only to display the regions of heat transfer. As is demonstrated in what follows, actual numerical values of these regimes are strongly dependent on geometry, temperature, pressure, and surface conditions. For small ΔT up to q^*, the

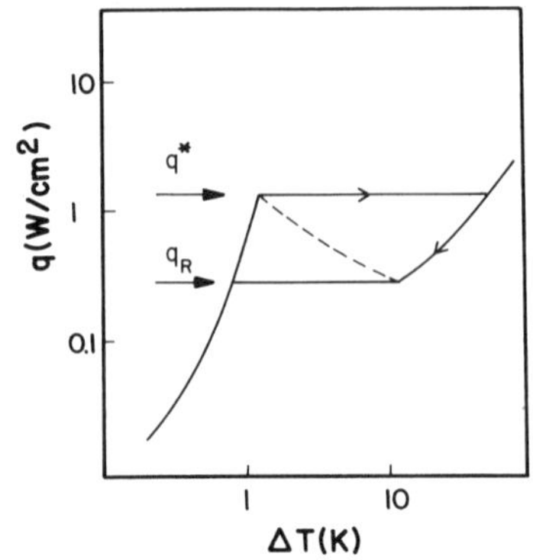

Fig. 5.1. A typical steady-state heat transfer curve for a metal surface at the end of a duct containing He II.

surface temperature difference is governed by interfacial phenomena having more to do with the character of the solid than that of the liquid helium. This is called the Kapitza regime. There is no boiling occurring, rather the temperature difference is a result of thermal impedance between the two dissimilar materials, the metal or insulating solid and liquid He II. The maximum heat flux q^* is strongly geometry and helium state dependent. It is characterized by the point where the helium adjacent to the interface exceeds the local boiling point. The maximum heat flux is also time dependent, achieving very high values for short-duration heat pulses. Once this maximum is exceeded, the heat transfer converts to a film boiling process where a film consisting of either He I, vapor, or both blankets the surface. Finally, in some configurations there is observed a hysteresis in the heat transfer curve exemplified by the requirement to reduce q below q^* to return to the Kapitza regime. This process is reasonably well understood in He I, being given by an engineering correlation. However, in He II the problem is more complex and has received less attention, owing to the experimental difficulty of achieving steady state and strong variations with configuration. It is the physical understanding of this heat transfer curve that is the goal of the present chapter. The description is based heavily on the heat transport modeling of He II contained in Chapter 4.

5.1. STEADY-STATE HEAT TRANSPORT

The first question to ask is: What are the limitations to heat transport in a channel containing He II? Since the heat transport equations for He II have already been developed, it should be straightforward to apply these equations to determine practical limitations. In doing so, it is assumed that the heat transport equations can be applied much the same as Fourier's law is applied to heat conduction in a solid. Note that He II cannot exist above the λ-transition, 2.172 K at SVP, which at least establishes temperature boundaries to the heat transfer problem. For a channel of finite length L, as shown in Fig. 5.2, subjected to a constant heat flux q, there is a temperature difference established across its length, that is $\Delta T = T_m - T_b$. In general, this temperature difference occurs because of two mechanisms, normal fluid viscous interactions with the boundaries and mutual friction between the two fluid components. Since most practical applications call for channel diameters and heat fluxes that are sufficiently large to allow the mutual friction term to dominate, the normal fluid viscous interaction will be neglected throughout this discussion. It is equally straightforward, in principle, to solve the steady-state heat transport problem including the normal fluid viscous term, although the mathematics are more cumbersome.

5.1.1. Peak Heat Flux in Wide Channels

The turbulent heat transport equation in one dimension is written in the form

$$\frac{dT}{dx} = f(T)\, q^m \tag{5.1}$$

where $f(T) = A\rho_n/(\rho_s^3 s^4 T^3)$ and m is a numerical coefficient which theory indicates should be equal to 3 but which experimentally has been shown to vary from below 3 to nearly 4 as the temperature approaches T_λ.[1-3] Recall that A is the Gorter–Mellink mutual friction parameter and ρ_n and ρ_s are the normal and superfluid densities, respectively. The origin of (5.1) is discussed extensively in Section 4.5. The quantity $f^{-1}(T)$ behaves much like a thermal conductivity in that it is a fluid property that controls the temperature gradient in the presence of a heat flux. It is therefore of interest to understand the variation of $f^{-1}(T)$ with the state of the helium. Plotted in Fig. 5.3 is this function as it depends on temperature and pressure between 1.4 K and T_λ and p = SVP and 2.5 MPa. Note that the temperature dependence is quite strong with a maximum occurring around T = 1.9 K at SVP. The pressure dependence is weaker and somewhat uncertain owing to the limited quantity of data available for the pressure dependence of the Gorter–Mellink parameter. In the data presented in Fig. 5.3, Vinen's[1] values for

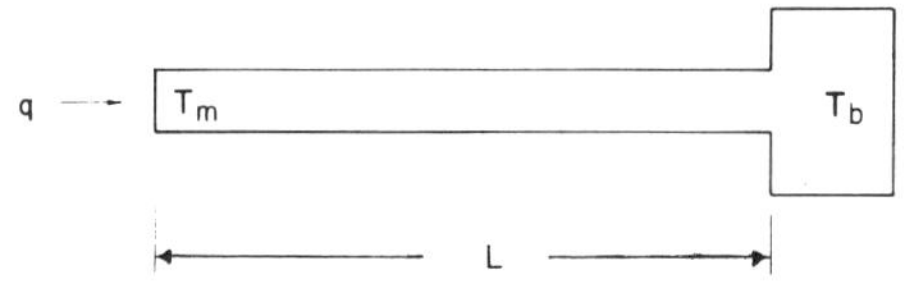

Fig. 5.2. Schematic of a channel containing He II.

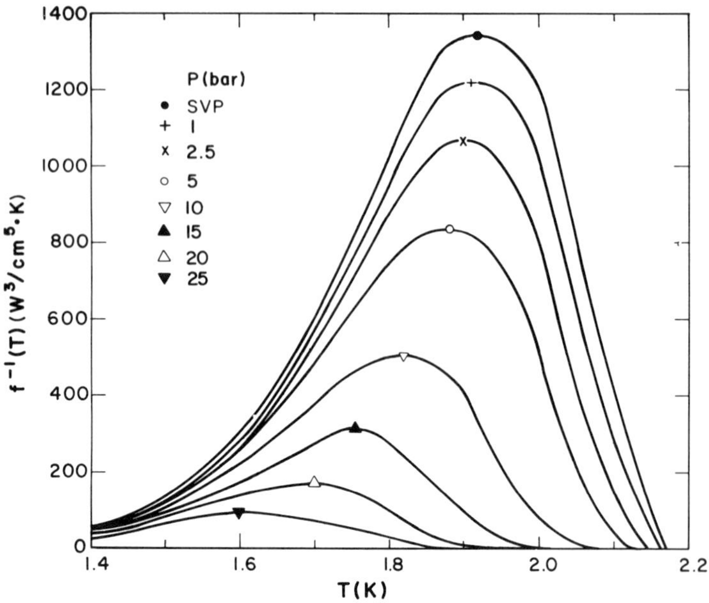

Fig. 5.3. Heat conductivity function for turbulent He II. Symbols indicate the location of the peak value.

the Gorter–Mellink parameter have been used. Furthermore, it has been assumed that $A\alpha\rho_n^2/\rho^3$, which is based on theory[4] and supported to some extent by experiment.[3] Based on an empirical fit to the Gorter–Mellink parameter, it is also possible to write an analytic expression for the heat conductivity function,

$$f^{-1}(T, p) = g(T_\lambda)[t^{5.7}(1 - t^{5.7})]^3 \tag{5.2}$$

where $g(T_\lambda) = \rho^2 s_\lambda^4 T_\lambda^3 / A_\lambda$, $t = T/T_\lambda$, $S_\lambda = 1.559$ J/g · K, and $A_\lambda \simeq 115$ cm · s/g. Note that at saturated vapor pressure, the maximum in (5.2) occurs at 1.923 K, which is generally consistent with experiment. The values presented in Fig. 5.3 are good to about $\pm 10\%$ at saturated vapor pressure and have been compared to experiment up to about 7 bars.[5] At this time, results are not available at higher pressures.

If the interest is in the heat transport in a finite-length channel with a sizable ΔT, it is possible to determine its value by integration of (5.1) and matching boundary conditions. The maximum heat flux q^* is then established according to the maximum allowable temperature difference across the channel, which for a given bath temperature T_b is $(T_\lambda - T_b)$. It follows that for a channel of length L

$$q^* = \left(\frac{1}{L} \int_{T_b}^{T_\lambda} \frac{dT}{f(T)} \right)^{1/m} \tag{5.3}$$

where for $m = 3$, the heat conductivity function is given in (5.2). This integral is mostly a function of T_b and only weakly dependent on other factors such as pressure. It is possible to define a function of the form

$$q^* L^{1/m} \equiv Z(T_b) = \left(\int_{T_b}^{T_\lambda} \frac{dT}{f(T)} \right)^{1/m} \tag{5.4}$$

which should be independent of channel length. Plotted in Fig. 5.4 are experimentally determined peak heat fluxes q^* for different channel lengths varying everywhere from 0.1 to 3 m. Two different correlations of the data are displayed: one for $m = 3$ and the other for $m = 3.4$. In either case, the agreement between data and correlation is quite acceptable.

A similar analysis is possible to determine the pressure dependence of the maximum heat flux q^*. Integration of the corresponding heat conductivity function $f^{-1}(T, p)$ predicts a decreasing maximum heat flux with elevated pressure. By analytic integration of (5.2), a prediction can be made for the behavior of q^* with pressure. The results of this analysis for four bath temperatures are displayed in Fig. 5.5. Also displayed are experimentally observed[3,6] maximum heat fluxes for short channels up to 0.3 MPa. The agreement is reasonable for the available data.

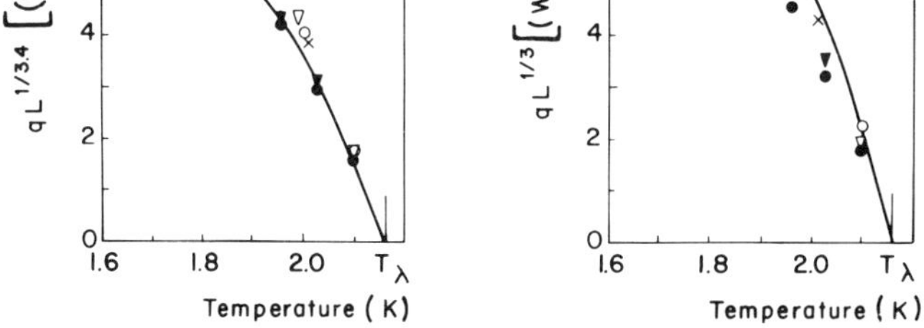

Fig. 5.4. Generalized steady-state limiting heat flux in He II (as compiled by Seyfert[60]).

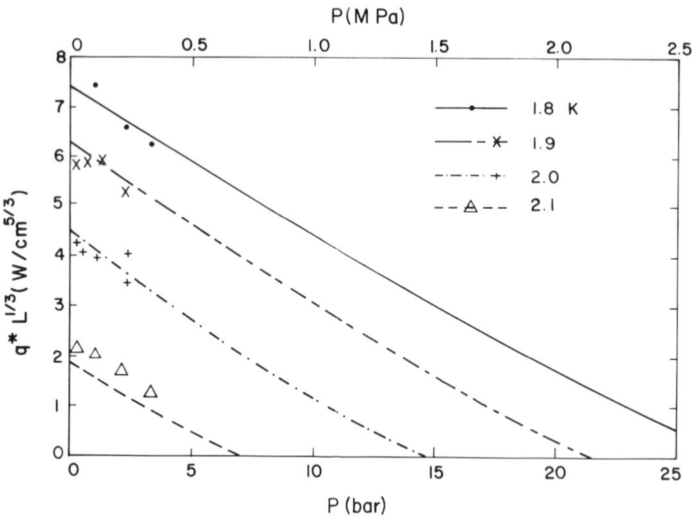

Fig. 5.5. Maximum heat flux in a He II-containing channel as a function of pressure (data from Ref. 3 and 6).

It should be kept in mind that the form and physical explanation for heat transport in He II place no fundamental limit to the maximum steady-state value for q^*. Everything depends on the allowable temperature difference. For example, with $T_b = 1.8$ K and $m = 3$, Fig. 5.4 predicts a product $q^* L^{1/3} = 7.4$ W/cm$^{5/3}$. Therefore, for a channel of length 10 μm, this analysis would predict a q^* in excess of 70 W/cm^2, truly a remarkable heat flux for liquid helium temperatures.

5.1.2. Peak Heat Flux in Cylindrical Geometries

Besides the simple linear geometry represented by a one-dimensional tube with constant heat flux, there has been considerable work carried out on cylindrical geometries consisting of a heated cylinder or wire immersed in a large bath of He II.[7–10] It is easy to show,[11] by assuming that the Gorter–Mellink equations apply in cylindrical geometry, that the steady-state heat transport equation can be written

$$\frac{dT}{dr} = f(T)\, q_0^m \left(\frac{r_0}{r}\right)^m \tag{5.5}$$

where q_0 is the heat flux per unit area of the heated cylinder of radius r_0. The difference in (5.5) occurs because the heat flux decreases as the radius increases. Comparison of (5.5) with experiment has given reasonable agreement, with essentially the same heat conductivity function as applies in linear geometries.[10,11] In a similar fashion to that applied to (5.3), integration of (5.5) leads to the maximum heat flux,

$$q_0^* = \left(\frac{m-1}{r_0}\int_{T_b}^{T_\lambda}\frac{dT}{f(T)}\right)^{1/m} \tag{5.6}$$

The important observation to make about (5.6) is that the peak heat flux q_0^* has as its scaling length r_0 rather than L as in the linear system. This means that, provided the radius of the container is much larger than that of the heater, the boundary conditions far from the heater should not affect q^* significantly. This is certainly contrary to the behavior in linear geometries.

Unlike the linear one-dimensional system, there have been fewer attempts to correlate the peak heat transport in cylindrical geometries with that of (5.6). This is due in part to the lack of reliable data, which are more difficult to obtain in cylindrical geometries. One such effort made use of data for small temperature differences $\Delta T \approx 10$ mK in the range of $T_b = 1.8$ K. It was found that the expression given by (5.6) is not entirely suitable to correlate the experimental values of q_0^* without introducing a radius-dependent quantity ψ defined by

$$q_0^* = \left(\frac{2\psi}{r_0}\int_{T_b}^{T'}\frac{dT}{f(T)}\right)^{1/3} \tag{5.7}$$

where $T' \simeq T_b + 0.01$ K. ψ was found empirically to depend on radius, being roughly proportional to $r_0^{1/2}$. The results of this correlation are shown in Fig. 5.6. Note that ψ is always less than 1, indicating that the peak heat flux is always less than that predicted by the idealized theory. This fact is somewhat surprising because the temperature gradients appear to be given accurately by (5.5).[11]

5.1.3. Peak Heat Flux in Saturated He II

Until now, it has been assumed arbitrarily that the peak heat flux q^* is determined by the condition that the helium adjacent to the heater surface reaches the λ-point. This condition does not always occur for reasons having to do with the helium temperature distribution and the phase diagram, displayed in Fig. 5.7. First of all, because of the high effective thermal conductivity of He II, it is reasonable to assume that the helium within the heat transfer region obeys equilibrium thermodynamics. This

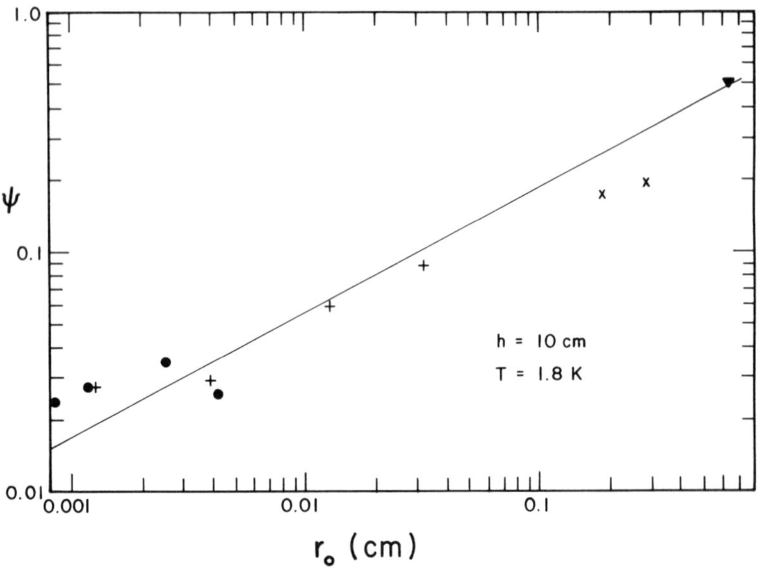

Fig. 5.6. Empirical correlating function for heat transfer in cylindrical geometries.

assumption allows the state of the helium everywhere in an experiment or engineering system to be described by a point on the phase diagram.

Now consider a simple example, that of a heat transfer experiment in a saturated bath of liquid helium boiling at 1.8 K, 1.6 kPa (12.5 torr). The process is occurring at a certain depth below the liquid–vapor interface; see Fig. 5.8. Thus, without any heat being applied, the state of the helium can be described by location ① on the phase diagram in Fig. 5.7. The pressure applied at the heat transfer surface is therefore $p = p_0 + \rho g h$, where h is the hydrostatic head of the liquid helium. If heat is applied to induce heat transfer in the system, there will be a local temperature excursion ΔT that is determined by heat flow mechanisms. The local temperature increases but the pressure is fixed, unless the experimental configuration is small enough for the thermomolecular effect to make a significant contribution. Avoiding this complication for the time being, we see that the temperature will follow a horizontal line as shown in the inset of Fig. 5.7 until it meets the liquid–vapor interface at which point boiling commences. The value of the maximum temperature excursion is determined by the slope of the vapor pressure curve:

$$\Delta T_m = \int_{p_0}^{p_0 + \rho g h} \left(\frac{dT}{dp}\right) dp \tag{5.8}$$

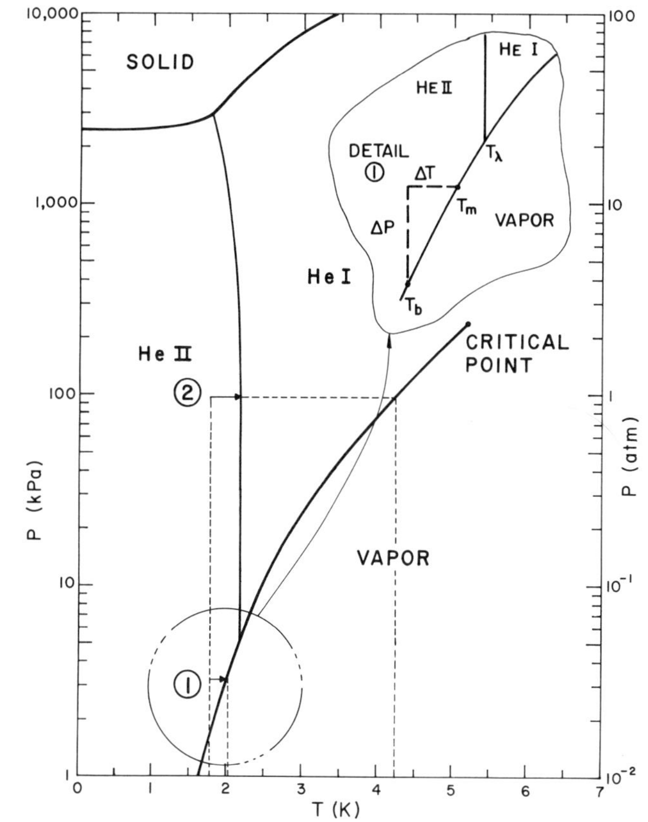

Fig. 5.7. Phase diagram of helium showing location ① of near saturation conditions and location ② of subcooled conditions.

The slope of the vapor pressure curve, dT/dp, is known through the Clausius–Clapeyron equation,

$$\frac{dp}{dT} = \frac{\lambda}{T\,\Delta v} \approx \frac{\lambda}{T v_g} \tag{5.9}$$

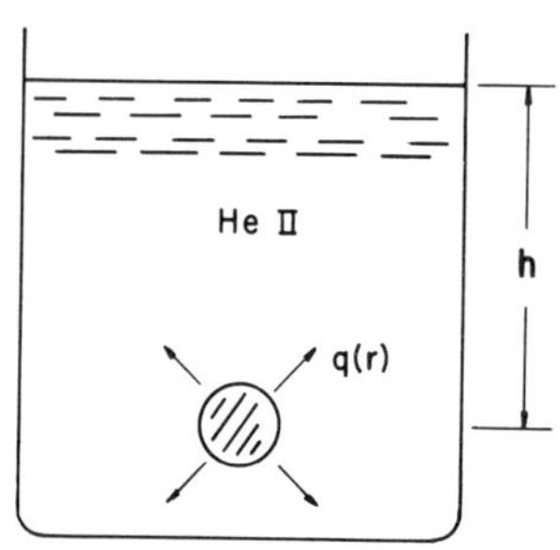

Fig. 5.8. Idealized heat transfer experiment in near-saturated He II.

and assuming the helium vapor to be an ideal gas, that is $v_g = RT/p$, we can approximate by

$$\Delta T_m \simeq \frac{RT^2}{\lambda} \ln\left(1 + \frac{\rho g h}{p_0}\right) \tag{5.10}$$

This expression is suitable for $\Delta T_m \ll T_b$. For larger values of ΔT it is better to evaluate the saturation temperature at the pressure corresponding to the given hydrostatic head. This can be achieved most easily by using tabulated values for vapor pressure.[12]

As an aside it is worth mentioning that the above description is not without controversy. In fact, a number of studies[13–15] have attempted to demonstrate experimentally the existence of superheated liquid in saturated He II. However, all these experiments have been performed using He II contained within a duct. The difficulty here is that the time constant for the steady state in this geometry can be quite long, leading to the belief that the observed superheat may be metastable. The resolution of this dilemma will have to wait for future work in this area.

To attach some quantitative understanding to the above discussion, it is worth noting that the pressure corresponding to the λ-point is $p_\lambda = 4.97$ kPa (37 torr), which is equivalent to a column of helium about 3.55 m high. This fact is important because if a He II system with vertical dimension larger than several meters is constructed, it would experience a heat transfer limitation determined by the He II–He I phase transition. Alternatively, it is possible to create a condition whereby the pressure at the heat transfer surface is arbitrarily high. This can occur in a closed-volume He II region cooled by a heat exchanger to a saturated bath. The pressure on the closed volume can therefore take on any value between saturation and 2.5 MPa.

The subcooled He II state is shown on th phase diagram by position ②. Here it is assumed, for example, that the applied pressure is 100 kPa (1 bar). A similar argument to that presented above applies when determining the temperature excursion; however, in the subcooled case the maximum temperature is governed by the λ-transition (at $p = 100$ kPa, $T_\lambda = 2.163$ K). Thus, the limits are relatively well fixed and only weakly dependent on applied pressure.

The conditions that exist once the maximum heat flux is exceeded are of great importance to understanding the heat transfer in this regime. Generally, there are two cases that can occur, each of which is associated with one of the two conditions indicated on the phase diagram in Fig. 5.7. For the saturation case which applies to position ①, q^* corresponds to the helium adjacent to the interface achieving saturation conditions. A schematic representation of the resulting physical condition for $q > q^*$ is shown in Fig. 5.9a. Displayed is a solid heat transfer surface blanketed by a vapor film which in turn is bounded by the He II at local saturation temperatures. This phase boundary is defined clearly because the He II–vapor transition is first order.

The alternative film boiling heat transfer situation occurs whenever q^* is exceeded under subcooled conditions, such as ② in Fig. 5.7. For this case the phase transition is between He II and normal liquid He I. Since, with rare exception, the maximum heat flux in He I is substantially less than that in He II, exceeding q^* under subcooled conditions invariably results in a double transition, first creating a film of liquid He I followed by boiling of the He I to form a vapor film. This triple-phase phenomenon brings all three helium states in close proximity to the heat transfer interface. A schematic representation of this process is shown in Fig. 5.9b. Through the He II–He I interface, shown as a dotted line in the figure, the density ρ and temperature T should be continuous. Recent visual experiments of boiling in saturated and subcooled He II have allowed observation of the interfaces between the vapor–He I and the He I–He II phases.[16] The latter observation is particularly significant since the physical properties of helium should be continuous through the interface.

5.1.4. Forced Convection Heat Transfer

The subject of steady-state heat transport would not be complete without some discussion of the effect of forced convection on the temperature gradient and peak heat flux. This subject is a somewhat more general heat transport problem than has been considered so far because it includes an additional variable, the net flow velocity **v**. Forced convection was introduced in Chapter 4 as part of the two-fluid model applied to He II heat transport.

Since there have been fewer experimental investigations relating to forced flow He II heat transport, the analysis of the problem is incomplete. A general configuration to consider is shown in Fig. 5.10. A channel of constant cross section and length L connects two reservoirs at temperatures T_1 and T_2. Steady-state heat is applied to one end of the channel and the temperature difference is established, ultimately leading to the peak heat flux q^*. In fact, this configuration is not easily obtainable because a tem-

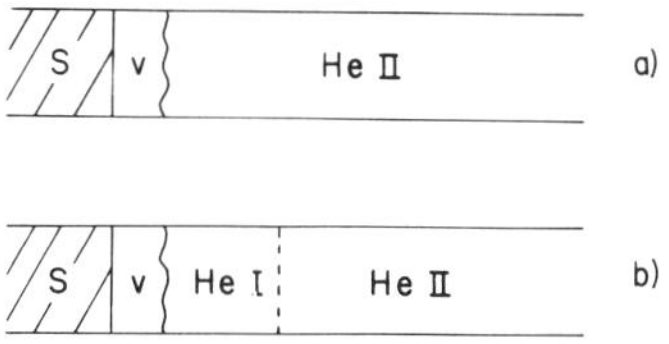

Fig. 5.9. Schematic of boiling in He II: (a) saturation condition and (b) subcooled condition.

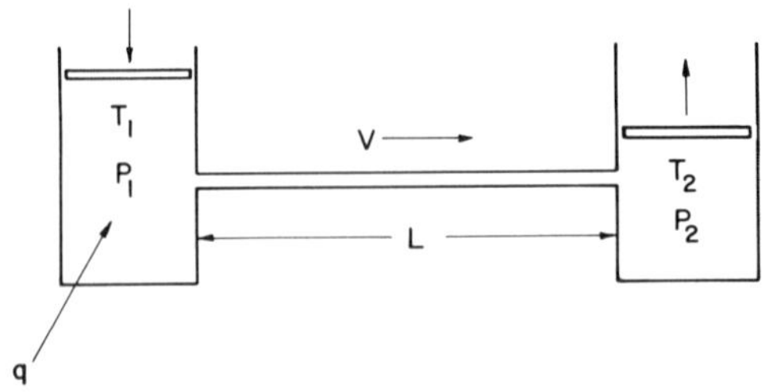

Fig. 5.10. Schematic representation of configuration required to obtain forced flow He II.

perature difference normally corresponds to a pressure difference under saturation conditions. However, it is possible to create the appropriate circumstances with a frictionless, massless piston that forces the liquid from volume 1 to volume 2 at velocity $\mathbf{v}$.

To treat the problem of forced flow He II analytically, it is necessary to solve the heat balance equation that describes the system of interest.[17] In developing this equation, two fundamental assumptions are made about the physical behavior of He II. These assumptions are not proved here but are justified by the analysis of experimental data.

The first assumption is that the heat flow by internal convection mechanisms is not affected by the net velocity of the fluid. This point has been discussed as part of the two-fluid model in Chapter 4. As an aside, it is possible to understand physically the invariance to velocity by analogy to an ordinary heat conduction mechanism. In the latter case, heat transported by conduction in a moving medium is no different from that of the medium at rest provided $\mathbf{v} \ll c$, the speed of sound. Furthermore, by making the normal set of simplifications to reduce the problem to one-dimensional heat flow in turbulent He II, the nonlinear Gorter–Mellink equation can be used to describe the heat conducted by internal convection,

$$q_{ic} = -\left(\frac{1}{f(T)}\frac{dT}{dx}\right)^{1/3} \tag{5.11}$$

where $f(T)$ is the same temperature-dependent heat conductivity function. The power law dependence of (5.11) has been assumed to be 1/3 although the analysis follows the same procedure if a different coefficient is assumed.

The second assumption is that the heat carried by ordinary convection mechanisms can be described by the flow of enthalpy between two points in the system,

$$q_{fc} = \rho \mathbf{v}\, \Delta h \tag{5.12}$$

where $\Delta h = h_1 - h_2$ represents the specific enthalpy difference between temperatures T_1 and T_2, that is, $\Delta h = \int_{T_1}^{T_2} C_p\, dT$. For simplicity (5.12) assumes the fluid density to be constant, which is a reasonably good approximation for He II.

The above two assumptions lead to an equation that is appropriate for analyzing the temperature profile in forced flow He II. This is achieved by combining differential forms of (5.11) and (5.12) and equating them to the time rate of change of the local enthalpy. The resultant equation is similar to the time-dependent heat equation for static He II except that it contains the extra convection term. In one dimension this expression takes the form

$$\frac{d}{dx}\left[\left(\frac{1}{f}\frac{dT}{dx}\right)^{1/3}\right]-\rho\mathbf{v}\frac{dh}{dx}=\rho\frac{dh}{dt} \tag{5.13}$$

Given the boundary conditions for a channel containing He II, it is a straightforward problem to integrate (5.13) and thus determine the temperature profile as a function of flow velocity and time. Unfortunately, this solution requires numerical methods because the equation is nonlinear and the functions such as h and f have rather strong temperature dependencies.

A good approximate solution to the steady-state problem, $dh/dt=0$, can be obtained by assuming constant properties with $\Delta h = C_p\,\Delta T$. This approximation leads to an exactly soluble differential equation. By making the following change of variables

$$\Theta^* = \frac{T-T_2}{T_1-T_2} \tag{5.14a}$$

$$x^* = \frac{x}{L} \tag{5.14b}$$

and

$$K' = 2\rho C_p \mathbf{v}(fL)^{1/3}(T_1-T_2)^{2/3} \tag{5.14c}$$

an exactly soluble form of the Bernoulli equation results,

$$-\frac{d}{dx^*}\left[\left(\frac{d^*\Theta}{dx^*}\right)^{1/3}\right]+K'\frac{d\Theta^*}{dx^*}=0 \tag{5.15}$$

which coupled with the appropriate boundary conditions can determine the steady-state temperature in a one-dimensional channel.

The results of this analysis for a channel of length $2L$ with its center at T_1 and ends fixed at T_2 are displayed in Fig. 5.11. The left-hand side of the figure can be interpreted as the case where the velocity of flow is opposite to the heat flow by conduction while the right-hand side refers to these quantities working in parallel. Note that the limitations of this solution are primarily in terms of the temperature dependence of the heat capacity C_p

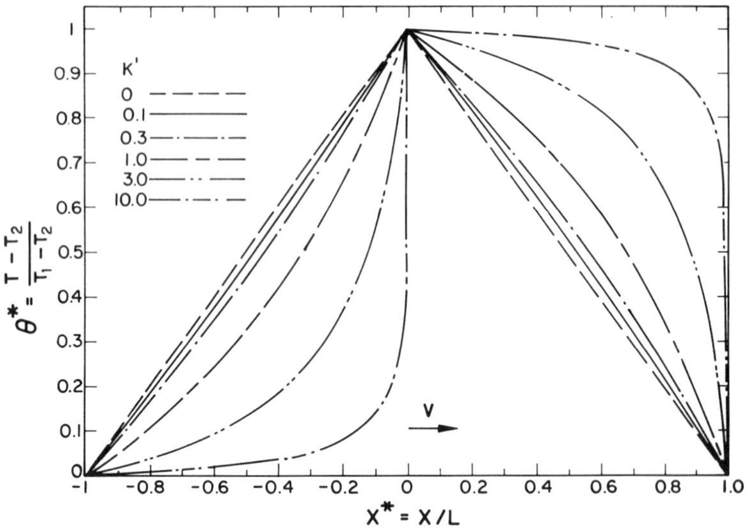

Fig. 5.11. Normalized temperature distribution in forced flow He II with fixed temperature boundary conditions $K' = 2\rho C_p \mathbf{v}(fL)^{1/3}\,(T_1 - T_2)^{2/3}$.

and heat conductivity function $f(T)$. The solution should be quite good for small $\Delta T = T_1 - T_2$ such that $\Delta T/T \ll 1$. The impact of this approximation is seen in the zero velocity profile ($K' = 0$) which is linear, while in fact for large ΔT the profile has considerable curvature. Also note that the nonzero velocity profiles for positive and negative K' are symmetric about the line corresponding to $K' = 0$.

There have been several reports of measured temperature profiles within forced flow He II which have been compared to numerical analyses based on the theory described above.[5,18] In general, these measurements have shown temperature profiles analogous to those displayed in Fig. 5.11. The numerical analysis is more exact because it includes the temperature dependence of the helium properties.

Given the solution to the temperature profile it is straightforward to determine the total heat transport, $q = q_{fc} + q_{ic}$, by integration of (5.15). This result can be normalized to the form

$$\frac{q}{q_0} = -\left(\frac{d\Theta^*}{dx^*}\right)^{1/3} + \frac{K'}{2}\Theta^* \tag{5.16}$$

where K' is defined above and

$$q_0 \equiv \left(\frac{T_1 - T_2}{fL}\right)^{1/3} \tag{5.17}$$

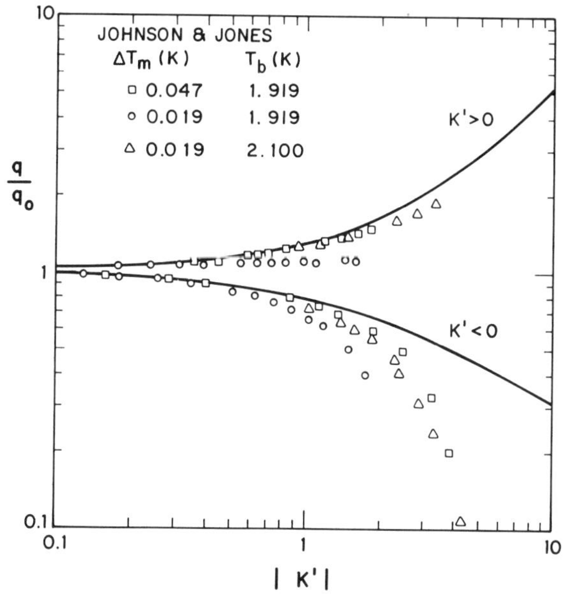

Fig. 5.12. Normalized peak heat flux for forced flow He II with $K' = 2\rho C_p v(fL)^{1/3} (T_1 - T_2)^{2/3}$.

which represents the heat carried by the internal convection mechanisms for He II having zero velocity ($K' = 0$). The results of this calculation are shown by the solid curves in Fig. 5.12. Although there is considerable enhancement of heat transport in the direction of flow even for small K', values of K' greater than unity are required before the total heat transport is enhanced significantly in a midpoint heated channel of length $2L$. This result occurs because forced flow suppresses the total heat transport when the velocity and heat flux are antiparallel.

Johnson and Jones[19] have carried out an extensive investigation of heat transport in saturated forced flow He II. In their experiment a pressure difference was maintained by adjusting the relative height of the two baths. Results obtained required sizable corrections because temperature differences in saturated He II produce a corresponding pressure difference. A comparison of data and analyses is displayed in Fig. 5.12. Although the general trends appear correct, there is considerable discrepancy between the experimental data and theory, particularly for the cases where ΔT is small. A more recent work by Kashani and Van Sciver[18] has shown much closer agreement between theory and the experimental maximum heat flux.

5.2. TRANSIENT HEAT TRANSPORT

A thorough understanding of the time evolution of the temperature distribution in He II is of the utmost importance to describing the heat transfer. Up until this point, the assumption has been made that steady-state heat transfer can be interpreted by using equilibrium thermodynamic models and the helium phase diagram. This assumption is certainly an oversimplification because there are a number of time-dependent phenomena that affect the heat transport properties of He II. To determine the relative significance of transient phenomena in a heat transfer problem, the following question must be asked: At what point in time does a system exposed to some change in the heat flux reach a steay state? To answer the question it is necessary to take account of the times required for the physical condition associated with steady-state heat transport to occur.

Begin by considering a one-dimensional channel of length L cooled by a constant temperature bath at one end. Initially, the temperature of this system is uniform at T_b. However, if a heat flux is applied at the end away from the bath, a temperature distribution will evolve until eventually a steady-state condition is achieved with the temperature at the heated end being $T_m > T_b$. Both transient and steady-state temperature distributions are shown in Fig. 5.13.

There are a number of energy inputs required to achieve the steady-state temperature distribution. The first of these is associated with the acceleration of the two fluid components, normal and superfluid, to $\mathbf{v}_n$ and $\mathbf{v}_s$, respectively. Since the fluids are initially at rest, it is necessary to apply

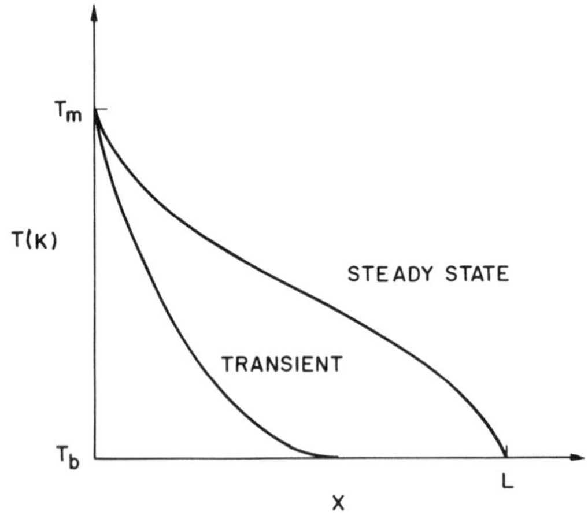

Fig. 5.13. Schematic temperature distributions in He II under steady-state and transient conditions.

sufficient kinetic energy for the steady-state distribution to be established. For typical heat fluxes, the normal fluid velocity is of the order of a few cm/s, which leads to a kinetic energy density of less than 1 erg/cm^3. This energy is very small on the scale of interest.

Since the He II is initially in the nonturbulent state, a second energy which must be applied to the system is that necessary to create the turbulence. This process has been described physically in Chapter 4. The result is that the time required to create turbulence in a channel is given by

$$\tau = aq^{-3/2} \tag{4.110}$$

where a is a temperature-dependent function of the order of 0.1 W$^{3/2}\cdot$s/cm^3. Equation (4.110) has no length dependence, presumably because the growth of turbulence travels at the velocity of second sound and therefore would be essentially uniform in a moderate-length channel. Using (4.110) we can determine the energy required to produce turbulence:

$$q\tau = aq^{-1/2} \tag{5.18}$$

which for heat fluxes in the range of a few W/cm^2 gives an estimated value of about 0.1 J/cm^2. For a channel having a length of the order of 1 m, the energy density associated with the creation of turbulence works out to be $q\tau \approx 1$ mJ/cm^3.

The third principal energy input required to create the steady-state temperature distribution is associated with the enthalpy content of the He II. The heat capacity of liquid helium is rather large, particularly near the λ-transition where on a volumetric basis it is of the order of 1 J/cm$^3\cdot$K. To establish the steady-state temperature distribution sufficient heat must be applied to increase the fluid temperature from that of the bath to its local steady-state value $T(x)$. In practical systems this increase is of the order of 0.1 K. The required energy density needed to achieve a given temperature distribution can be written

$$\varepsilon_C = \frac{\rho}{A_0 L} \int_0^L dx \int_{T_b}^{T(x)} C_p(T)\, dT \tag{5.19}$$

which is of the order of 100 mJ/cm^3. Furthermore, as the length of the channel increases the dominance of this term increases, particularly for long channels with $L \gtrsim 1$ m. Because the enthalpy profile dominates the transient heat input, other energies associated with this problem rarely are considered. Therefore, it is assumed that the transient temperature distribution is controlled by heat transport and enthalpy considerations. As a result, a diffusionlike model has been shown to describe the problem effectively.

In a solid, the main mechanism of heat transport is conduction. This mechanism is described by a well-known differential equation,

$$\frac{\partial T}{\partial t} = D\,\nabla^2 T \tag{5.20}$$

where $D = k/\rho C$ is the thermal diffusivity having units of m^2/s. The form of (5.20) has the inherent assumption of constant properties, k, ρ, and C. Otherwise, the equation is somewhat more complex. Equation (5.20) has been solved for a plethora of boundary conditions with dimensionalized results applicable to a variety of problems.[20,21] Without solving (5.20) for a particular set of conditions, some physical discussion of its implications can be made. In particular, it is possible to construct a characteristic time, called the diffusion time τ_D, which has the form

$$\tau_D = \frac{L^2}{D} \tag{5.21}$$

where the length L is the relevant length in the problem at hand. For the case of the one-dimensional channel heated at one end, L is the overall channel length. The diffusion time is a measure of the thermal relaxation of the system. Since the conduction is dominated by exponential terms, τ_D is not the actual relaxation time but is proportional to the time required to reach the steady state. To achieve a high rate of thermal diffusion it is necessary to have a large thermal diffusivity. Solids possessing the largest values of D are high-conductivity metals at low temperatures, where their specific heat is small. For example, high-purity copper near 4 K has $\rho C \approx 1\ \mathrm{kJ/m^3}$ and a thermal conductivity $k \approx 1\ \mathrm{kW/m \cdot K}$ which gives $D \approx 1\ \mathrm{m^2/s}$. For high-purity metals at low temperatures the diffusion time, assuming a characteristic length $L \approx 1$ m, is of the order of seconds.

Although He II clearly does not obey the simple diffusion equation (5.20), it is useful nonetheless for physical understanding to make some order-of-magnitude comparisons. Since turbulent He II in one dimension obeys the nonlinear heat flow equation given by (5.1), it is possible to define an effective thermal conductivity, albeit dependent on heat flux,

$$k_{\text{eff}} = \frac{1}{f(T)\,q^2} \tag{5.22}$$

As can be seen in Fig. 5.3., the quantity $f^{-1}(T)$ typically has values around 1000 $\mathrm{W^3/cm^5 \cdot K}$. Therefore, for a heat flux $q = 1\ \mathrm{W/cm^2}$, the effective thermal conductivity is around 1000 $\mathrm{W/cm \cdot K}$ or 100 $\mathrm{kW/m \cdot K}$, which is about two orders of magnitude larger than for high-purity metals at low temperatures. Of course, the heat flux dependence of k_{eff} works against the transport of large heat fluxes. At 10 W/cm^2, k_{eff} is reduced by two orders of magnitude to around 1 kW/m · K, which is comparable to that of copper in the same temperature range.

The effective thermal diffusivity of He II, D_{eff}, can also be defined according to the analogy with solid conduction. Around 1.9 K, $\rho C(1.9\ \mathrm{K}) \approx 0.5\ \mathrm{MJ/m^3 \cdot K}$, which gives a value for $D_{\mathrm{eff}} \approx 0.2\ \mathrm{m^2/s}$. For a characteristic length of 1 m, this effective thermal diffusivity leads to a characteristic time constant $\tau_D \approx 5$ s. Furthermore, since the diffusion time goes as the square of the characteristic dimension, larger systems have time constants that can be long compared to other characteristic times in the problem.

To treat a variety of transient heat transfer problems in He II with a heat diffusion model, it is necessary to develop a general equation. The derivation of the equation used to describe heat diffusion is analogous to that of the ordinary diffusion equation although the thermal conduction relationship must be replaced with the nonlinear Gorter–Mellink expression. In the usual way, an energy balance is established for an element of space, as is shown in Fig. 5.14. The difference between the heat flux in and that exiting the element is just the time rate of change of the enthalpy contained within the volume. The result of this analysis, written in the form of a one-dimensional heat diffusion problem, is

$$\rho C_p \frac{\partial T}{\partial t} = \frac{\partial}{\partial x}\left(\frac{1}{f(T)}\frac{\partial T}{\partial x}\right)^{1/m} \tag{5.23}$$

where $m \approx 3$ as determined by the choice of heat conduction equation. This expression has a very similar appearance to the ordinary diffusion equation, with the one exception that it involves an unusual power of the temperature gradient. Because of the nonlinear character of (5.23), it is apparent that extraordinary efforts are needed to solve the equation. There are in fact two methods available to treat this equation, both of which lead to approximate solutions.

The most elegant approach to the solution of this nonlinear heat transport equation employs the method of similarity solutions.[22–24] This

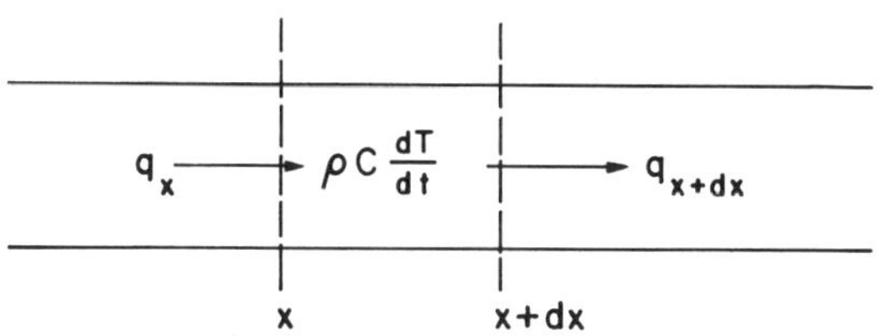

Fig. 5.14. Heat balance within a channel containing He II.

approach makes use of changes of variables which reduce the nonlinear partial differential equation (5.23) to a nonlinear ordinary differential equation. Assuming that the heat conductivity function and specific heat are constant over the range of interest and making the appropriate change of variables leads to a partial differential equation of the form

$$\frac{\partial \Theta}{\partial \tau} = \frac{\partial}{\partial x}\left[\left(\frac{\partial \Theta}{\partial x}\right)^{1/3}\right] \tag{5.24a}$$

where

$$\Theta = \frac{T - T_b}{T_\lambda - T_b} \tag{5.24b}$$

and

$$\tau \equiv \frac{t}{f^{1/3}\rho C(T_\lambda - T_b)^{2/3}} \tag{5.24c}$$

Note that this assumption of constant properties is a poor approximation for He II when $\Theta \approx 1$.

The solution of (5.24) is determined by choice of boundary conditions. Considered here are two problems of interest. The first problem concerns the application of a constant heat flux q at $x = 0$, which is referred to as the clamped flux problem.[23] It leads to boundary conditions of the form

$$\left.\frac{\partial \Theta}{\partial x}\right|_{x=0} = -\frac{q^3 f}{T_\lambda - T_b} \quad \text{for all } t \tag{5.25a}$$

and

$$\Theta(x, 0) = 0 \tag{5.25b}$$

$$\Theta(\infty, \tau) = 0 \tag{5.25c}$$

The first of these conditions originates from the requirement that the heat flux q obey the nonlinear heat conductivity equation at $x = 0$.

The similarity solution method uses what are termed stretching transformations which leave the partial differential equation unchanged. For the case at hand, the transformations can be written

$$x' = \lambda x, \quad \tau' = \lambda^\beta \tau, \quad \text{and} \quad \Theta' = \lambda^\alpha \Theta \tag{5.26}$$

where λ is a positive quantity and the parameters α and β obey the relationship $2\alpha - 3\beta + 4 = 0$. The boundary conditions for the clamped flux problem are also invariant if $\alpha = 1$ and $\beta = 2$. Invariance of the solution to the transformations (5.26) requires that it take the form

$$\Theta/\tau^{1/2} = y(x/\tau^{1/2}) \tag{5.27}$$

where y is an undetermined function. Substituting (5.27) into the original partial differential equation (5.24a) reduces it to an ordinary differential equation,

$$\tau \frac{d}{dx}\left(\frac{dy}{dx}\right)^{1/3} + x\frac{dy}{dx} - y = 0 \tag{5.28}$$

This equation has been solved by a combination of numerical and graphical methods.[22]

The first observation gleaned from the above analysis is that the general solution for the temperature distribution should take a form consistent with (5.27). Therefore, a plot of experimental data in this form should provide a universal curve representing the function y. Displayed in Fig. 5.15 are interpolated results from the long one-dimensional channel described earlier. The data do in fact follow a universal curve. In addition, the magnitude of the general solution can be determined from the exact solution to (5.28). This solution has a simple form when evaluated at $x = 0$:

$$\left.\frac{\Delta T}{\tau^{1/2}}\right|_{x=0} = a^2 q^2 \left(\frac{f}{\rho c}\right)^{1/2} \tag{5.29}$$

where a is a proportionality constant of order unity. At large values of x, the result approaches an asymptotic dependence such that

$$\frac{\Delta T}{\tau^{1/2}} = \left(\frac{4\sqrt{3}}{9}\right)\left(\frac{1}{\rho c}\right)^{3/2}\left(\frac{1}{f}\right)^{1/2} \frac{\tau}{x^2} \tag{5.30}$$

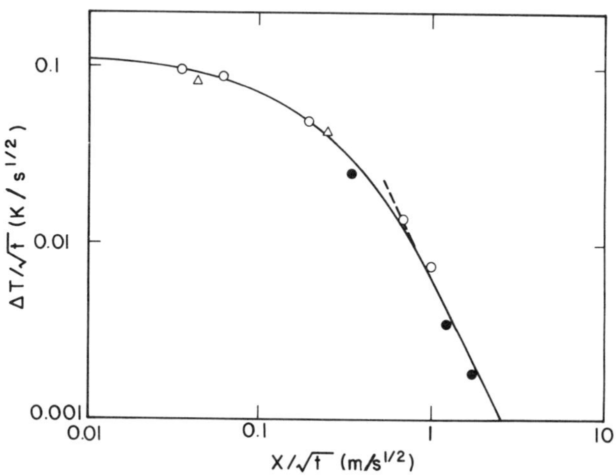

Fig. 5.15. Normalized temperature distribution for the clamped flux problem. Experimental data from several workers are displayed (as compiled by Dresner[22]).

This expression is exemplified by the linear region on a log–log plot for large $x/\tau^{1/2}$

An additional success of the theory is in its ability to predict the critical energy flux ΔE^* beyond which the heat transfer enters film boiling. For the clamped flux problem, this quantity has been measured experimentally by detecting the time to film boiling, Δt^*, for a given heat flux. A relationship that fits the experimental data has the simple form

$$\Delta t^* = \kappa q^{-4} \tag{5.31}$$

where κ is a temperature-dependent function.[25] The same relationship also follows from the above theoretical analysis, and κ is predicted to take the form

$$\kappa = \frac{\rho c(T_\lambda - T_b)^2}{f a^4} \tag{5.32}$$

where a is the same numerical coefficient as given in (5.29). By inserting average values for the heat conductivity function and the volumetric specific heat into (5.32), the predicted coefficient κ agrees with experiment to within 20%.

The other problem of interest in one-dimensional transient heat transfer is the determination of the temperature profile resulting from a given energy deposition. This pulsed-source problem, which can also be solved by the similarity solution method,[21] must obey the boundary conditions

$$\int_{-\infty}^{\infty} \rho C(T - T_b)\, dx = \Delta E \quad \text{for all time} \tag{5.33a}$$

$$\Theta(x, 0) = 0 \tag{5.33b}$$

$$\Theta(\infty, \tau) = 0 \tag{5.33c}$$

where $\Delta E = q\, \Delta t$ is the total thermal energy applied per unit channel area. It follows that the pulsed-source problem must have a solution of the form

$$\Theta \tau^{3/2} = y'(x/\tau^{3/2}) \tag{5.34}$$

where y' is another undetermined function. As in the clamped flux problem, (5.34) is a powerful result because it indicates the form in which to plot data. Plotted in terms of these variables, the experimental data should follow a universal curve.

By substitution of (5.34) in (5.24a), a solution to the temperature distribution can be obtained with the simplifying assumption of constant properties. The details of this analysis are available in Ref. 24. The resulting equation takes the form

$$\Delta T\, t^{3/2} = \frac{4}{3\sqrt{3}} \Delta E \left(\frac{f}{\rho c}\right)^{1/2} \left(\frac{1}{Z^4 + b^4}\right)^{1/2} \tag{5.35a}$$

where

$$Z = x\, \Delta E(\rho C f)^{1/2}\, t^{-3/2} \tag{5.35b}$$

and

$$b = \frac{2[\Gamma(\frac{1}{4})]^2}{3\sqrt{3\pi}} = 2.855 \tag{5.35c}$$

That this solution fits the original heat conductivity equation can be verified by the reader.

Experimental verification of the above analysis is displayed in Fig. 5.16. The inset shows the time variation of the temperature distribution

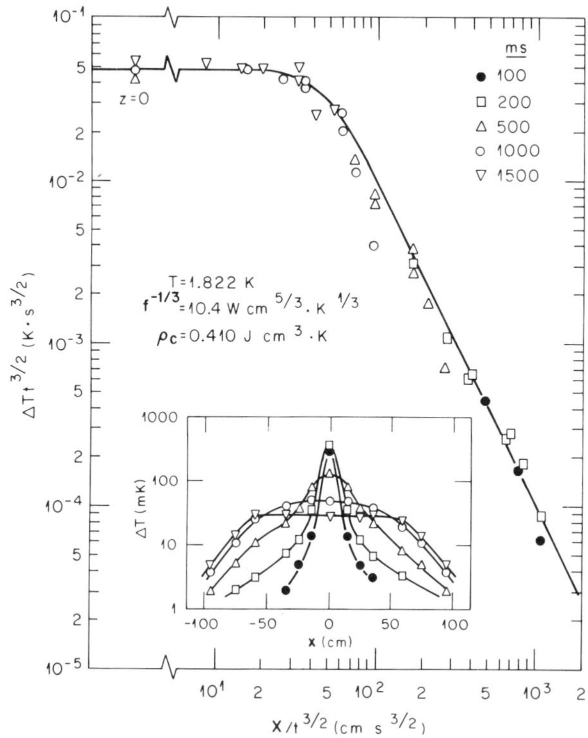

Fig. 5.16. Normalized temperature distribution for the helium channel subjected to a pulsed heat source (from Dresner[23]).

after a discrete energy pulse is applied. These data, plotted in the form suggested by the similarity solution, map out the universal curve given by (5.35). The universal curve plotted in the figure has as its necessary input average values for the heat conductivity function and heat capacity, but when ΔT is small, these properties do not vary substantially over the range of interest. The agreement between experiment and theory is entirely adequate for engineering applications.

The above analytical treatment is useful for providing a physical description of the problem as well as developing scaling relationships to correlate data. However, a complete solution including the temperature dependence of the physical parameters is only possible by numerical methods. The usual approach is to apply finite difference methods. Here the exact differentials in T, x, and t are replaced by finite differences with an appropriate choice of mesh size. These in turn lead to a set of simultaneous equations for the temperature at the nodes in the mesh. Solutions of this type have been carried out for two different boundary conditions.[26] In both cases the clamped heat flux condition at $x=0$ is assumed. However, different boundary conditions are applied at $x=L$ where the channel contacts the reservoir. These conditions have two forms,

$$\left.\frac{d\Theta}{dx}\right|_{x=L}=0 \quad \text{adiabatic end} \tag{5.36a}$$

and

$$\Theta|_{x=L}=0 \quad \text{isothermal end} \tag{5.36b}$$

Particularly for the first case, it is useful to scale the data in terms of a ratio of energies ($\Delta E^*/\Delta E_0$), where these quantities are defined individually as

$$\Delta E^* = q\,\Delta t^* \tag{5.37}$$

and

$$\Delta E_0 = L\int_{T_b}^{T_\lambda} \rho C_p\, dT \tag{5.38}$$

Therefore, (5.37) represents the energy applied through the heat transfer surface which is necessary to reach film boiling at Δt^*. The other energy, ΔE_0, represents the total enthalpy available within the constant cross section channel of length L. Obviously, for the adiabatic end condition, the ratio $\Delta E^*/\Delta E_0 \leqslant 1$. For short, high-energy heat pulses such that the heat diffusion length is much shorter than the total channel length L, the solution is independent of choice of boundary condition established at $x=L$.

Plotted in Fig. 5.17 are the numerical solutions to the one-dimensional

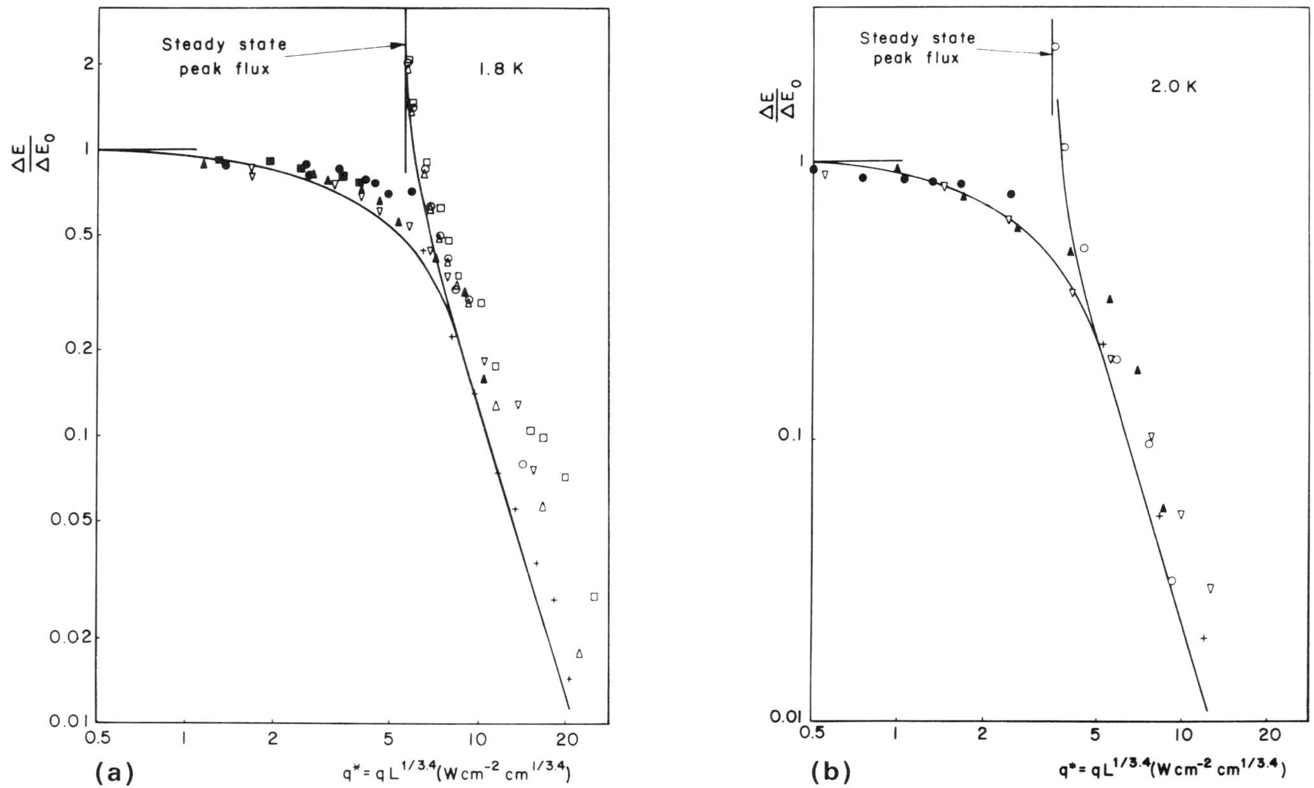

Fig. 5.17. Maximum energy flux for a step-function heat pulse. Steady-state peak heat flux represents the open-channel solution. $\Delta E/\Delta E_0 \leqslant 1$ is the closed-channel solution: (a) 1.8 K and (b) 2.0 K (as compiled by Seyfert et al.[26]).

heat transfer problem for two different bath temperatures, 1.8 and 2.0 K. In either case the ratio of energy is plotted versus the scaling parameter $qL^{1/m}$, where in this case $m = 3.4$ by selection of the authors. The data give comparable agreement for $m = 3$. Note that for values of $qL^{1/3.4}$ in excess of 10 at 1.8 K and 5 at 2.0 K, there is no significant difference between the two boundary conditions at $x = L$. The steady-state limit corresponds to the results presented in Section 5.1.1.

In conclusion, a general comment is in order concerning transient heat transfer in He II. As noted above the time constant to establish the steady state can be quite large, particularly in long one-dimensional system. Although this fact may be surprising considering the extremely high effective thermal conductivity of He II, it is due to the large heat capacity of He II resulting in a finite thermal diffusivity. Therefore, care should be exercised when applying steady-state heat transfer models. It is best first to estimate the thermal diffusion time to determine whether the system is actually in the steady state. For transient problems, the heat diffusion model is generally suitable. Unfortunately, it is not known whether there are limits of this model, and extrapolation beyond the regions where experimental data exist should be avoided.

5.3. KAPITZA CONDUCTANCE

A totally different problem of heat transfer in He II is that which occurs at an interface between a solid and the liquid. This process is in contrast to heat transport in the bulk fluid, which has been the subject so far. Surface heat transfer is more controlled by the interfacial character, including the properties of the solid state rather than that of the bulk He II. In general, there are two regimes of surface heat transfer in He II as exemplified by the two positive slope portions of the heat transfer curve; see Fig. 5.1. At low ΔT, no boiling occurs and the heat transfer is controlled by a phenomenon called Kapitza conductance. At high ΔT and for heat fluxes greater than q^*, the surface is blanketed by a film of He I or vapor or both. In this region the heat tranfer is controlled primarily by the character of this film. The present section concerns itself with the first problem, that of heat transfer directly from the solid surface into the liquid He II or Kapitza conductance. Section 5.4 reviews the subject of film boiling.

Thermal boundary conductance occurring at the interface between a solid and liquid He II was first discovered by Kapitza[27] in 1941 during an experiment to study the flow of heat around a copper block immersed in the liquid. Within the liquid helium the temperature gradients were seen to be negligible; however, a sizable temperature difference did occur between the copper block and the He II. This discovery has spawned a considerable quantity of fundamental and applied research. Kapitza conductance is of great technical interest because it often results in the largest temperature differences in a heat transfer problem. For an order of magnitude comparison, a heat flux of 0.1 W/cm^2 can lead to a temperature difference of about 100 mK across an interface due to this effect. Within turbulent He II the same heat flux of 0.1 W/cm^2 would require about 1000 m of one-dimensional channel to produce the same temperature difference!

The general term Kapitza conductance has taken on much broader connotation in recent years. In particular, it now refers to the interfacial thermal boundary conductance which occurs between any two dissimilar materials where electronic transport does not contribute. Thus, Kapitza conductance occurs at the interface between a metal and water at room temperature. However, since the effect is strongly temperature dependent it makes a negligible contribution to the heat transfer coefficient except at low temperatures.

The measurement of Kapitza conductance is achieved by a method shown schematically in Fig. 5.18a. A solid of some finite thermal conductivity is in intimate contact with He II contained in a one-dimensional configuration. The temperature at various points within the solid and He II are measured as they vary with applied heat flux q. In the steady state, a temperature profile is obtained as shown schematically in Fig. 5.18b. The profile can be extrapolated to the He II-solid interface to determine the surface temperature difference ΔT_s. By this method, the bulk properties of the two media can be eliminated from the measured temperature differences.

Kapitza conductance usually is given an empirical definition. Ideally, the Kapitza conductance is defined in the limit where q and ΔT_s are vanishingly small, that is,

$$h_{K0} = \lim_{\Delta T_s \to 0} \frac{q}{\Delta T_s} \tag{5.39}$$

where the 0 subscript refers to the limit as $\Delta T_s \to 0$. This quantity has a

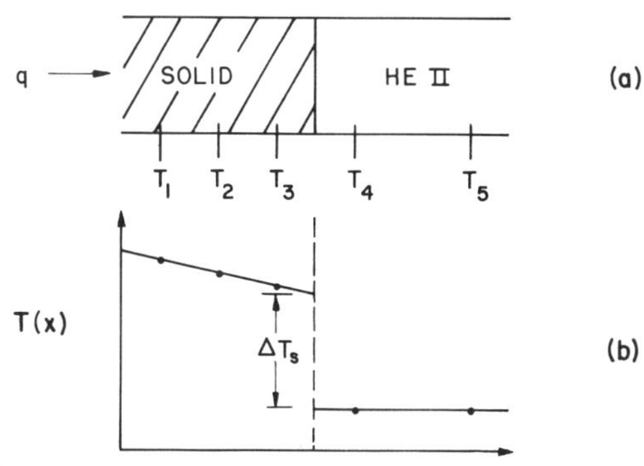

Fig. 5.18. Schematic of Kapitza conductance experiment: (a) temperature sensors located in the vicinity of a solid–He II interface and (b) temperature profile.

fairly strong temperature dependence going as T^n with n varying anywhere from around 2 to 4. A more general definition of Kapitza conductance simply relates it to finite values of q and ΔT_s:

$$h_K = \frac{q}{\Delta T_s} \tag{5.40}$$

Because of its nonlinear nature, definition (5.40) is of more practical interest to engineering applications.

There are a number of applications for He II where knowledge of the Kapitza conductance is of substantial importance. In refrigeration involving He II, its value strongly impacts the proper design of components, particularly heat exchangers. Because of the strong temperature dependence, the importance of Kapitza conductance to heat transfer problems increases with decreasing temperature. In very-low-temperature dilution refrigeration the Kapitza conductance becomes the dominating heat transfer process. As a result, the majority of the recent measurements[28] have been performed in the temperature range relevant to this mode of cooling. Knowledge of the Kapitza conductance of materials at higher temperatures, $T > 1$ K, is also important. Here the desire is to cool large devices such as superconducting magnets or space instruments. For proper design of these devices, it is necessary to have an estimate for the effective heat transfer coefficient. In nonboiling He II the heat transfer coefficient is all the result of Kapitza conductance. Although Kapitza conductance is an experimentally defined quantity, there has been considerable theoretical work aimed at understanding this complex phenomenon. Therefore, before discussing the empirical behavior of h_K any further, a review of the physical theories used to explain its behavior is presented.

5.3.1. Phonon Radiation Limit

The first theory to successfully characterize the qualitative features of Kapitza conductance is referred to simply as the phonon radiation limit.[29–31] The model is an overestimate of the true Kapitza conductance because it includes too many energy transport mechanisms. However, the theory does show the proper temperature dependence of h_K.

Any body above absolute zero has thermal energy, which in the case of insulators is in the form of a phonon spectrum while for conductors it may be due partially to the electron density of states. Phonons are quantized lattice vibrations and are analogous to photons, which are quantized electromagnetic radiation. Ignoring for the moment any effect electrons may have on these concepts, it is reasonably straightforward to identify the phonon energy spectrum (see Chapter 2). Since Kapitza conductance is mainly of interest at low temperatures, it is not a bad approximation to use the Debye theory to describe this energy spectrum. In the Debye model, the internal energy may be written as a temperature-dependent quantity,

$$E_{\text{ph}} = aT^4 \tag{5.41}$$

where $a = \frac{3}{5}\pi^4 (N/V) k_B/\Theta_D^3$ and $T \ll \Theta_D$, the Debye temperature. For most solids, the Debye temperature is in the range of several hundred kelvins, making this approximation quite reasonable for the Kapitza conductance at helium temperatures.

To quantify the problem of phonon radiation between two media, assume there is a unit interfacial area dA on which phonons are incident at velocity c. A schematic of the hemispherical region surrounding this elemental area is shown in Fig. 5.19. The angle of incidence of the phonon is given in spherical coordinates by θ and ψ. Now assume that the perpendicular component of the incident phonon transmits the energy. More detailed theory[28] includes the coupling of transverse phonons at the interface. For simplicity, the present treatment will neglect this effect. The perpendicular component of the velocity can be written in terms of the angle θ such that

$$c_\perp = c \cos\theta \tag{5.42}$$

The heat flux into the area dA is the product of the energy density and perpendicular velocity component such that $dq = c_\perp E_{ph}$. It follows that the net heat flux per unit area is obtained by integration of dq over the hemisphere in θ and ψ. This procedure yields

$$\begin{aligned} q &= \frac{1}{4\pi}\int_0^{2\pi} d\psi \int_0^{\pi/2} cE_{\text{ph}} \sin\theta\cos\theta\, d\theta \\ &= \tfrac{1}{4}cE_{\text{ph}} \end{aligned} \tag{5.43}$$

In the Debye approximation, the speed of sound is defined as

$$c = \frac{k_B\Theta_D}{\hbar}\left(\frac{6\pi^2 N}{V}\right)^{-1/3} \tag{5.44}$$

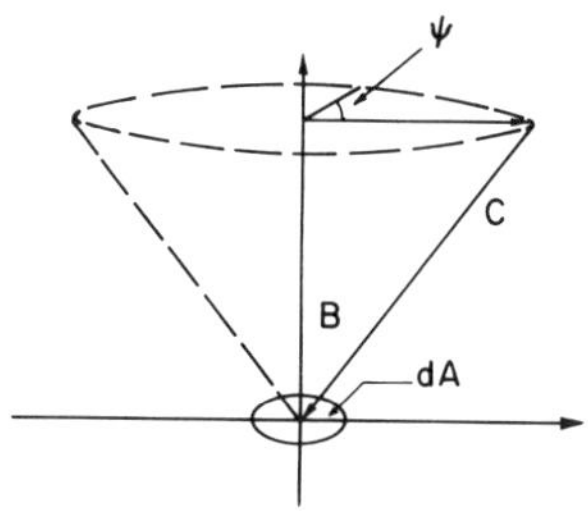

Fig. 5.19. Hemispherical region surrounding an elemental surface area for phonon heat transfer.

where for solids c is of the order of 3 km/s while for He II the speed of sound never exceeds 240 m/s. Substituting (5.44) and (5.41) into (5.43), we obtain an equation for the total heat flux carried by phonon radiation:

$$q = \sigma T^4 \tag{5.45a}$$

and

$$\sigma = \frac{\pi^4}{10\,\hbar}\left(\frac{k_B}{\Theta_D}\right)^2\left(\frac{3N}{4\pi V}\right)^{2/3} \tag{5.45b}$$

This form is analogous to the heat transported by photon radiation. Note that (5.45) includes the variable material properties through the molar volume and Debye temperature. The quantity σ can vary considerably between materials, which according to the theory leads to a quite different heat transport character. For example, since σ is inversely proportional to c^2, it follows that the ratio of heat fluxes by phonon radiation should differ by a factor of 100 between He II and solids at low temperatures. This is one of the major weaknesses of the phonon radiation theory in that it neglects the differing thermal character of the two media.

Now consider the radiation of phonons between two different media, between which there exists an interface. For the sake of discussion, assume there is no appreciable temperature gradient occuring in either bulk material and that the flow of heat from one side to the other produces an interfacial temperature difference ΔT_s. In order for this interface to be defined as a boundary between two bulk media, it must be confined to a thickness that is small compared to the characteristic phonon wavelength, $\lambda_D = hc/k_B\Theta_D$. For solids $\lambda_D \approx 100\ \mu$m, which is large compared to most interfacial dimensions.

The net heat flux through the interface is actually a difference between two values, the radient energy incident on the high-temperature side, $q(T+\Delta T)$, minus that incident from the low-temperature side, $q(T)$. Explicitly, the net heat flux may be written

$$q_{\text{net}} = q(T+\Delta T) - q(T) \tag{5.46}$$

Assume that these quantities can be described in terms of the phonon heat flux expression (5.45) and also that the coefficient σ is the same in either media. It follows that the net heat flux is a difference between the phonon radiation expressions of the two media:

$$q_{\text{net}} = \sigma(T+\Delta T)^4 - \sigma T^4 \tag{5.47a}$$

$$= 4\sigma T^3\,\Delta T\left[1 + \frac{3}{2}\frac{\Delta T}{T} + \left(\frac{\Delta T}{T}\right)^2 + \frac{1}{4}\left(\frac{\Delta T}{T}\right)^3\right] \tag{5.47b}$$

In the limit of small ΔT, (5.47) gives an explicit definition for the phonon radiation limit Kapitza conductance in terms of σ:

$$h_K^p = 4\sigma T^3 \tag{5.48}$$

where the superscript p refers to the phonon radiation limit. The assumption that σ is independent of media clearly is not accurate, particularly when dealing with interfaces between solids and He II. However, it is argued that the smaller value of σ—that of the solid—controls the heat transfer because the net heat flux q_{net} must be in the direction from high to low temperature. As a reference point, the phonon radiation limit applied to copper ($\Theta_D = 343$ K, $N/V = 0.86 \times 10^{23}$ cm^{-3}) predicts a Kapitza conductance of

$$h_K^p = 4.4\ T^3\ \text{kW}/m^2 \cdot \text{K} \tag{5.49}$$

or 30 kW/m$^2 \cdot$ K at 1.9 K. It is of interest to compare the Kapitza conductance predicted by the phonon radiation limit with experiment.

Listed in Table 5.1 are calculated values from the phonon radiation limit and the highest values obtained experimentally for typical metals and nonmetals. It is important to note that the experimental values vary considerably, in some cases by as much as an order of magnitude. However, since the phonon radiation limit should represent an overestimate of h_K, it

Table 5.1. Comparison of Highest Experimental Values for the Kapitza Conductance with the Phonon Radiation Limit[a]

Solid	Θ_D (K)	h_K^p(1.9 K) (kW/m$^2 \cdot$ K)	h_K(1.9 K) (kW/m$^2 \cdot$ K)
Hg	72	440	30
Pb	100	190	32
In	111	171	11
Au	162	155	8.8
Ag	226	55	6
Sn	195	54	12.5
Cu	343	30	7.5
Ni	440	19	4.0
W	405	18	2.5
KCl	230	22	6.9
SiO_2 (quartz)	290	19	5.7
Si	636	6.4	4.2
LiF	750	5.1	4.5
Al_2O_3	1000	1.5	1.6

[a] Compiled by Snyder.[31]

is most appropriate to compare it to the highest experimental values. The first result gleaned from this comparison is that the phonon radiation calculation always gives values for h_K that are higher than experiment, sometimes by as much as an order of magnitude. This fact is consistent with the perception that the phonon radiation limit overestimates h_K. Second, although it is not apparent from the tabular data, the general temperature dependence of h_K predicted by the phonon radiation limit is borne out. Recall that experiment yielded a form for $h_K \approx T^n$ with $2<n<4$. Finally, as with the phonon radiation limit, there is a measurable dependence of h_K on the value of Θ_D for the particular solid. This fact can be seen most clearly in Fig. 5.20, which is a log plot of h_K versus Θ_D^{-1}. The linear interpolation indicates there is a colleration of the form

$$h_K \propto \Theta_D^{-n} \tag{5.50}$$

where the phonon radiation limit predicts $n=2$ and experiment, as shown in Fig. 5.20, yields values of $n \approx 1$.

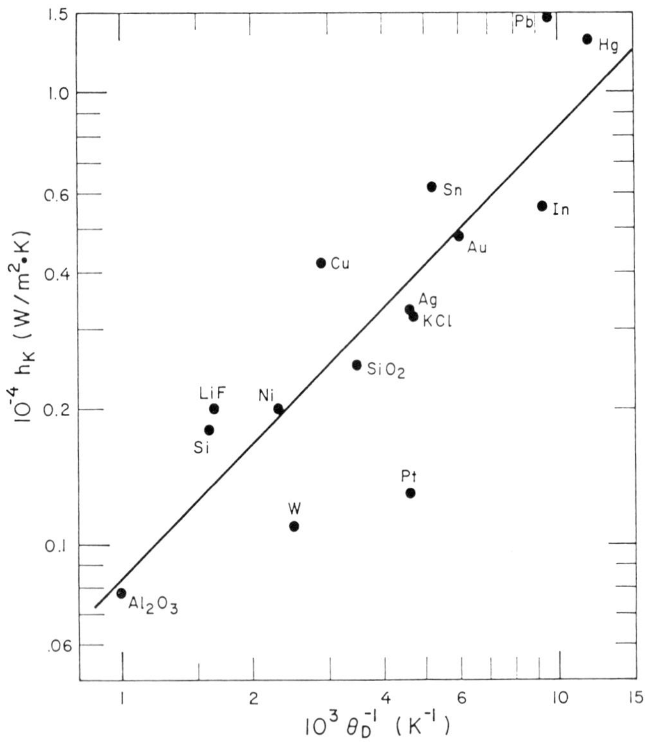

Fig. 5.20. Kapitza conductance at 1.5 K—largest values observed for each solid (compiled by Challis[32]).

There are a number of problems with the phonon radiation limit, particularly when it applies to solid–He II interfaces. It is clearly a crude approximation to a complex problem and is limited by the numerous factors incorrectly accounted for in the theory. These factors include:

1. A failure to distinguish adequately between the different media on either side of the interface. The Debye temperatures that enter the problems are associated with the solid media. Any correction to the theory should include the characteristics of both media.
2. An assumed coupling between both longitudinal and transverse phonon modes. This is particularly a problem for He II where transverse phonons cannot be sustained.
3. Reflections at boundaries are not considered, implying a perfect transmission coefficient. This assumption clearly overestimates the heat transport. A finite reflection coefficient at the interface would be expected, particularly for solid–He II interfaces where the phonon spectra are so different.
4. Crystal structure at the interface is ignored. This is potentially an important factor owing to the solid having long-range order while the liquid is not periodic.
5. Phonons are assumed to be the only heat transport mechanism. However, some experiments have shown that electrons in metals play a significant role in the heat transfer at the interface.
6. The existence of interfacial films or impurities is not considered. Layers of adsorbed impurities can have substantial effect on the heat transfer from practical surfaces. However, even for clean surfaces the detailed character of the helium adjacent to the interface must be considered.

The above factors are part of improved theories of Kapitza conductances. The first such improvement was due to Khalatnikov in 1952.[33] It basically addresses the first three objections to the phonon radiation limit as listed above.

5.3.2. Acoustic Mismatch Theory

The first real advance in the theory of Kapitza conductance was made by the development of the acoustic mismatch theory of Khalatnikov.[33] This theory is based on an analogy with classical acoustics or boundary scattering in optics. In some ways the theory is similar to the phonon radiation limit except that it includes a very important additional mechanism—that of the finite reflection coefficient at the boundary between the two media. The basic approach involves the derivation of the phonon transmission coefficient using conservation of momentum and energy at the boundary. Because the acoustic mismatch theory explicitly accounts for the difference between the two media, it is expected to be a better description for the solid–He II case where the speed of sound differs by an order of magnitude on either side of the interface.

Consider an interfacial region between two media, for example, a solid and He II. This example is of current interest so it will be emphasized throughout the following discussion. The liquid is able to transmit only longitudinal phonons. A schematic of the interface is shown in Fig. 5.21. By analogy with optics, these regions can be thought of as a low-refractive-index solid adjacent to a high-refractive-index liquid. Because of the difference in refractive index, a phonon that is incident on the boundary from the liquid side would be reflected off the interface unless its angle of incidence is less than a critical value, θ_L^c. Within this angle the transmitted phonons are diffracted to an angle θ_s on the solid side. The locus of maximum angles θ_L^c forms a cone of transmission which is determined solely by the speed of sound ratio in the two media:

$$\theta_L^c = \arcsin\left(\frac{c_L}{c_s}\right) \tag{5.51}$$

For interfaces between metals and He II, the ratio c_L/c_s is about 0.1, which corresponds to a critical angle of about 6°. The acoustic mismatch theory assumes that phonons strike the interface between the two media at an angle θ_s but only those that are scattered into an angle less than θ_L^c can be transmitted. All other phonons are reflected internally.

The above argument can be extended to describe the phonon heat transport across an interface. The total heat transport is determined by integration over the cone of transmission, much as was accomplished

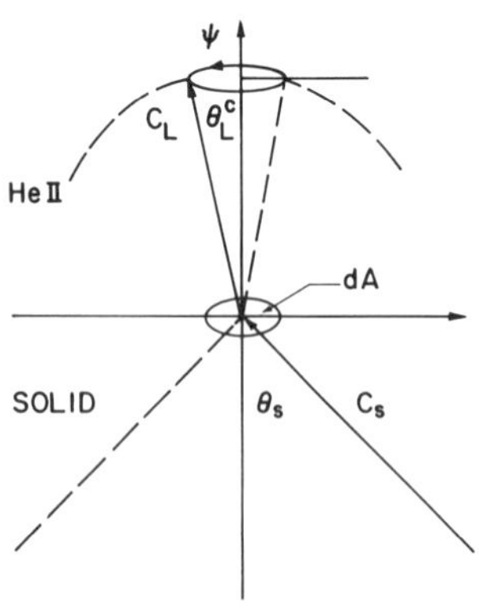

Fig. 5.21. Schematic of interface between solid and He II.

above for phonon radiation although now only angles less than θ_L^c are allowed:

$$q = \frac{1}{4\pi} \int_0^{2\pi} d\psi \int_0^{\theta_L^c} c_L E_{\text{ph}} \sin\theta \cos\theta \, d\theta$$

$$= \frac{1}{4} c_s E_{\text{ph}} \left(\frac{c_L}{c_s}\right)^3 \tag{5.52}$$

Note that (5.52) is similar to the phonon radiation energy transport except that it is modified by the ratio $(c_L/c_s)^3$, which for solid–He II interfaces is of the order of 10^{-3}.

The above expression theoretically predicts the heat transport carried by phonons that are refracted into the angle θ_L^c. However, not all phonons that fall within this cone are actually transmitted, because there is a finite transmission coefficient for phonons incident on the boundary. This transmission coefficient t is given in terms of the acoustic impedance Z of each medium:

$$t = \frac{4 Z_L Z_s}{(Z_L + Z_s)^2} \tag{5.53}$$

where $Z_L = \rho_L c_L$ and $Z_s = \rho_s c_s$. For the case concerning the solid–He II interface it is apparent that $Z_s \gg Z_L$ and (5.53) can be simplified such that

$$t = 4 \frac{Z_L}{Z_s} = 4 \frac{\rho_L c_L}{\rho_s c_s} \tag{5.54}$$

As an example, consider the interface between He II and copper. In this case $Z_s \approx 10^3 Z_L$, which corresponds to a transmission coefficient of approximately 0.5%.

Combining the equations for the heat flux (5.52) and the transmission coefficient (5.54), we find that an expression for the transmitted heat flux is

$$q_t = qt \tag{5.55a}$$

$$= \left(\frac{\rho_L c_L^4}{\rho_s c_s^4}\right) c_s E_{\text{ph}} \tag{5.55b}$$

which is similar to the phonon radiation limit except for the extra term in the parentheses.

An additional aspect to the acoustic mismatch theory concerns the fact that the fluid can sustain only longitudinal phonons. Instead of the expression for the energy density used in the phonon radiation limit, in the acoustic mismath theory the expression must include only longitudinal phonons. Thus, the corresponding form for the energy density is

$$E_{\text{ph}}^{L} = \frac{4\pi^5 k_B^4 T^4}{15\,\hbar^3 c_L^3} \tag{5.56}$$

Combining the energy density equation for longitudinal phonons (5.56) with the transmitted heat flux, we obtain q_t given by (5.55), a temperature-dependent expression for the transmitted heat flux:

$$q_t = \frac{4\pi^5 k_B^4 \rho_L c_L}{15\,\hbar^3 \rho_s c_s^3} T^4 \tag{5.57}$$

As with the phonon radiation limit it is possible to replace the prefactor in (5.57) by σ.

The net heat flux across the interface is obtained by subtracting the incident flux on either boundary. By suitably redefining σ it is possible to use (5.47) and (5.48) to predict the Kapitza conductance in the acoustic mismatch theory. For a small interfacial temperature difference, the result is

$$h_K^A = \frac{16\pi^5 k_B^4 \rho_L c_L}{15\,\hbar^3 \rho_s c_s^3} T^3 \tag{5.58}$$

where the superscript A refers to the acoustic mismatch theory. Note that the expression for the Kapitza conductance derived from the acoustic mismatch theory is determined by the properties of both media, a dependence left out of the phonon radiation limit. A more rigorous calculation in the acoustic mismatch theory[33] replaces the sound speed in the solid, c_s, by its transverse component, c_t. Also a multiplicative factor F, which is a function of the ratio c_t/c, is included. For most solids, F is of the order of unity and so is not a particularly important factor in (5.58).

It is more conveninent to have an expression for the Kapitza conductance in terms of the Debye temperature and other properties of the media. By replacing the transverse speed of sound c_t by its expression in terms of the Debye temperature Θ_D, a simplified expression is obtained:

$$h = \left(\frac{6\pi^4}{5}\right) \frac{RF\rho_L c_L T^3}{M\Theta_D^3} \tag{5.59}$$

where R is the universal gas constant equal to 8.31 J/mol · K and M is the molecular weight. For helium properties at saturation pressure and assuming $F = 1.6$, we can simplify to a useful form,

$$h = 5.5 \times 10^7 \left(\frac{T^3}{M\Theta_D^3} \right) \quad (\mathrm{kW/m^2 \cdot K}) \tag{5.60}$$

where M is in units of g/mol. As is the case with the phonon radiation limit, the acoustic mismatch theory predicts a variation of h_K^A with the Debye temperature as $h_K^A \approx \Theta_D^{-3}$. This dependence is similar to that of the phonon radiation limit, $h_K^P \approx \Theta_D^{-2}$. However, both theoretical approaches overestimate the experimentally determined Debye temperature variation, $h_K \approx \Theta_D^{-n}$ where $n \approx 1$.

5.3.3. Small Heat Flux Kapitza Conductance ($\Delta T \ll T$)

Now consider the relationship between the above theoretical treatments and actual experimental data. Plotted in Fig. 5.22 are model calculations and experimental data for the Kapitza conductance of a He II–copper interface at temperatures above 1.3 K.[31] As can be seen in the figure, the phonon radiation limit forms an upper bound to experimental data consistent with the model excluding boundary scattering effects. Similarly, the Khalatnikov acoustic mismatch theory predicts a Kapitza

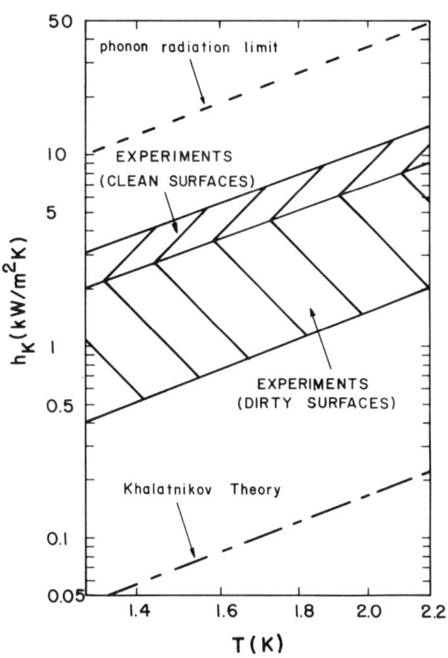

Fig. 5.22. Experimental values for the Kapitza conductance of copper between 1.3 K and T_λ (compiled by Snyder[31]).

conductance about 200 times smaller than the phonon radiation limit. Insertion of numerical values for copper into (5.60) results in the expression for the Kapitza conductance,

$$h_K^A = 0.021\ T^3\ \mathrm{kW/m^2 \cdot K} \tag{5.61}$$

The experimental data displayed in Fig. 5.22 show basically the same temperature dependence with some variations. The best fit to these data indicate h_K proportional to T^n with n ranging between 2 and 4. Perhaps more surprisingly from the experimental viewpoint is that the magnitude of h_K at a given temperature varies by at least an order of magnitude among samples. Part of this variation can be attributed to surface morphology. The upper shaded region in the figure is for copper surfaces that are cleaned carefully either chemically or mechanically and perhaps recrystallized at room temperature to reduce surface strain. On the other hand, lower values are generally obtained for dirty surfaces, indicated by the lower shaded region, for which less effort was made to maintain surface cleanliness. Based on the available experimental data, approximate forms for the Kapitza conductance in this temperature range are suggested:

$$h_K \simeq 0.9\ T^3\ \mathrm{kW/m^2 \cdot K} \quad \text{for clean surface} \tag{5.62}$$

$$\simeq 0.4\ T^3\ \mathrm{kW/m^2 \cdot K} \quad \text{for dirty surfaces} \tag{5.63}$$

with potentially as much as a factor of 2 variation in value.

It can be seen from Fig. 5.22 that although the theoretical treatments bracket the experimental data, neither does a particularly good job of predicting the results. Nevertheless, the physical interpretation contained in the acoustic mismatch theory generally is believed to be correct. In fact, the theory does a much better job of interpreting the magnitude of h_K in the very-low-temperature regime, $T < 0.3$ K.[28,34] Therefore, the problem with the theory appears to be that it does not include additional thermal coupling mechanisms which can make a large contribution, particularly at relatively high temperatures, $T > 1$ K.

A number of improvements to the acoustic mismatch theory have attempted to bring the calculations in closer agreement with experiment. One such improvement adds to the model a high-density helium layer at the interface between the solid and bulk liquid.[34–37] The existence of this layer has been demonstrated in helium adsorption studies.[38] It occurs because the helium molecules are bound tightly to the surface by van der Waal interactions. Since the interfacial region consists of several components—the solid, perhaps two high-density atomic layers of solid helium, and then the liquid—it is possible to have boundary scattering occur at each of these interfaces. Finite phonon transmission and reflection coefficients can be assumed to occur at each boundary. Defining the phonon absorption coefficient ν as the fraction of incident phonons that are absorbed, we can make this an adjustable parameter and fit the data to the best choice of $0 < \nu < 1$. This approach allows for a good fit to experimental data above 0.5 K.[36]

There are several additional features to the Kapitza conductance which are worth noting, partly because they prove the existence of additional physical processes but also because they relate to problems of applied interest. The first feature to consider is the dependence of the Kapitza conductance on externally applied pressure. Plotted in Fig. 5.23 are the ratio of the measured Kapitza resistance ($1/h_K$) at 20 bars to that at saturated vapor pressure.[35,39] Also displayed in the figure are two theoretical treatments, the Khalatnikov theory and the same theory modified to include the dense helium layer without finite reflection coefficients. Note first that the experimentally determined ratio is not much different from unity, particularly in the range of technical interest, above 1 K. The unmodified Khalatnikov theory on the contrary predicts a sizable effect for all temperatures, mostly due to the variation of the fluid properties with pressure. For the modified theory a smaller ratio is predicted, particularly at high temperatures, although it is still above the experimental results. Because the pressure dependence of the Kapitza conductance is not a large effect, it is generally not considered in practical calculations.

A second factor that leads to variations in h_K is the application of a magnetic field. This is an important physical observation for it indicates that there must be other heat transport mechanisms contributing at the interface. In particular, since electrons in the solid are affected by a magnetic field, there must be a coupling between electrons in the metal and phonons in the helium adding to the heat transport. Two types of experiment have been performed to investigate this effect. The first has shown a larger Kapitza conductance for a Type I superconductor for fields above B_C. For lead the ratio h_K (normal)/h_K (superconductor) has been shown to vary between 1.3 and 3 for different samples.[31] This observation

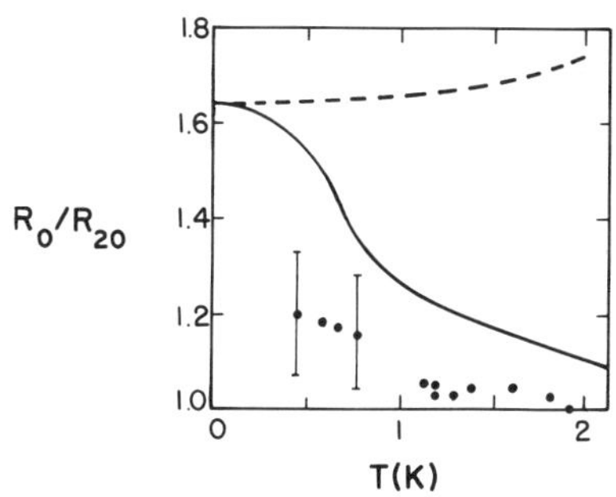

Fig. 5.23. The ratio of the Kapitza resistance under saturated vapor pressure and under a pressure of 20 bars (R_0/R_{20}): ●, experimental points; - - -, Khalatnikov theory; —, dense-layer theory (from Wilks[39]).

is probably the most direct evidence of some type of electron–phonon coupling at the interface. For normal metals such as copper, there have been fewer investigations of the effect of magnetic field on Kapitza conductance. Some reports have indicated h_K for normal metals increases by about 10% in a 1 T magnetic field.[31] However, there is insufficient data available to predict this effect to higher magnetic fields.

Beyond the above two experimental features, there is evidence that the Kapitza conductance is affected strongly by the application of interfacial coatings to the solid, either in the form of applied materials or adsorbed gases. Generally, coatings decrease the apparent Kapitza conductance because they tend to the insulating materials with low thermal conductivities. The effect of a few monolayers of adsorbed gas has not been investigated thoroughly, although it has little importance for practical applications where the surfaces are not kept very clean.

5.3.4. Large Heat Flux Kapitza Conductance ($\Delta T \approx T$)

Up until this point, the theory and experiments associated with the Kapitza conductance for small ΔT have been emphasized. However, in applications of He II large heat fluxes can occur, which in turn result in large temperature differences across the interface such that $\Delta T \approx T$. Because of this occurrence, it is desirable to develop a method of handling the heat transfer process for finite ΔT. Returning to the simplest theories, either phonon radiation limit or acoustic mismatch for finite ΔT, the heat flux through the interface may be written as a sum of terms involving the ratio $(\Delta T/T)$, (5.47). It follows that the Kapitza conductance for finite ΔT is larger than h_{K0} by the magnitude of this expansion,

$$h_K(\Delta T) = h_{K0}\left[1 + \frac{3}{2}\frac{\Delta T}{T} + \left(\frac{\Delta T}{T}\right)^2 + \frac{1}{4}\left(\frac{\Delta T}{T}\right)^3\right] \tag{5.64}$$

where for $\Delta T/T \approx 0.5$, the bracketed quantity is approximately 2. Note that the expansion given by (5.64) makes the initial assumption of an explicit T^3 dependence of h_{K0} consistent with theory. However, experimental measurements vary considerably from this exact form, obeying power laws varying between T^2 and T^4. Some additional characteristic to the Kapitza conductance may also be expected to occur when the surface temperature exceeds T_λ.

There have been several attempts[39–41] to correlate the Kapitza conductance for finite ΔT with the form of Eq. (5.64). These have not been entirely successful largely because of the deviation between the theory and experimental temperature dependence of this effect. Some of this work[41] has suggested a change in the temperature dependence of the Kapitza conductance as the surface temperature exceeds T_λ. This effect may be of importance to understanding the fundamental processes involved in Kapitza heat transfer for large ΔT. However, because the effect is not large, it has little practical importance.

For practical applications it is necessary to have a fairly simple form to describe the heat transfer rate in the Kapitza regime. Based on experimental data and consistent with theory, a general form for the surface heat flux can be written:

$$q_s = a(T_s^n - T_b^n) \tag{5.65}$$

where a and n are adjustable parameters. Note that if one equation is able to fit the experimental data for one sample over the whole temperature difference range, then it should be possible to expand (5.65) consistent with the low heat flux temperature dependence. Similar to the behavior of the experimental data for small ΔT, the high ΔT Kapitza conductance also varies considerably with sample. Plotted in Fig. 5.24 are measurements of T_s as it depends on q_s for different materials. A range of results for copper are shown which indicate the variation with surface preparation.[42] Listed in Table 5.2 is a summary of published data for Kapitza conductance of pure metals at high ΔT. Displayed are the surface preparation, surface temperature at a heat flux $q = 1$ W/cm^2, and the best-fit functional form to these data. Note that most of the fits give values of $n = 3 \pm 0.5$, which is systematically lower than the theoretical value of $n = 4$. Also, the best fit to the coefficient in (5.65) is for $a \simeq 0.05$ W/cm$^2 \cdot$ K^{n+1} but with substantial

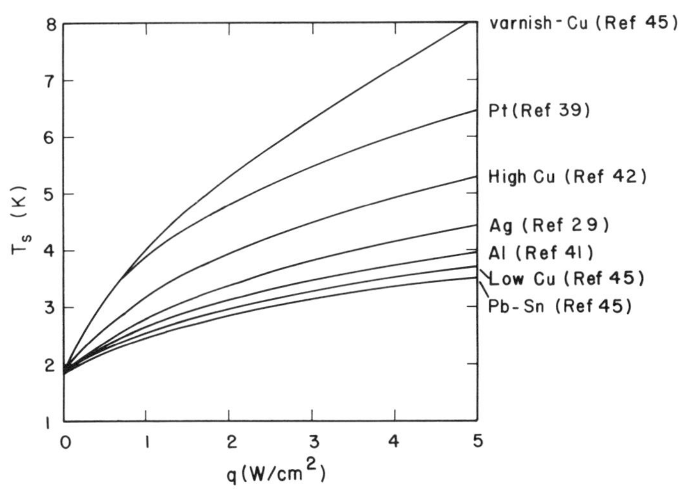

Fig. 5.24. Surface temperature versus heat flux for large ΔT Kapitza conductance.

Table 5.2. High Heat Flux Kapitza Conductance Fitting Parameters for Metals at 1.8 K

Metal	Surface condition	T_s at 1 W/cm^2	α (W/cm$^2 \cdot$ K)	n	Reference
Cu	As received	3.1	0.0486	2.8	43
	Brushed and baked	2.85	to		
	Annealed	2.95	0.02	3.8	43
	Polished	2.67	0.0455	3.45	45
	Oxidized in air for 1 month	2.68	0.046	3.46	45
	oxidized in air at 200° C for 40 min	2.46	0.052	3.7	45
	50–50 PbSn solder coated	2.43	0.076	3.4	45
	Varnish coated	4.0	0.0735	2.05	45
Pt	Machined	3.9	0.019	3.0	39
Ag	Polished	2.8	0.06	3.0	29
Al	Polished	2.66	0.049	3.4	41

variation. Finally, it is interesting to note that the variation in Kapitza conductances at large ΔT is not nearly as great as is obtained in the limit $\Delta T \rightarrow 0$, where an order-of-magnitude deviation in h_K is seen.

5.4. FILM BOILING AND RECOVERY—He II

Until now the existence of boiling phenomena in He II has been mentioned only as the condition reached when the critical heat flux q^* is exceeded. However, there are a number of complicated heat transfer processes that can occur above q^*. This regime of heat transfer is of significant technical importance for its occurrence can lead to catastrophic events in cryogenic systems where good heat transfer must be maintained continuously. Unfortunately, boiling heat transfer in He II is the least understood process, subject to engineering correlation rather than direct calculation.

In He II above the peak heat flux, the fluid in the region of heat transfer can contain several phases in coexistence. Consequently, the physical interpretation of the heat transfer processes is more difficult than in single-phase He II. For heat fluxes above q^* there occurs a discontinuous jump in the surface temperature. This transition marks the formation of a stable film of helium vapor, liquid He I, or both blanketing the heat transfer surface. These general characteristics of the transition to film boiling are best couched in the context of the surface heat transfer curve such as Fig. 5.1. In the film boiling state, the heat transfer is much less effective because of the low-thermal-conductivity film insulating the surface from the bulk He II. Typical values for the film boiling heat tranfer coefficient $h_{fb} = q/\Delta T_s$ are 10–100 times smaller than in the Kapitza regime. However, these values are strongly dependent on a number of physical parameters including heater configuration, bath temperature, and pressure, either saturated or subcooled. An additional feature in the heat transfer curve, also commonly observed in He I, is the occurrence of a recovery heat flux q_R that is less than q^*. The existence of $q_R < q^*$ causes hysteresis in the heat transfer curve. Unlike pool boiling He I, this hysteresis is not observed universally in He II.

There are three possible film boiling conditions that can exist in He II above q^*. To establish which condition is expected for a given set of externally imposed factors, it is necessary to consider the helium state in the vicinity of the heat transfer surface. The first condition corresponds to the local pressure at the heated surface being less than the saturation pressure at the λ-point, that is $p_\lambda = 0.05$ bar. This is referred to as the saturation boiling condition. It is achieved by having a local temperature excursion above T_{sat}, the saturation temperature at the local pressure. The resulting boiling state is a coexistence of two phases—vapor and He II. The vapor film is quite thin as has been shown experimentally by high-speed photography as well as by calculation.[46] The saturation boiling condition has been investigated primarily for thin heater wires in an open bath of He II.

The second condition occurs when the local pressure at the heat transfer surface is large enough to exceed p_λ at the heat transfer surface. A consequence of exceeding q^* is the production of a film of low-thermal-conductivity He I which blankets the surface. If the heat flux q is not greater than the corresponding critical heat flux in He I for that configuration, the heat transfer process will be stable, allowing nucleate boiling to occur in the He I film and heat conduction in He II. Note that the boundary between the He II and He I is not clearly defined because the phase transition from He I to He II is second order, allowing for a continuous density profile. It is important to be aware that, because q^* in He II is under most circumstances much higher than that in He I, the limiting of this process to two phases usually only occurs for temperatures near T_λ.

The third boiling condition occurs at the solid–helium interface when q^* is exceeded in subcooled He II for temperatures well below T_λ, that is, $T_b < 2.1$ K. In this regime, the critical heat flux is greater than that in He I and the He I film becomes unstable and enters film boiling. The result is a triple-phase process consisting of vapor film blanketing the surface, a He I film, and bulk He II.[16] As with the second condition, the He I–He II boundary is not sharply defined. It is apparent that this third film boiling condition, although occurring commonly in engineering systems employing subcooled He II, is the most complex to understand because of the existence of multiphase processes.

There exists one additional type of boiling in He II which does not fall in any of the three above categories. This type of boiling occurs in the bulk fluid rather than at the interface. It can be achieved only in special configurations[47] where the surface heat flux does not exceed q^* but the channel heat flux surpasses the local boiling condition. As a result, He I and vapor are nucleated in the bulk and the heat transport properties of the fluid are modified. This process is analogous to bulk boiling in ordinary fluids.

The fundamental description of the film boiling heat transfer clearly requires a more complex theoretical description than considered so far. Furthermore, it appears that there is no broadly applicable theory capable of handling the multiphase boiling processes. As a result, most research on this problem is of the category of engineering correlations combined with empirical evidence. Since the understanding of the process relies heavily on suitable experimental data, a review of measurements of film boiling heat transfer coefficients is presented first. Subsequently, a comparison is made between the available theories and experimental results. Finally, a description is included of some of the less understood heat transfer phenomena such as recovery from the film boiling state and time-dependent effects.

5.4.1. Film Boiling Heat Transfer Coefficient

The experimental measurements of the film boiling heat transfer coefficient have been extensive; however, they have also been rather restrictive in regime of investigation. The most obvious restriction is associated with the total allowable heat transfer, which is a system limitation. Therefore, film boiling heat transfer surfaces are usually much less than 1 cm^2 in area, which places rather strict constraints on the sample configuration. Additionally, most experiments have been carried out in saturated He II, leading to a boiling state consisting of vapor–He II coexistence. The results in turn are strongly dependent on the hydrostatic head of the helium bath.

To begin, consider film boiling heat transfer from horizontal cylinders in saturated He II. The observed heat transfer coefficient is seen to depend on diameter in a significant way. Plotted in Fig. 5.25 are typical film boiling heat transfer coefficients h_{fb} for fixed bath temperature and hydrostatic head.[48,49] The general tendency is for h_{fb} to increase with decreasing diameter, such that there occurs approximately a factor of 2 increase in value with an order-of-magnitude decrease in diameter. As is discussed below, this behavior can be interpreted in terms of a fairly simple vapor film conduction model. In addition to the diameter dependence, h_{fb} for cylindrical heaters as well as for flat plates is also a strong function of

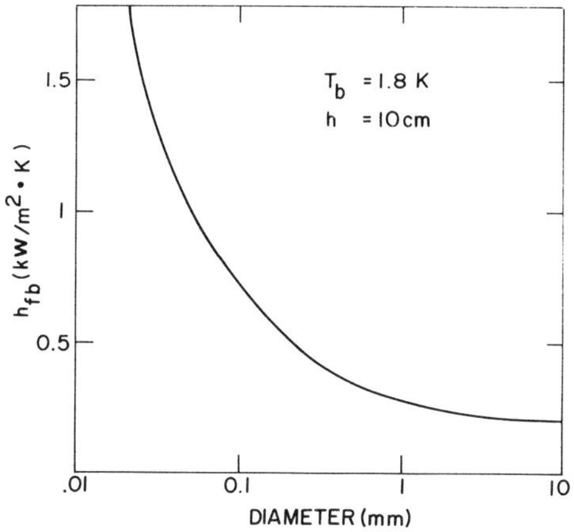

Fig. 5.25. Film boiling heat transfer coefficients as a function of heater diameter for constant hydrostatic head ($h = 10$ cm).

depth of immersion in the saturated helium. Plotted in Fig. 5.26 are typical results for a heated wire as it depends on hydrostatic head h and surface temperature. At small depths, the relationship is roughly linear with h.[49]

There exist two different regimes of film boiling heat transfer in saturated He II. These are referred to as noisy and silent boiling. Noisy boiling is apparent by the existence of audible sound emanating from the heat transfer region. Silent boiling is film boiling without this audible sound. Frequencies observed in noisy boiling can vary anywhere from a few Hz to tens of kHz. The frequency is generally a function of heat flux although in no well-established pattern. The regions of noisy and silent boiling are seen to depend on bath temperature and depth of immersion.

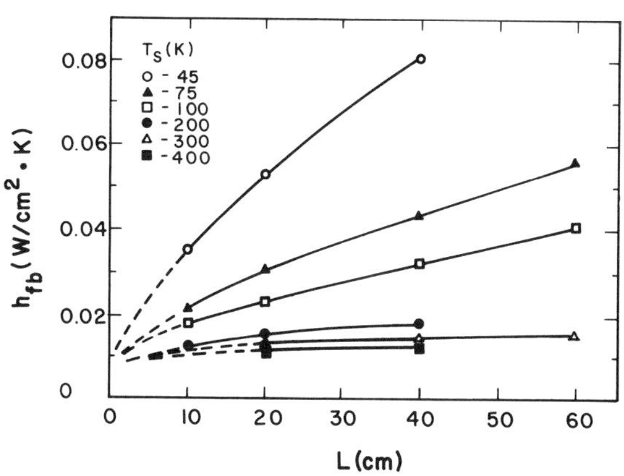

Fig. 5.26. Film boiling heat transfer coefficients versus immersion depth (from Betts and Leonard[49]).

Plotted in Fig. 5.27 is a map of these regimes based on one set of data on wire heaters.[50] This map should not be construed as universally applicable. It appears from these results that noisy boiling occurs for larger immersion depths and lower temperatures. The occurrence of noisy or silent boiling also has a significant effect on the heat transfer coefficient h_{fb}. In regimes of overlap where either noisy or silent boiling is seen to occur, it is usually the case that a slightly higher heat transfer coefficient is measured during silent boiling. This effect, which appears systematic in published experiments, is nonetheless only of the order of 10–20% in the majority of reported results.[49–51]

The situation that occurs in subcooled He II, where the boiling is associated with multiphase phenomena, is generally more complex. In addition, there have been fewer experiments performed under these conditions so data are less prevalent. For the limited results reported to date it is observed generally that the film boiling heat transfer coefficient is higher in the subcooled condition compared to that at saturation. Furthermore, higher pressures on subcooled He II tend to increase h_{fb}.[48] This latter effect may be caused by the increased thermal conductivity of the helium vapor film for higher pressures. Few investigations of the dependence of h_{fb} on heater diameter have been reported for subcooled He II.

To give a quantitative idea of the relative magnitude of the measured film boiling heat transfer coefficients, Table 5.3 list some typical values for different configurations. The wide range of values of h_{fb} is generally consistent with the above qualitative description. It is also worth mentioning that the measured surface temperatures in these experiments can be quite high, exceeding 100 K in several cases. This high surface temperature occurs because of the relatively high value of peak heat flux, where for fine wires q^* occasionally exceeds 10 W/cm^2. As a result, it is difficult to model the film boiling state because of the very large temperature gradients that

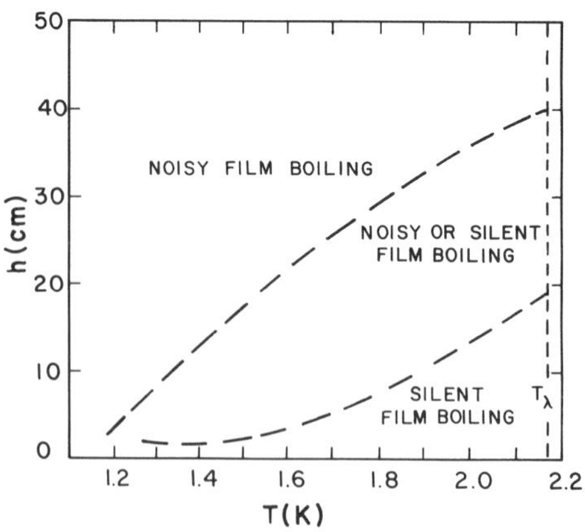

Fig. 5.27. Regimes of noisy and silent boiling from wire heaters in saturated He II (from Leonard[50]).

Table 5.3. Typical Film Boiling Heat Transfer Coefficients

Sample	T_b (K)	T_s (K)	Δp (kPa)[a]	h(kW/m$^2\cdot$K)	Reference
Wire ($d = 76$ μm)	1.8	150	0.42	1.1	55
Wire ($d = 25$ μm)	1.8 K	150	0.56	2.2	
Flat rectangular plate	1.8	75	0.14	0.22	49
(39 mm × 11 mm)	1.8	75	0.28	0.3	
	1.8	75	0.84	0.55	
Flat surface ($d = 13.7$ mm)	2.01	40	0.13	0.69	3
	2.01	25	0.237	0.98	
Horizontal cylinder	1.88	40	0.10	0.2	51
($d = 14.6$ mm)	2.14	40	0.10	0.2	
Wire ($d = 200$ μm)	2.05	150	0.14	0.66	52
Cylinder ($d = 1.45$ mm)	1.78	80	0.06	0.22	53

[a] 1 kPa = 7.5 torr = 70.3 cm He.

occur in the vapor film. Theoretical efforts to model this problem have met with a reasonable degree of success. Generally, the analyzed problem consists of a stable film boiling in saturated He II.

5.4.2. Theoretical Models for Film Boiling Heat Transfer

A simple model can be constructed for film boiling heat transfer based on an assumption that heat is transported through the vapor film by thermal conduction only. The conceptual picture therefore would represent a stable vapor film of constant thickness δ surrounding the heat transfer surface. Consider a cylindrical heater of radius r as shown in Fig. 5.28. An estimate of the vapor film thickness δ can be obtained from the relationship

$$\delta = \frac{\bar{k}}{h_{fb}} \tag{5.66}$$

where the mean thermal conductivity $\bar{k}$ is taken over the temperature range between the bath T_b and surface T_s. As an example, if the heat transfer coefficient is about 0.015 W/cm$^2\cdot$K, obtained for flat plates in saturated He II with a surface temperature of 150 K, δ takes on a value of about 300 μm. This thickness is fairly large compared to typical heater wire diameters. Note that the mean free path in the vapor film is of the order of 1 μm, which is much shorter than the film thickness.

A simple heat conduction model can be employed to interpret the diameter dependence of h_{fb}, shown in Fig. 5.25. Making one further assumption that the vapor film thickness does not vary appreciably with

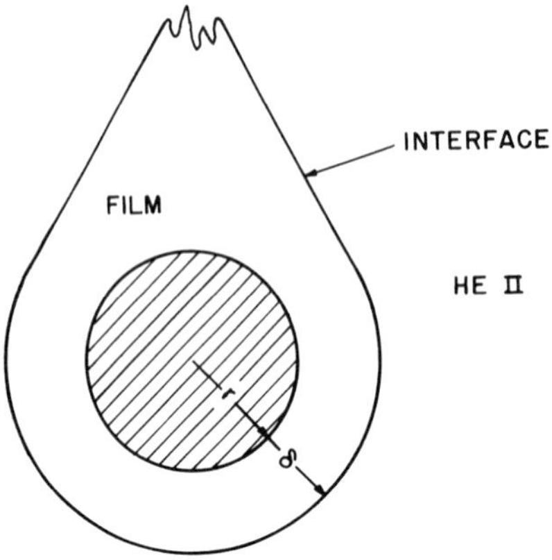

Fig. 5.28. Schematic of cylindrical heater surrounded by a vapor film.

heater radius, a relationship can be derived for the ratio of heat transfer coefficients:

$$\left.\frac{h}{h_0}\right|_{fb} = \frac{\delta/r}{\ln(1+\delta/r)} \tag{5.67}$$

where h_0 refers to the film boiling heat transfer coefficients obtained for flat plates. Thus, as the radius of the heater decreases the film boiling heat transfer coefficient is expected to increase purely as a result of radial heat conduction. Generally, such a result is borne out by experiment.

A more detailed theoretical attempt to correlate film boiling heat transfer is due to Rivers and McFadden.[54] This work treats film boiling heat transfer in saturated He II in terms of a boundary layer model. The equations that describe the problem are conservation of mass, momentum, and energy for two-dimensional, steady-state heat transfer. In rectangular coordinates these equations may be written

$$\frac{\partial u}{\partial x} + \frac{\partial v}{\partial y} = 0 \tag{5.68}$$

$$u\frac{\partial u}{\partial x} + v\frac{\partial u}{\partial y} = \frac{1}{\rho_f}F_b + \frac{\mu_f}{\rho_f}\frac{\partial^2 u}{\partial x^2} \tag{5.69}$$

and

$$u\frac{\partial T}{\partial x} + v\frac{\partial T}{\partial y} = \frac{k_f}{\rho_f C_{pf}}\frac{\partial^2 T}{\partial x^2} \tag{5.70}$$

where the body force F_b represents the buoyancy of the vapor film. The solution of this set of equations has been accomplished[54] by assuming fourth-order polynomials for the temperature and velocity profiled and matching boundary conditions at the wall and the vapor–He II interface. The result is a dimensionless form to the Nusselt number,

$$\mathrm{Nu}\,\mathrm{Gr}^{-1/4} = f(Q_b, \mathrm{Gr}, H_i, \mathrm{Pr}) \tag{5.71}$$

where the Prandtl number for the film is

$$\mathrm{Pr} = \frac{\eta_f C_{pf}}{k_f} \tag{5.72}$$

and the Grashof number,

$$\mathrm{Gr} = \frac{g\,D^3 \rho_f(\rho_b - \rho_f)}{\eta_f^2} \tag{5.73}$$

describes the heat transfer process. To determine the exact form of (5.71) for a given problem it is necessary to apply numerical integration. The end product of this analysis is a solution for the Nusselt number and steady-state film thickness as a function of the interfacial heat flux. Two regimes become evident in this result. The transition between these two regimes occurs at a value of $Q_b\,\mathrm{Gr}^{-1/4} \approx 1$, where Q_b is the dimensionless interfacial heat flux,

$$Q_b = \frac{Dq_b}{k_f \Delta T_f} \tag{5.74}$$

For values of $Q_b\,\mathrm{Gr}^{-1/4} < 1$, the heat transfer is dominated by convection. In this regime, the heat transfer coefficient is a function of the integral enthalpy which can be normalized to the form

$$H_i = \frac{\Delta h_i}{C_{pf} \Delta T_f} \tag{5.75}$$

For values of $Q_b\,\mathrm{Gr}^{-1/4} > 1$, the heat transfer is via conduction mechanisms. In this regime, the heat transfer coefficient can be approximated by (5.66). Comparison between experiment and the theory described above is only partially successful.[46]

The above theory of film boiling heat transfer is by no means a complete description of the problem. An alternative approach suggested by Labuntzov and Ametistov[55] is based on the idea that film boiling is a non-equilibrium process involving heat and mass transfer at the vapor–He II interface. Thus, it is essential to account for the processes of vaporization and condensation. These processes have been investigated theoretically in detail; the following relationships for the dimensionless fluxes of mass, momentum, and heat have been obtained:

$$\Delta\tilde{p} - 2\sqrt{\pi}\,\frac{1-0.4\,\beta}{\beta}\,\tilde{j} = 0.44\,\tilde{q} \tag{5.76a}$$

where

$$\Delta\tilde{p} = \frac{p'' - p_s}{p_s} \tag{5.76b}$$

$$\bar{q} = \frac{q}{p_s(2R'T_i)^{1/2}} \tag{5.76c}$$

$$\tilde{j} = \frac{j}{\rho_s(2R'T_i)^{1/2}} \tag{5.76d}$$

In the above equations R' is the gas constant for helium ($R' = R/M = 2079$ J/kg · K), T_i is the temperature of the liquid helium at the interface, β is the condensation coefficient, ρ_s and p_s are the equilibrium density and pressure corresponding to T_i, and p'' is the vapor pressure corresponding to the helium vapor film at the interface. The derivation of (5.76) is beyond the scope of the present treatment although it is obtained analytically from the Boltzmann kinetic equation.[56] Equation (5.76) can be applied to the solution of interface mass and heat transfer for ordinary liquids as well as He II. However, for ordinary liquids heat transfer is controlled by convective processes in the bulk. For He II these convective processes are enhanced by two-fluid internal convection which dominates the heat transfer in most cases. The form of (5.76) is approximate since it is assumed that $\Delta\tilde{p} \ll 1$, $\tilde{j} \ll 1$. If this is not the case, it is necessary to use the full nonlinear solutions.

Now consider the existence of a film boiling state as shown schematically in Fig. 5.29. In the steady state the vapor film is of constant thickness δ, and consequently $j = 0$. There is no net mass flow across the He II–vapor interface. Under these conditions, (5.76) can be simplified to

$$\Delta\tilde{p} = 0.44\,\tilde{q} \tag{5.77}$$

or for a given hydrostatic head

$$q = 2.27\,\rho g h(2RT_i)^{1/2} \tag{5.78}$$

The physical meaning of (5.78) is that it represents the heat flux necessary to establish a stable vapor film in He II. Note that this is a considerably different interpretation for the critical heat flux than that based on the peak temperature difference within the bulk fluid. Thus, there is some contradic-

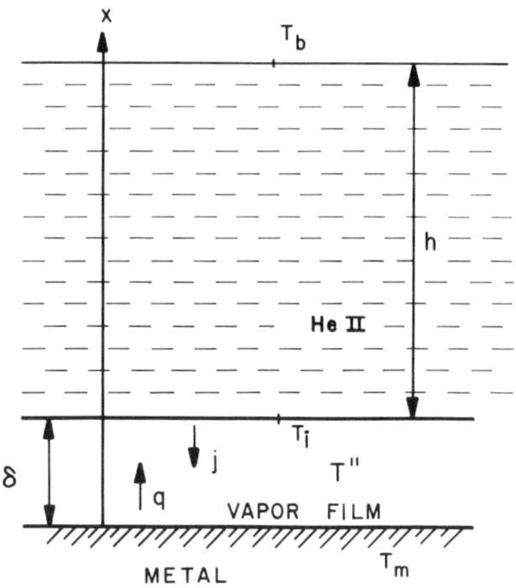

Fig. 5.29. Heat mass transfer process in film boiling He II.

tion between the results of Section 5.1 and Eq. (5.78). There are several possible resolutions for this contradiction.

The first explanation is to imagine conceptually that the film boiling state can be obtained spontaneously anytime the condition described by (5.78) is satisfied. However, this argument is contradictory to experiment, particularly for critical heat fluxes in one-dimensional channels. On the other hand, there is some disagreement between experiment and the He II peak heat flux predicted by turbulent heat transport in cylindrical geometries. It is possible that the condition described by (5.78) is a clue to this discrepancy; however, this point has yet to be analyzed.

As an alternate explanation,[57] it is suggested that the film boiling state once established obeys the kinetic relationship derived above. This idea is not contradictory to the peak heat flux being the point where the helium temperature near the interface reaches T_{sat} or T_λ. However, once this film is established it will remain stable against collapse until the heat flux falls below the value given by (5.78), provided it is less than the applied heat flux. Thought of in this manner, the molecular kinetic theory provides a mechanism of interpreting the minimum film boiling heat flux in He II, that is q_R.

The connection between these two alternate concepts for recovery from film boiling has been investigated analytically for heated cylinders in saturated He II. The comparison of this theoretical concept with experiment has been carried out for the few configurations where minimum film boiling heat flux data are available.[58] The best agreement occurs with data acquired on small-diameter wires. Plotted in Fig. 5.30 are the experimentally measured peak and minimum film boiling heat fluxes versus temperature for a cylindrical wire of diameter 76.2 μm. The theoretical plot

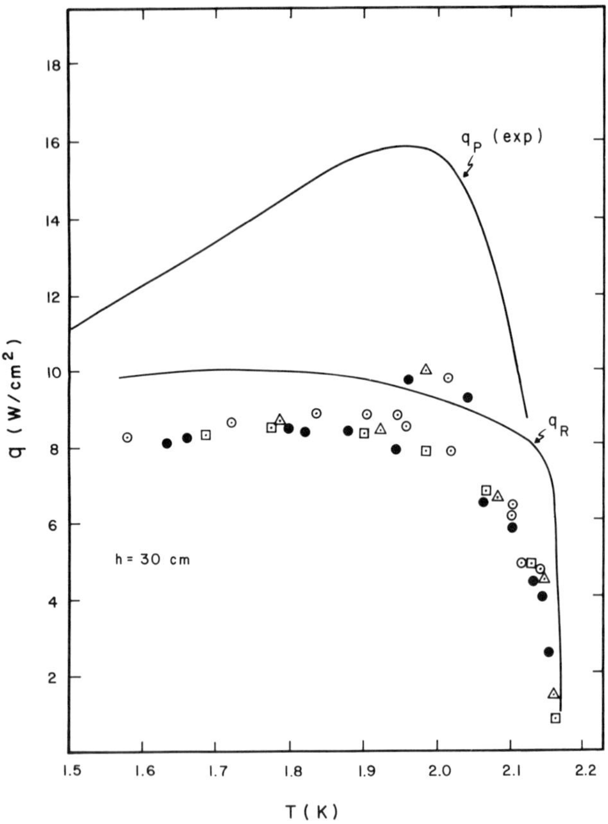

Fig. 5.30. Comparison of calculations using the kinetic theory with experimental results (from Kryukov and Van Sciver[57]).

for q_R is also shown to provide quite close agreement with experiment. Unfortunately, the correspondence between theory and experiment for other configurations of larger dimension is not nearly as close. Consequently, there is still considerable disagreement as to the correct mechanism needed to describe the film boiling state in He II.

An alternate approach for the recovery from the film boiling state has been suggested by experiments[3] that have shown a correlation between the minimum film boiling heat flux and the film boiling heat transfer coefficient. For a specific configuration consisting of a flat plate at the end of a duct, the value of q_R was shown to increase monotonically with h_{fb}. This condition was observed to previal in both saturated and subcooled He II. Furthermore, the ratio q_R/h_{fb} was found to take on an essentially constant value of about 23 K, suggesting the existence of a critical temperature difference for recovery from film boiling. This critical temperature difference should be related in some fundamental way to the collapse of the vapor film. Unfortunately, no further understanding of this effect is currently available.

5.4.3. Transient Film Boiling Heat Transfer

Once the heat flux in a film boiling process is reduced below q_R, recovery to the nonboiling state begins. This dynamic process is governed by transient heat transfer within the system. Although rather complex, there have been two successful attempts to correlate this process with known physical phenomena. The first relates recovery from film boiling to a time-dependent heat exchange through the vapor film, that is, independent of the He II properties.[59] This approach seems to correlate best with transient recovery in saturated film boiling. The alternate approach, which has been applied to the understanding of transient recovery in subcooled He II,[21] treats the problem in terms of time-dependent heat transport in the bulk fluid.

The first of the above two approaches is considered in terms of conventional time-dependent heat transfer. It is assumed that at time $t=0$ the film boiling state is initiated and that heat is applied at a rate Q for a time Δt_f, after which the heat generation ceases ($Q=0$) and the recovery process begins. Empirically, the recovery process is found to take a length of time Δt_R which is a function of Δt_f before the vapor film collapses. It is assumed further that the controlling mechanism for recovery is the enthalpy stored within the heated sample. The fluid simply acts as a constant temperature bath. Using the Debye approxmation to the specific heat, we can show that the above assumptions lead to a correlation between Δt_f and Δt_R which can be written explicitly as

$$Q\,\Delta t_f = 0.176\,\Theta_D\left(\frac{M}{mR}\right)^{1/3}(hA\,\Delta t_R)^{4/3}\left(1+\frac{4}{5}\alpha+\frac{4}{6}\alpha^2+\cdots\right) \quad (5.79)$$

where $\alpha = hT_m/Q \leqslant 1$ with T_m being the maximum temperature of the heat transfer surface. The total mass of the heat transfer sample is m and its cooled surface has area A. The properties of the heat transfer sample enter through its molecular weight M and Debye temperature Θ_D.

The correlation suggested by (5.79) has shown reasonable agreement with experiment.[59] By allowing the heat transfer coefficient h to be an adjustable parameter, we can fit experimental data for the relationship between Δt_f and Δt_R. This fit is shown in Fig. 5.31. By establishing the correlation based on only the leading term in (5.79), the best-fit heat transfer coefficients are $h=0.018$ W/cm$^2\cdot$K at SVP and $h=0.062$ W/cm$^2\cdot$K at

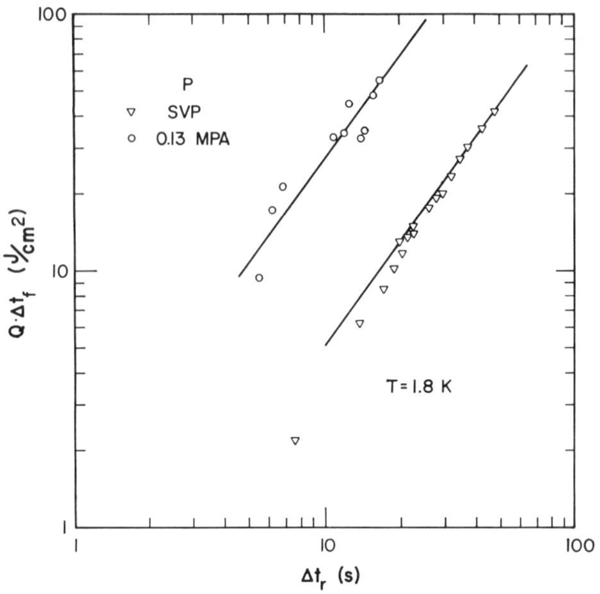

Fig. 5.31. Correlation between the energy applied to a film boiling heat transfer sample and the time to recovery.

0.13 MPa. It is interesting to note that these values of h are roughly 60% of typical steady-state film boiling heat transfer coefficients.

The other transient film boiling heat transfer problem concerns recovery in subcooled He II.[21,60] Rather than correlating the time to recovery, the approach is to determine the maximum steady-state heat flux which allows recovery after a large short-duration heat pulse is applied to a heat transfer sample. The short-duration heat pulse is assumed to be larger than the maximum energy flux needed to break down the He II. Consequently, the heat transfer at the solid–helium interface breaks down and further heat is stored in the solid. After the heat pulse, the excess heat can be transferred through the He II by the established transient heat transfer mechanisms. Now if the steady-state heat flux is not too large, the combination of it with the remaining excess energy being released from the heat transfer sample will not exceed the maximum transient energy flux in the He II. The result is a temporary recovery to the nonboiling state. Recovery is temporary only if the steady-state heat flux is larger than the maximum steady-state heat flux in He II for that particular configuration.

The above set of conditions are illustrated graphically in Fig. 5.32, which is a normalized transient heat transfer plot for He II at 1.8 K. The solid curve represents the maximum heat flux that can be applied for a given time before breakdown of He II heat transfer occurs. This condition

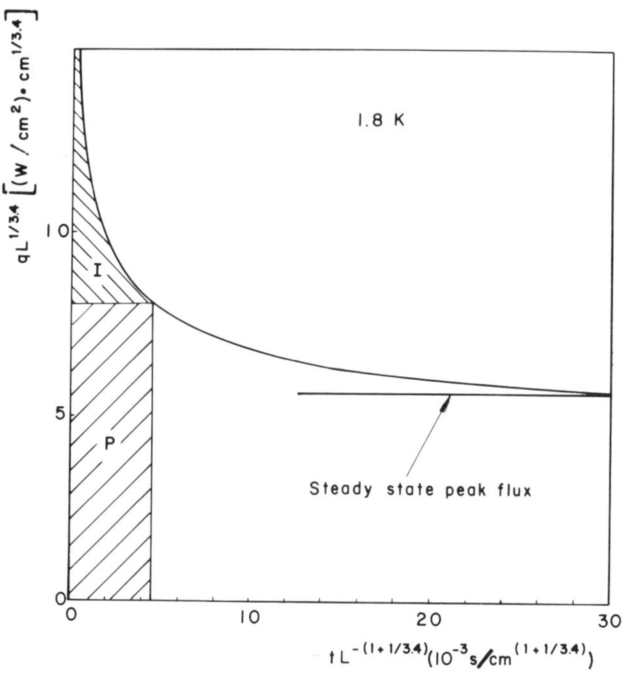

Fig. 5.32. Schematic description of the transient recovery process in He II with postheating (from Seyfert[21,60]).

is established according to Section 5.2. Now consider a short-duration, high-level heat pulse applied to the heat transfer sample. Since this energy is larger than that transferable by the He II, the excess is stored in the heat transfer sample. The total energy contained in the pulse is shown by region I in the figure. Subsequent to the pulse, the heat flux is dropped to a lower level q_p which is still above the maximum steady-state heat flux q^*. However, temporary recovery will occur because the transient heat transfer mechanism can continue. Thus, the sum of the two areas I and P determines the length of time before unrecoverable thermal breakdown occurs. This model has been compared successfully to transient heat transfer experiments in He II.[21] The problem is of significant practical interest to superconducting magnet technology.

PROBLEMS

5.1. Consider a sphere of radius r_0 in a large bath of He II. Calculate the steady-state temperature gradient as a function of radial coordinate r. Determine the peak heat flux as it depends on bath temperature. Assume that mutual friction is the only important interaction and the ambient pressure is 1 bar. [Hint: This problem is analogous to that of heat transfer in cylindrical geometries discussed in Section 5.1.2.]

5.2. Estimate the maximum surface heat flux q_0 for a cylindrical wire of diameter 1 mm in He II at 1.8 K, 1 bar. What would be the effect of increasing the external pressure to 10 bars?

5.3. Show that the transient heat transfer solution given by (5.35) for the fixed energy deposition obeys the heat conductivity equation for He II.

5.4. Estimate the Kapitza conductance at 1.9 K of aluminum, for small ΔT, by each of the following methods:
   a. Phonon radiation limit.
   b. Acoustic mismatch theory.
   c. Experimental results listed in Table 5.2.

5.5. Show that the variation of the film boiling heat transfer coefficient with heater radius may be written in the form given by (5.67). For the data listed in Table 5.3 from Ref. 55, estimate the film thickness δ and the limiting heat transfer coefficient h_0 for large radii.

5.6. A heater is located at the bottom of a 50 cm long vertical channel containing He II at 1.9 K. Assume the pressure at the top of the channel is saturation and that heat flow is governed by mutual friction.
   a. Find the peak heat flux q^*. [Note: Although Eq. (5.10) is valid for this case, it is more accurate to use tabulated vapor pressure of helium.]
   b. Estimate the heater surface temperature at q^*.

REFERENCES

1. W. F. Vinen, Mutual Friction in a Heat Current in Liquid Helium II. I. Experiments on Steady State Heat Currents, *Proc. R. Soc. London* **A240**, 114 (1957).
2. V. Arp, Heat Transport Through Helium II, *Cryogenics* **10**, 96 (1970).
3. S. W. Van Sciver, Kapitza Conductance of Aluminum and Heat Transport Through Subcooled He II, *Cryogenics* **18**, 521 (1978).
4. W. F. Vinen, Mutual Friction in a Heat Current in Liquid Helium II. III. Theory of Mutual Friction, *Proc. R. Soc. London* **A242**, 493 (1957).
5. R. Srinvasan and A. Hofmann, Investigations on Cooling with Forced Flow of He II, *Cryogenics* **25**, 641 (1985).
6. C. Linnet and T. H. K. Frederking, Thermal Conditions at the Gorter–Mellink Counterflow Limit between 0.01 and 3 Bar, *J. Low Temp. Phys.* **21**, 447 (1975).
7. A. C. Leonard and M. A. Clermont, Correlation of the Vaporization Onset Heat Flux for Cylinders in Saturated Liquid Helium II, in *Proceedings of the 4th International Cryogenics Engineering Conference*, pp. 301–306, IPC Science and Technology Press, London, 1972.
8. J. S. Goodlang and R. K. Irey, Nonboiling and Film Boiling Heat Transfer to a Saturated Bath of Liquid Helium, *Adv. Cryog. Eng.* **14**, 159 (1969).
9. T. H. K. Frederking and R. L. Haben, Maximum Low Temperature Dissipation Rates of Single Horizontal Cylinders in Liquid Helium II, *Cryogenics* **8**, 32 (1968).
10. S. W. Van Sciver and R. L. Lee, Heat Transfer to He II in Cylindrical Geometries, *Adv. Cryog. Eng.* **25**, 363 (1980).
11. S. W. Van Sciver and R. L. Lee, Heat Transfer from Circular Cylinders in He II, in *Cryogenic Processes and Equipment in Energy Systems*, ASME Publication No. H00164, 1981, pp. 147–154.
12. H. Van Dijk, M. Durieux, J. R. Clement, and J. K. Logan, *The 1958 ^{4}He Scale of Temperatures*, National Bureau of Standards Monograph 10, U.S. Government Prining Office, Washington, DC, June 17, 1960.

13. G. Krafft, Superheating and Bubble Formation in Helium II, *J. Low Temp. Phys.* **31**, 441 (1978).
14. L. J. Rybarcyk and J. T. Tough, Superheating in He II and the Extension of the Lambda Line, *J. Low Temp. Phys.* **43**, 197 (1981).
15. S. W. Van Sciver and O. Christianson, Heat Transport in a Long Tube of He II, in *Proceedings of the 7th International Cryogenics Engineering Conference*, pp. 228–234, IPC Science and Technology Press, London, 1978.
16. S. Breon, Boiling Phenomena in Saturated and Subcooled He II, Ph.D. thesis, University of Wisconsin–Madison (1986).
17. S. W. Van Sciver, Heat Transport in Forced Flow He II: Analytic Solution, *Adv. Cryog. Eng.* **29**, 315 (1984).
18. A. Kashani and S. W. Van Sciver, Steady State Forced Convection Heat Transfer in He II, *Adv. Cryog. Eng.* **31**, (1986).
19. W. W. Johnson and M. C. Jones, Measurements of Axial Heat Transport in Helium II with Forced Convection, *Adv. Cryog. Eng.* **23**, 363 (1978).
20. F. P. Incropera and D. P. Dewitt, Fundamentals of Heat Transfer, Chap. 5, Wiley, New York, 1981.
21. G. E. Myers, *Analytical Methods in Conduction Heat Transfer*, McGraw–Hill, New York, 1971.
22. L. Dresner, Transient Heat Transfer in Superfluid Helium, *Adv. Cryog. Eng.* **27**, 411 (1982).
23. L. Dresner, Transient Heat Transfer in Superfluid Helium—Part II, *Adv. Cryog. Eng.* **29**, 323 (1984).
24. L. Dresner, *Similarity Solution of Non-Linear Partial Differential Equations*, Pitman Publishing, Boston, MA, 1983.
25. S. W. Van Sciver, Transient Heat Transport in He II, *Cryogenics* **19**, 385 (1979).
26. P. Seyfert, J. Lafferranderie, and G. Claudet, Time Dependent Heat Transport in Subcooled Superfluid Helium, *Cryogenics* **22**, 401 (1982).
27. P. L. Kapitza, The Study of Heat Transfer on Helium II, *J. Phys.* (*USSR*) **4**, 181 (1941).
28. O. V. Lounasmaa, *Experimental Principles and Method Below 1 K*, Academic Press, London, 1974.
29. T. H. K. Frederking, Thermal Transport Phenomena at Liquid Helium II Temperatures, *Adv. Cryog. Heat Transfer* **64**, 21 (1968).
30. L. J. Challis, Kapitza Resistance and Acoustic Transmission Across Boundaries at High Frequencies, *J. Phys. C. Solid State Phys.* **7**, 481 (1974).
31. N. S. Snyder, *Thermal Conductance at the Interface of a Solid and Helium* II (*Kapitza Conductance*). NBS Technical Note 385, U.S. Government Printing Office, Washington, DC, 1969; Heat Transport through Helium II: Kapitza Conductance, *Cryogenics* **10**, 89 (1970).
32. L. J. Challis, Experimental Evidence for a Dependence of the Kapitza Conductance on the Debye Temperature of a Solid, *Phys. Lett.* **26A**, 105 (1968).
33. I. M. Khalatnikov, *Introduction to the Theory of Superfluidity*, Chap. III, W. A. Benjamin, New York, 1965.
34. R. E. Peterson and A. C. Anderson, The Kapitza Thermal Boundary Resistance, *J. Low Temp. Phys.* **11**, 639 (1973).
35. L. J. Challis, K. Dransfeld, and J. Wilks, Heat Transfer Between Solids and Liquid Helium II, *Proc. R. Soc. London* **A260**, 31 (1961).
36. D. Cheeke and H. Ettinger, Macroscopic Calculation of the Kapitza Resistance Between Solids and Liquid ^{4}He, *Phys. Rev. Lett.* **37**, 1625 (1976).
37. P. H. E. Meijer and J. S. R. Peri, New Kapitza Heat Transfer Model for Liquid ^{4}He, *Phys. Rev. B* **22**, 195 (1980).
38. J. G. Dash, *Films on Solid Surfaces*, Academic Press, New York, 1974.
39. J. Wilks, *The Properties of Liquid and Solid Helium*, Chap. 14, Clarendon Press, Oxford, 1967.

40. B. W. Clement and T. H. K. Frederking, Thermal Boundary Resistance and Related Peak Flux During Supercritical Heat Transport from a Horizontal Surface Through a Short Tube to a Saturated Bath of Liquid Helium II, *Liquid Helium Technology, Proceedings of the International Institute of Refrigeration*, Commission I, Boulder, CO, Pergamon Press, Oxford, 1966, pp. 49–59 (see also Ref. 29).
41. K. Mittag, Kapitza Conductance and Thermal Conductivity of Copper, Niobium, and Aluminum in the Range from 1.3 to 2.1 K, *Cryogenics* **13**, 94 (1973).
42. S. W. Van Sciver, Kapitza Conductance of Aluminum and Heat Transport from a Flat Surface through a Large Diameter Tube to Saturated He II, *Adv. Cryog. Eng.* **23**, 340 (1977).
43. G. Claudet and P. Seyfert, Bath Cooling with Subcooled Superfluid Helium, *Adv. Cryog. Eng.* **27**, 441 (1981).
44. S. W. Van Sciver, Developments in He II Heat Transfer and Applications to Superconducting Magnets, *Adv. Cryog. Eng.* **27**, 375 (1981).
45. A. Kashani and S. W. Van Sciver, Kapitza Conductance of Technical Copper with Several Different Surface Preparations, *Cryogenics* **25**, 238 (1985).
46. R. K. Irey, Heat Transport in Liquid Helium II, in *Heat Transfer at Low Temperatures*, W. Frost (Ed.), Plenum Press, New York, 1975.
47. D. Gentile and M. X. Francois, Heat Transfer Properties in a Vertical Channel Filled with Saturated and Pressurized Helium II, *Cryogenics* **21**, 234 (1981).
48. S. W. Van Sciver, Heat Transfer in Superfluid Helium II, in *Proceedings of the 8th International Cryogenics Engineering Conference*, pp. 228–237, Butterworth Scientific, London, 1980.
49. K. R. Betts and A. C. Leonard, Free Convection Film Boiling from a Flat, Horizontal Surface in Saturated He II, *Adv. Cryog. Eng.* **21**, 282 (1975).
50. A. C. Leonard, Helium II Noise Film Boiling and Silent Film Boiling Heat Transfer Coefficient Values, in *Proceedings of the 3rd International Cryogenics Engineering Conference*, pp. 109–114, ILIFFE Science and Tech. Publications, Guilderford, Surrey, U.K., 1970.
51. J. S. Goodlang and R. K. Irey, Non-Boiling and Film Boiling Heat Transfer to a Saturated Bath of Liquid Helium, *Adv. Cryog. Eng.* **14**, 159 (1969).
52. R. C. Steed and R. K. Irey, Correlation of the Depth Effect on Film Boiling Heat Transfer in Liquid Helium II, *Adv. Cryog. Eng.* **15**, 299 (1970).
53. R. M. Holdredge and P. W. McFadden, Boiling Heat Transfer from Cylinders in Superfluid Liquid Helium II Bath, *Adv. Cryog. Eng.* **11**, 507 (1966).
54. W. J. Rivers and P. W. McFadden, Film-Free Convection in Helium II, *Trans. ASME, J. Heat Transfer*, **88C**, 343 (1966).
55. D. A. Labuntzov and Ye. V. Ametistov, Analysis of Helium II Film Boiling, *Cryogenics* **19**, 401 (1979).
56. P. L. Bhatnager, E. P. Gross, and M. Krook, A Model for Collision Processes in Gases. I. Small-Amplitude Processes in Charged and Neutral One-Component Systems, *Phys. Rev.* **94**, 511 (1954).
57. A. P. Kryukov and S. W. Van Sciver, Calculation of the Recovery Heat Flux from Film Boiling in Superfluid Helium, *Cryogenics* **21**, 525 (1981).
58. G. D. Lemieux and A. C. Leonard, Maximum and Minimum Heat Flux in Helium II for a 76.2 μm Diameter Horizontal Wire at Depths of Immersion Up to 70 cm, *Adv. Cryog. Eng.* **13**, 624 (1968).
59. S. W. Van Sciver, Correlation of Time Dependent Recovery from Film Boiling Heat Transfer in He II, *Cryogenics* **21**, 529 (1981).
60. P. Seyfert, Practical Results on Heat Transfer to Superfluid Helium, in *Stability of Superconductors*, pp. 53–62, International Institute of Refrigeration Commission A 1/2, Saclay, France, 1981.

Chapter 6

Pool Boiling He I Heat Transfer

Normal helium is a simple liquid with state properties that can be described reasonably well by statistical models of the type introduced in Chapter 3. However, the dynamics of heat transfer are of particular interest to engineering applications. Heat transfer, which is a nonequilibrium process, is probably the most important single characteristic of cryogenic fluids. The subject has considerable physical basis, and the models used to describe the phenomena are a combination of fundamental physics and engineering correlations. Pool boiling He I heat transfer is probably the most studied engineering problem related to liquid helium. Pool boiling is a common term used to describe an experimental configuration consisting of a heater, either a plate or wire, immersed in a large bath of the fluid. Normally, the bath has such an extent that it is possible to assume it to be infinite in size relative to the heater sample. This problem is a classic in heat transfer research; although more complex configurations are needed to model true engineering systems. Heat transfer to forced flow helium, which is a wholly different problem, is the subject of Chapter 7.

There are a number of general characteristics of He I which are worth noting. It has a rather small thermal conductivity and large specific heat, suggesting that conduction heat transport is of little significance to the overall heat transfer picture. Particularly in the steady state, the heat transport is dominated by convection mechanisms. This behavior is clearly quite different from that seen in He II, where two-fluid internal convection allows extremely high effective thermal conductivity. Because of the difference in heat transport properties it is not possible to use the models developed for He II in the characterization of He I. Rather, the behavior of He I is cast in terms of heat transfer models based on engineering correlations.

The traditional approach to the interpretation of heat transfer is best suited for engineering applications. The general philosophy is to assume that the heat transfer process is too complicated to understand from first principles. A specific problem requires solution of a complex set of equations which are only treatable in the simplest geometries. Therefore, engineering problems are scaled on the basis of dimensionless variables, which are functions of the properties of the system. It is possible to construct nondimensional relationships which when fit to experimental data can be applied universally to other systems. The strength of this approach is in its relative ease of application. These dimensionless relationships have been explored extensively and their forms are available in the literature.[1,2] Furthermore, the computation of the parameters for a given set of conditions allows straightforward predictions for experimental data. When carried out correctly, the correspondence between experiment and correlations is quite satisfactory. The essential ingredient to this approach is sufficient quantity of experimental data, not only for the particular fluid in question but also for other fluids with widely varying properties. This need must be satisfied to instill reasonable confidence in the particular correlation at hand. Fortunately, for most liquids this kind of ground work has been laid already. In general, the behavior of He I is in satisfactory agreement with the accepted correlations. The quality of the agreement is in part the subject of the present chapter.

For problems of heat transfer, the most important dimensionless quantity to consider is the Nusselt number, Nu. It represents a dimensionless heat transfer coefficient defined by the relationship

$$\mathrm{Nu} = \frac{hL}{k} \tag{6.1}$$

where $h = q/(T_s - T_b)$, the heat transfer coefficient of the surface, T_s and T_b are the local surface and bath temperatures, k is the thermal conductivity, and L is a characteristic length in the problem. In pool boiling, L is the characteristic dimension of the heater, that is, its diameter or width. The Nusselt number normally appears in correlations used to describe both free convection and forced convection heat transfer.

In the case of pool boiling heat transfer the two relevant dimensionless numbers are the Grashof number (Gr) and Prandtl number (Pr). The Grashof number indicates the ratio of buoyancy forces relative to viscous forces. It is represented by the relationship

$$\mathrm{Gr} = \frac{g\beta(T_s - T_b)\,L^3}{\nu^2} \tag{6.2}$$

where g is the acceleration of gravity, β is the bulk expansivity, and ν is the kinematic viscosity. The Prandtl number contains most of the fluid properties. It is written as a ratio of the mass and thermal diffusivities of the fluid.

$$\mathrm{Pr} = \frac{\nu}{D} = \frac{\eta C_p}{k} \tag{6.3}$$

where D is the thermal diffusivity of the fluid. For systems that are dominated by natural convection mechanisms, that is, with negligible forced flow, the Nusselt number can be written as a function of these two number,

$$\mathrm{Nu} = \phi(\mathrm{Gr})\,\psi(\mathrm{Pr}) \tag{6.4}$$

where ϕ and ψ are functions determined by the empirical correlation of data.

Most empirical correlations for natural convection are given in terms of the Rayleigh number, which is simply the product of the Grashof and Prandtl numbers,

$$\mathrm{Ra} \equiv \mathrm{Gr}\,\mathrm{Pr} = \frac{g\beta(T_s - T_b)}{D\nu} \tag{6.5}$$

Simplified correlations can then be written in the form

$$\mathrm{Nu} = C\,\mathrm{Ra}^n \tag{6.6}$$

where C is an empirically determined parameter. The coefficient n is dependent mostly on the geometrical and flow conditions. For a vertically oriented plate in an open bath, $n = \frac{1}{4}$ when the flow is laminar and $n = \frac{1}{3}$ in the turbulent regime. The type of heat transfer condition that exists in a particular system can be described by the corresponding value of the Rayleigh number. The critical Rayleigh number Ra_c defines the transition between these regimes. For flat plates, the transition between pure conduction and convection occurs for $\mathrm{Ra}_c \approx 10^3$, while the transition between laminar and turbulent convection heat transfer usually occurs for $\mathrm{Ra}_c \approx 10^9$. These concepts assume single-phase heat transfer and consequently are not applicable in heat transfer processes that involve change of phase.

6.1. REGIMES OF HEAT TRANSFER

To obtain a better physical feeling for pool boiling heat transfer, it is helpful to consider a hypothetical experimental system. Such an experiment, shown in Fig. 6.1, consists of a flat heated plane with some arbitrary orientation exposed to an effectively infinite bath of liquid helium. The experiment consists of heating the plane from inside the insulated region and measuring the temperature difference between the bath and surface, ΔT_s, as it varies with heat flux q. There are a number of variables that affect the results in this experiment. Among these are bath temperature and pressure, surface orientation, physical characteristics of the heated surface

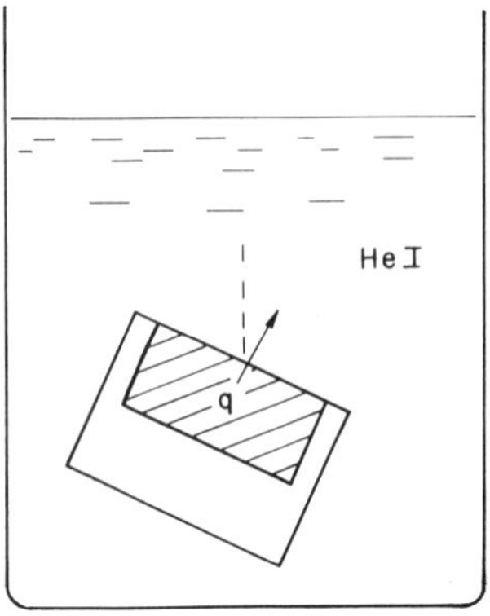

Fig. 6.1. Schematic of pool boiling heat transfer process from a planar surface of arbitrary orientation.

including coatings, and frequency of heat flux. The general impact of these variables is described further below.

Given this experimental configuration, a measurement consists of determining a relationship between the heat flux q and ΔT_s. A typical example of such a relationship is shown in Fig. 6.2. There are principally three regions of heat transfer as indicated in the figure: (1) natural convection, (2) nucleate boiling, and (3) film boiling. Each has a characteristically different physical description. A schematic representation of these regimes is shown in Fig. 6.3. At the lowest heat fluxes up to a few mW/cm^2, heat is transferred by natural convection; see Fig. 6.3a. No phase change is

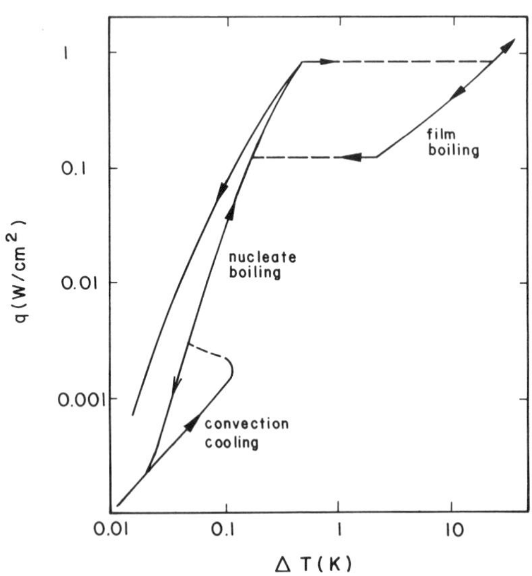

Fig. 6.2. Typical heat transfer relationship for pool boiling liquid.

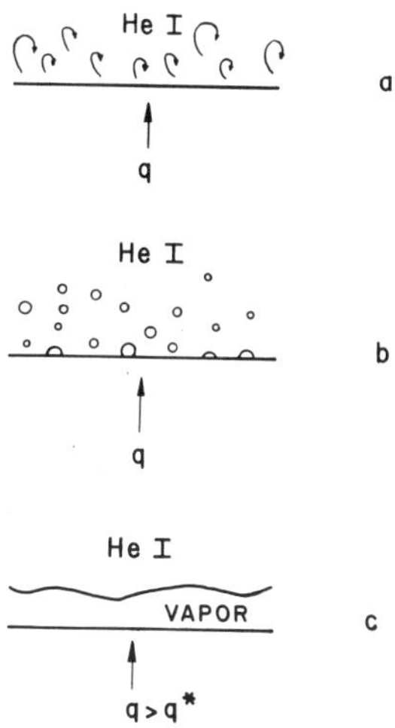

Fig. 6.3. Schematic representation of regimes of heat transfer: (a) natural convection, (b) nucleate boiling, and (c) film boiling.

evident. This mechanism is characterized by density-driven convection currents near the heated surface. Surface temperature differences can be determined by the type of correlation given in (6.6). As the heat flux is increased, bubbles of helium vapor begin to form at preferred sites on the surface. These are typically surface imperfections. In the natural convection region, a certain amount of hysteresis in the heat transfer curve results from the activation and deactivation of these nucleation sites.

As the heat flux is increased further, the nucleation sites get fully activated, meaning that each site contains one bubble. At this point the surface is referred to as being "educated" since now increasing the heat flux only serves to accelerate the rate of bubble growth and detachment. In the nucleate boiling regime, Fig. 6.3b, there is a layer of superheated liquid adjacent to the heater surface. As a bubble detaches, cold liquid from above rushes down to cool the surface. This bubble growth and detachment causes macroscopic turbulence.

At still higher heat fluxes, the nucleate boiling bubbles get so large and are detaching at such a great rate that they become unstable and coalesce into a continuous vapor film; see Fig. 6.3c. The heat flux at which this occurs is referred to as the peak nucleate boiling heat flux q^*. This regime is called film boiling. The condition is unstable and causes hysteresis in the heat transfer curve, as shown by the upper region in Fig. 6.2. On decreasing the heat flux, it is necessary to go to a value lower than q^* for recovery to the nucleate boiling regime. This recovery value is referred to as the minimum film boiling heat flux q_{mfb} or q_R. In the film boiling regime, the surface temperature difference is an order of magnitude higher than with nucleate boiling. The hysteresis in this regime of heat transfer is associated with the stability of a vapor film below a higher-density liquid.

Numerous factors affect these heat transfer characteristics. For example, the surface condition of the heat transfer sample can affect both the peak heat flux q and ΔT_s in the nucleate boiling regime. It is possible to obtain variations in these quantities by as much as a factor of 2–3 between samples. The mechanism by which surface preparation affects the heat transfer characteristics is believed to be associated with the number of available nucleation sites.

Surface orientation has a profound effect on the heat transfer behavior. Variation of the surface orientation with respect to the gravitational force can cause significant changes in the heat flux and minimum film boiling heat flux. The highest values for both these quantities occur with the surface facing upward, because the buoyancy force aids bubble detachment. This argument supports the observed result that q^* and q_{mfb} are minimum with the surface facing downward.

The thermodynamic state of the liquid helium bath is also an important parameter in the heat transfer process. The bath temperature has a significant effect on various heat transfer parameters, particularly the peak nucleate boiling heat flux q^*. Similarly, the bath pressure affects these values, particularly when considering the subcooled or supercritical state. These variables can be taken into account through the changes in the helium properties with temperature and pressure.

The frequency with which the heat transfer event occurs is also important for both the peak heat flux and temperature difference. At low frequencies up to perhaps 10 Hz, the behavior does not deviate significantly from that of the steady-state process. Heat transfer is controlled largely by convection mechanisms. However, at higher frequencies approaching the kilohertz range there is insufficient time for the bubble nucleation to occur. Consequently, the behavior becomes dominated by simple heat diffusion in the liquid adjacent to the solid. Then temperature differences are caused by two physical mechanisms—the thermal conductivity of the helium and interfacial conductance (Kapitza conductance).

Finally, variations in geometry can have a profound effect on the heat transfer. Many engineering systems consist of channels, tubes, or other complex geometries, which are vastly different from the open infinite bath configuration. Such factors can cause differences in the heat transfer at least in part caused by the limited coolant volume. Some of the physical phenomena that can occur include heat-induced natural circulation and vapor locking in narrow channels.

6.2. CONVECTIVE HEAT TRANSFER

At very low heat fluxes, $q \approx 1\ \mathrm{mW/cm^2}$, heat is transferred by a combination of conduction and convection. It is described by a heat transfer coefficient $h = q/\Delta T_s$, where h is only weakly dependent on ΔT_s. This regime of heat transfer has only limited technological application in liquid helium because the heat fluxes are quite small. However, the problems of low heat flux heat transfer and of transitions between conduction and convection are of great interest for their fundamental significance. Certain special cases of heat transfer fall in the general area of exactly soluble classical physics problems.

A good example of such a special interest is the problem of convection onset in a layer of fluid that is heated from below. The main difference between this problem and similar ones concerning pool boiling heat transfer is that the fluid layer is to have a thickness dimension d that enters into the problem in one of the dimensionalized parameters. This problem is referred to as Bénard convection and the instability associated with the transition is called the Rayleigh–Bénard instability. The transition is between one stationary state and another. As the heat flux is increased, the condition where the fluid is at rest carrying heat by conduction is transformed to that where a polygonal convective cell structure occurs. This type of structure has been observed in visual experiments with room temperature fluids. These flows exhibit regularity and structure that have inspired considerable theoretical research into the dynamics of small perturbations in fluids heated from below. Theoretical modeling is achievable because the disturbances are assumed to be sufficiently small that their description, at most, adds linear terms to the fluid equations.

The theoretical description of Bénard convection begins with the continuity equation and the equations for conservation of momentum and energy. The growth or decay of perturbations in the velocity and temperature fields is governed by the following linearized equations[3]:

$$\nabla \cdot \mathbf{v} = 0 \tag{6.7a}$$

$$\frac{\partial \mathbf{v}}{\partial t} = -\frac{1}{\rho_0} \nabla p + \nu \nabla^2 \mathbf{v} - \mathbf{g}\beta T \tag{6.7b}$$

$$\frac{\partial T}{\partial t} = D \nabla^2 T - w\gamma \tag{6.7c}$$

Solutions to these equations establish the regions of convection growth or decay. The boundary between these regions is defined by a "critical" Rayleigh number which effectively is a nondimensional temperature difference. The most interesting of the three equations is (6.7c) which describes the effect of motion on the temperature gradient. Without the second term, $w\gamma$, the expression is simply the heat diffusion equation. The parameter γ is defined as the undisturbed temperature gradient due only to conduction ($\gamma = q/k$). The physical meaning of the term $w\gamma$ is that of the

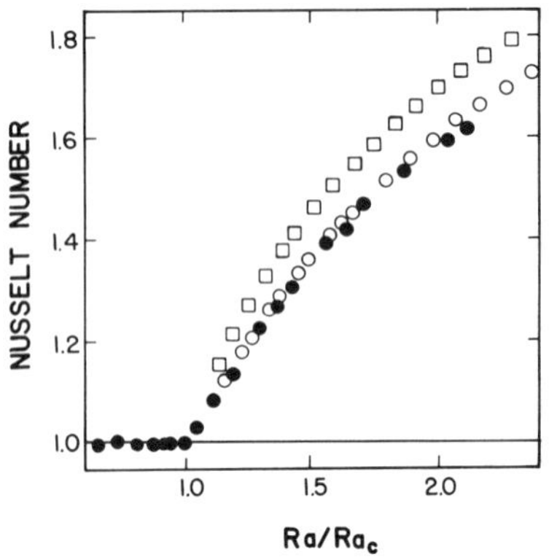

Fig. 6.4. Nusselt number as a function of Ra/Ra_c for Rayleigh–Bénard instability (from Behringer and Ahlers[4]).

motion generator. Heat is swept upward while cold returns. Heat conduction is necessary to generate the initial temperature gradient, but since mass flow is involved, viscosity enters to resist the growth of the perturbation.

The problem of Bénard convection in He I has been studied by a number of workers.[4,5] Experimental measurements normally consist of determining the variation of the Nusselt number with the normalized Rayleigh number (Ra/Ra_c). For any fluid, the Nusselt number represents the ratio of the effective thermal conductivity to the actual thermal conductivity obtained without convection. Plotted in Fig. 6.4 is the normalized Nusselt number for one set of experiments on He I.[4] A number of interesting facts can be gleaned from these data. For example, above Ra_c the Nusselt number increases quite strongly with Ra. This is to be expected because the convection currents improve the heat transport. However, it is worth noting that the behavior of the Nu versus Ra/Ra_c plot appears to be somewhat universal in form. Slight differences in the data displayed in Fig. 6.4 are attributed to geometrical factors in the experiment.

The Rayleigh–Bénard instability is an interesting classical fluids problem. Its connection with helium heat transfer in practical configurations is limited, yet it does give fundamental insight into the fluid flow problem. As the heat flux is increased above a few mW/cm^2 bubbles begin to nucleate on the surface and simple convection is no longer the generally applicable solution. Then the problem is referred to as nucleate boiling because it involves a phase change.

6.3. NUCLEATE BOILING HEAT TRANSFER

Above a heat flux of a few mW/cm^2 in liquid helium the heat transfer surface becomes "educated" and the nucleation sites are essentially all active. This condition is quite different from that of natural convection because it is controlled mostly by the hydrodynamics of bubble growth and detachment rather than convection in the liquid.

Two conditions must exist at or near the heat transfer interface before there can be activation of bubble nucleation sites. First, there must be a boundary layer of liquid adjacent to the surface which is in the superheated condition. The thickness of this layer is determined by the thermal conductivity of the liquid, k, and the allowable superheat, $\Delta T_s = T_s - T_b$, where T_s is the maximum superheat temperature and T_b is the bath temperature. The boundary layer thickness can therefore be written

$$\delta \approx \frac{k \, \Delta T_s}{q} \tag{6.8}$$

Typically, δ is of the order of 1–10 μm for liquid helium near its normal boiling point. Second, local surface imperfections must exist. These imperfections are necessary to provide preferential regions where bubbles can form. Shown schematically in Fig. 6.5, these imperfections usually are envisioned to be grooves or slots which allow a bubble to form with negative curvature, thus taking advantage of surface tension to stimulate the bubble nucleation. There are two principal reasons why the superheated boundary layer must form near the interface before substantial vapor nucleation can occur. First, the liquid near the interface is subcooled by the hydrostatic head ($\Delta p = \rho g h$). The local temperature must increase above ambient before the saturation condition can be attained. Second, and probably more important, in order to have vapor bubbles of positive radius of curvature, the surface tension of the liquid must be overcome.

6.3.1. Nucleation Theory

So far the discussion of nucleate boiling has been quite general and qualitative. However, it is worth considering two specific questions of

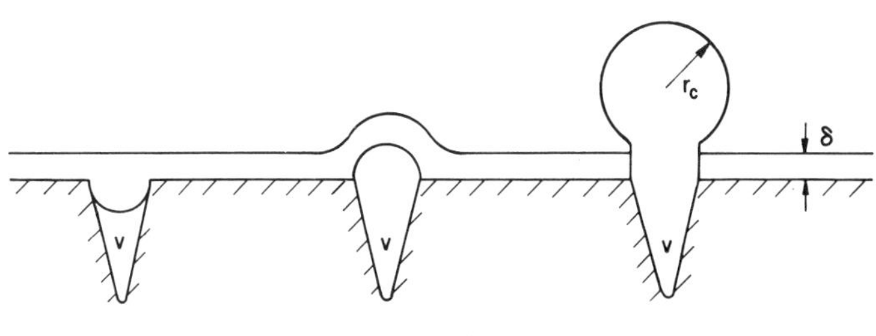

Fig. 6.5. Bubble nucleation on an imperfect surface: (a) negative radius of curvature, (b) positive radius of curvature, and (c) critical radius.

quantitative nature. These concern the general nucleation problem but are worked out as examples for the case of liquid helium. The first question pertains to the growth of a vapor bubble on a nucleation site or actually anywhere in the bulk fluid. For a given amount of superheat, a bubble will be stable against the surface tension which is trying to collapse it. This problem involves consideration of the stability of a vapor bubble immersed in the bulk liquid—a case similar to that shown schematically in Fig. 6.5c.

The stability of a vapor bubble in the liquid can be evaluated in terms of the Clausius—Clapeyron equation. Considering the change of state between the liquid and vapor, the vapor pressure derivative is written

$$\frac{dp}{dT} = \frac{\lambda}{T\,\Delta v} \tag{6.9}$$

where λ is the latent heat of the liquid and $\Delta v = v_g - v_l$ is the difference between the specific volumes of the gas and liquid. For the interface of the bubble to be stable, it must have a pressure inside which exceeds the local saturation pressure by an amount related to the surface tension σ. For a spherical bubble, the expression has a standard form given by

$$p_v - p_s = \frac{2\sigma}{r} \tag{6.10}$$

where r is the local bubble radius. Obviously, the smaller the bubble the larger must be the pressure difference.

To get a feel for the order of magnitude of the quantities involved, assume that helium vapor obeys the ideal gas law and that the specific volume of the vapor is much greater than that of the liquid. These assumptions lead to the following approximation:

$$\Delta v \approx v_g \approx \frac{RT}{p} \tag{6.11}$$

Substituting (6.11) into the Clausius–Clapeyron equation leads to a differential relationship

$$\frac{dp}{p} = \frac{\lambda}{R}\frac{dT}{T^2} \tag{6.12}$$

This expression must be integrated between saturation (T_s, p_s) and the condition inside the bubble (T_v, p_v) as determined by the stability relationship (6.10). Such a procedure yields a common relationship for the required vapor pressure within a bubble.

$$p_v = p_s e^{\lambda \Delta T_s / R T_s^2} \tag{6.13}$$

where a further approximation has been made that the temperature difference, $\Delta T_s = T_v - T_s$, is small compared to T_s.

The present discussion is aimed at determining the minimum radius of a stable vapor bubble in the bulk liquid. Substituting the expression for equilibrium of a vapor bubble (6.10) and allowing the radius to be undetermined, we obtain an expression for the critical radius,

$$r_c = \frac{2\sigma}{p_s}\left(e^{\lambda \Delta T_s / R T_s^2} - 1\right)^{-1} \tag{6.14}$$

which subsequently can be solved to determine the approximate value of r_c for any fluid.

As an example of how (6.14) can improve the understanding of liquid helium boiling, consider the value of r_c for helium at 4.2 K and 1 bar (100 kPa). Under these conditions, $\lambda = 82$ J/mol · K and $\sigma = 0.15$ mJ/m^2. For a critical temperature difference assume $\Delta T_s \approx 0.3$ K, which is typical for superheat obtained on smooth heat transfer surfaces. Rough surfaces with many imperfections have been shown to have lower values of ΔT_s. Inserting these numerical values into (6.14) yields a critical radius $r_c \approx 0.0164\ \mu$m. This value for a radius defines a sphere containing approximately 10^4 atoms. Thus, it is reasonable to assume that the bubble represents a thermodynamic system. A bubble larger than r_c would be expected to grow while one smaller would collapse. Note that the above calculation is limited by the assumptions that $v_g \gg v_l$ and the ideal gas behavior for the vapor phase. These assumptions can lead to considerable inaccuracies in calculations of both the critical radii and the nucleation temperature.

The above calculation contains an assumed value for the superheat required to initiate the nucleation process, ΔT_s. Experimentally determined superheats actually vary by as much as half an order of magnitude. The highest values are obtained for the most ideal surfaces where nucleation is assumed to be homogeneous. These systems give nucleation superheats around 0.35 K at 4.2 K, 1 bar. In fact, homogeneous nucleation superheat has been measured over the entire He I range and shown to agree with the empirical relationship[6]

$$T_v = T_s + 4.322\left(1 - \frac{T_s}{T_c}\right)^{1.534} \tag{6.15}$$

An expression can be derived for the homogeneous nucleation temperature based on a model suggested by Frenkel.[7] The analysis yields the rate of formation of bubbles having the critical radius r_c as defined by (6.10). As a function of the fluid properties, the rate is given by

$$R = n_l \left(\frac{\sigma}{m}\right)^{1/2} \exp\left(-\frac{4\pi}{3}\frac{\sigma r_c^2}{k_B T}\right) \tag{6.16}$$

where n_l is the number density in the liquid and m is the mass of a helium atom. The critical radius r_c is a function of the superheat ΔT_s, as well as other parameters such as absolute temperature T_s.

The expression given by (6.16) can be used to calculate the critical radius r_c or preferably the superheat ΔT_s. Such a calculation requires one arbitrary assumption: that of the reaction rate for the onset of nucleation. However, since the critical parameters enter into the exponential, the results are only weakly dependent on the choice of R. Typical values for this quantity are assumed to be $1\ \text{cm}^{-3} \cdot \text{sec}^{-1}$. Furthermore, it is not possible to use the form developed above for the critical radius because the nonideality of the vapor phase plays an important role. Flint and Van Cleve[8] were able to obtain good agreement between experiment and theory if they used the actual behavior of the helium vapor pressure curve. The results of their calculation are shown in Fig. 6.6 along with experimental values for the nucleation superheat. These results were obtained on polished silicon chips oriented vertically in a bath of saturated liquid helium. The critical temperature differences are determined by noting the point where the hysteresis ended in a q versus ΔT curve; see Fig. 6.2. These data indicate a close correspondence between homogeneous nucleation theory and experiment. Typically, nucleation temperatures on real roughened surfaces are lower than those indicated in Fig. 6.6.

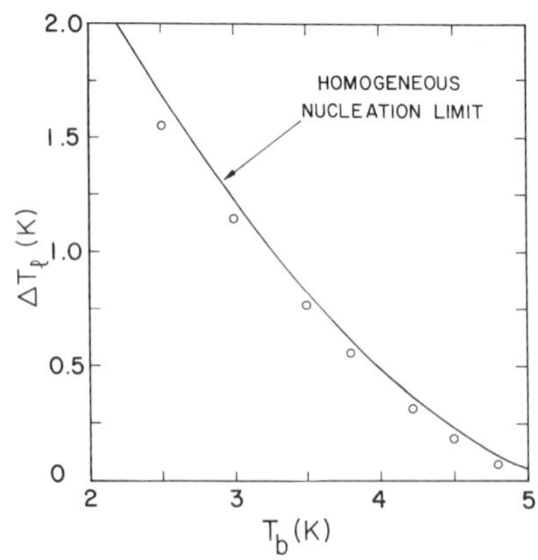

Fig. 6.6. Homogeneous nucleation limit for liquid helium heat transfer (from Flint and Van Cleve[8]).

6.3.2. Heat Transfer Correlations

Once nucleation has occurred and the bubbles are large enough to be stable against collapse in the bulk fluid, the heat transfer becomes dependent on the hydrodynamics of bubble detachment and growth. To model the nucleate boiling heat transfer in this regime, it is necessary to know a number of quantities including the rate of bubble growth, frequency of detachment, and something of the hydrodynamics of the two-phase fluid consisting of the liquid in the vicinity of the bubble. A conceptual picture of the hydrodynamics of an individual bubble is beneficial in understanding the heat transfer mechanisms.

In the vicinity of the nucleation site, it was observed by Hsu and Graham[9] that a departing bubble took with it an area of superheated liquid equal to twice the projected area of the bubble. Based on this hypothesis, shown schematically in Fig. 6.7, it is possible to account for the heat removed by one bubble as a sum of two quantities,[10]

$$q_b = \frac{4\pi}{3} \lambda \rho v \int_A nfr_b^3 \, dA + 2\pi \, \Delta T_s C_l \rho_l \delta \int_A nfr_b^2 dA \qquad (6.17)$$

where n is the number of nuclei per unit area of surface and the integrals are over the entire heat transfer surface area. The first quantity on the right-hand side is that due to the latent heat of the helium within the bubble, while the second is the heat required to superheat a new layer of liquid that replenishes the layer taken away with the departing bubble. The difficulties associated with applying (6.17) to real problems are multifold. First, the frequency of detachment f is involved with both terms in (6.17). The heat flux is proportional to f, which is largely an experimental quantity. The quantity is dependent on n, the number of nucleation sites per area, and r_c, the critical size of a departing bubble. The amount of superheat ΔT_s also enters (6.17) in the second term. In principle, this quantity can be determined from (6.16); however, for real surfaces it can vary considerably.

An additional complication enters when attempting to determine the total heat transferred in the nucleate boiling regime. In an actual process,

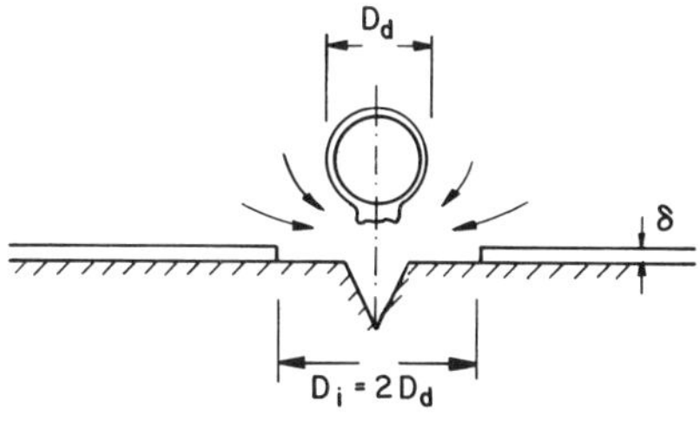

Fig. 6.7. Schematic of departing bubble and area of superheated liquid (from Bald[10]).

there are two heat transfer terms: one due to bubble hydrodynamics, q_b, and one due to natural convection in the bulk fluid, q_{nc}:

$$q_{nb} = q_b + q_{nc} \tag{6.18}$$

There may be an additional contribution due to the interaction between natural convection and boiling, but is is unclear what form it would take. It is tempting to neglect the natural convection term, assuming it is small; however, this is not always possible. Except in very special geometries, it is not possible to determine q_{nc} exactly. Therefore, engineering correlations are needed to describe the second term in (6.18). In general, the natural convection heat transfer can be written as a function of the dimensionless Rayleigh number as given by (6.6).

The above analysis has the potential of being able to describe the heat transfer in the nucleate boiling region. However, difficulty arises when determining the variables that enter (6.17) and (6.18). These variables mostly include the functional form of the natural convection function, the amount of superheat ΔT_s, the bubble density n, and detachement frequency f. Since this analysis has limited practical usefulness, the preferred approach is to characterize the total heat transfer in terms of an engineering correlation.

The most popular and probably the best correlation used to describe the nucleate boiling regime is due to Kutateladze.[11] It is based on theory and experimental scaling of heat transfer to many different fluids:

$$\frac{h}{k_l}\left(\frac{\sigma}{g\rho_l}\right)^{1/2} = 3.25 \times 10^{-4}\left[\frac{qC_{pl}\rho_l}{\lambda\rho_v k_l}\left(\frac{\sigma}{g\rho_l}\right)^{1/2}\right]^{0.6} \times\left[g\left(\frac{\rho_l}{\eta_l}\right)^2\left(\frac{\sigma}{g\rho_l}\right)^{3/2}\right]^{0.125}\left(\frac{p}{(\sigma g\rho_l)^{1/2}}\right)^{0.7} \tag{6.19}$$

where g is the acceleration of gravity and λ is the latent heat of vaporization. Although (6.19) is a complex expression, it does predict a reasonably correct functional dependence for the nucleate boiling heat transfer. Rearranging (6.19) into a more manageable form leads to the relationship

$$q = 1.90 \times 10^{-9}\left[g\left(\frac{\rho_l}{\eta_l}\right)^2\chi^3\right]^{0.3125}\left(\frac{p\chi}{\sigma}\right)^{1.75}\left(\frac{\rho_l}{\rho_v}\right)^{1.5} \times\left(\frac{C_p}{\lambda}\right)^{1.5}\left(\frac{k_l}{\chi}\right)(T_s - T_b)^{2.5} \tag{6.20a}$$

where

$$\chi = \left(\frac{\sigma}{g\rho_l}\right)^{1/2} \tag{6.20b}$$

and T_s and T_b are the surface and bath temperatures, respectively. The expression given by (6.20) is evaluated more easily. For the case of He I at 4.2 K, 1 bar, the coefficient of proportionality can be calculated to equal 5.80 $W/cm^2 \cdot K^{2.5}$.

The Kutateladze correlation is in reasonable accord with experimental measurements of heat transfer in He I. However, there is a wide variation in experimental data owing to the differences in sample preparation and surface material and orientation.[12] It is not too surprising, based on the hydrodynamic arguments above, that different surface preparations would yield much different results. Figure 6.8 is a compilation of nucleate pool boiling data for flat copper surfaces facing upward and experiencing an increasing heat flux. The data vary over at least half an order of magnitude in ΔT with the smoothest surfaces apparently allowing a larger superheat. The larger superheat seen in smooth surfaces is consistent with the homogeneous nucleation arguments presented at the beginning of this section. Finally, note that an empirical fit used by Schmidt[13] to describe the data is also plotted in Fig. 6.8. This fit, which is quite close to the Kutateladze correlation with the exception of the coefficient of proportionality, is a conservative form useful for engineering applications.

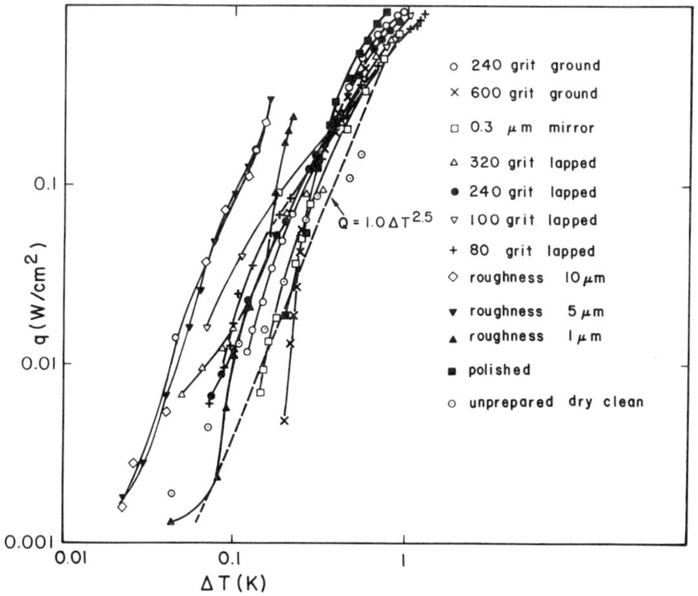

Fig. 6.8. Nucleate boiling heat transfer to He I (compilation of data and suggested correlation from Schmidt[13]).

6.3.3. Maximum Nucleate Boiling Heat Flux

The qualitative picture applied to the understanding of the maximum heat flux in pool boiling He I is as follows. Imagine a surface populated with a number of nucleation sites. At high heat fluxes, these sites are actively nucleating bubbles that grow to a stable size and detach at a frequency f controlled by gravitational considerations. With increasing heat flux, the number and size of these bubbles grow until a point is reached where they cover a sizable fraction of the heater surface. At this point, the individual bubbles are no longer the lowest-energy condition. They will prefer to coalesce into a continuous vapor film which will blanket the entire heat transfer surface. This condition usually is referred to as the onset of film boiling. The difficulty is in being able to understand and predict the value of heat flux at which this event occurs, q^*.

To obtain some physical feeling for the occurrence of the maximum heat flux, consider an idealized surface facing upward as in Fig. 6.3b. The vapor bubbles are departing at velocity $\mathbf{v}_v$, while the replenishing liquid moves in the opposite direction at velocity $\mathbf{v}_l$. These velocities are not independent variables because the heat flow is determined by the rate of growth and departure of the vapor bubbles.

The hydrodynamics of this process is described by the Helmholtz instability,[14] which pertains to the critical velocity of immiscible fluids moving relative to each other. Assume the liquid and vapor phases represent these two immiscible fluids. Then the boundary separating the two fluids would show an upward-moving vapor and downward-moving liquid. For these two fluids to pass each other undisturbed, the boundary that separates them must remain stable. The stability of this boundary is a function of a number of parameters including the relative velocities and densities of the two fluids. This is believed to be the condition that imposes the peak heat flux limit in classical liquids including He I.

Because of the relative motion of each fluid, there can be a surface wave set up at the interface. The velocity of this wave, c_s, is dependent on a number of factors including the surface tension and properties of the individual phases. The relationship for the surface wave velocity is

$$c_s^2 = \frac{\sigma m}{\rho_l + \rho_v} - \frac{\rho_l \rho_v}{(\rho_l + \rho_v)^2} (\mathbf{v}_v - \mathbf{v}_l)^2 \tag{6.21}$$

where $m = 2\pi/\lambda$ is the wave number of the surface wave. Since (6.21) consists of a difference between two positive quantities, it is possible for c_s^2 to have either a positive or negative value. For $c_s^2 > 0$ the surface wave can exist. If $c_s^2 < 0$, the surface wave velocity is imaginary, implying an instability in the interface. Therefore, the condition for maximum vapor velocity is obtained by equating c_s^2 to zero. Using conservation of mass flow, that is, $\rho_v \mathbf{v}_v + \rho_l \mathbf{v}_l = 0$, a simple expression is obtained:

$$\mathbf{v}_v^* = \left(\frac{\rho_v \sigma m}{\rho_l(\rho_l + \rho_v)} \right)^{1/2} \tag{6.22}$$

If the vapor velocity exceeds the value given by (6.22) there should be an unstable two-phase flow. The result is destruction of the interface between the two phases, which in turn leads to a condition where the vapor film blankets the heat transfer surface. This condition occurs at the maximum heat flux.

Zuber et al.[15] used the above reasoning to predict analytically the maximum heat flux q^*. Assuming that the heat is transported primarily by the vapor velocity and that the latent heat of the liquid that goes into the formation of the bubble is the dominant energy, Zuber and coworkers argued that the maximum heat flux can be written as a product of these quantities, that is,

$$q^* = K\lambda\rho_v \left(\frac{\sigma(\rho_l - \rho_v)\, g}{\rho_2^2} \right)^{1/4} \left(\frac{\rho_l}{\rho_l + \rho_v} \right)^{1/2} \tag{6.23}$$

where K is a numerical factor which in the case under consideration has a value of $\pi/24$. It is possible to derive a similar relationship for q^* from dimensionless group analysis. Based on experimental evidence, a similar relationship has developed acceptance,[16,17] that is,

$$q^* = 0.16\, \lambda\rho_v^{1/2}[\sigma g(\rho_l - \rho_v)]^{1/4} \tag{6.24}$$

which is in close agreement with (6.23) with the exception of the last factor. The numerical factor in (6.24) is determined by correlation of numerous experimental results. For a numerical comparison, (6.23) predicts for liquid helium at 4.2 K, 1 bar a value of $q^* = 0.85$ W/cm^2, which is in reasonable agreement with experimental results ranging from 0.5 W/cm^2 to around 1.5 W/cm^2. However, to test the relationship represented by (6.24) for a particular fluid, it is necessary to make measurements over a wide range of vapor densities or temperatures. Furthermore, to determine whether the correlation is universally acceptable, measurements of q^* for a variety of fluids are required. These experimental investigations determine the empirical constant of proportionality. Plotted in Fig. 6.9 are normalized measurements of $q^*/\lambda\rho_v$ versus the density function given in (6.24).[17] Reasonable agreement with the correlation is seen for the three cryogenic fluids considered.

It is worthwhile considering the temperature and orientation dependencies of q^* in comparison to (6.24). At high temperatures, q^* would be

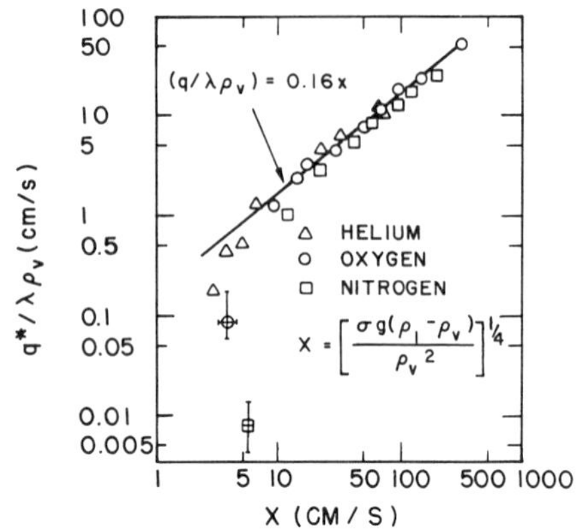

Fig. 6.9. Comparison of peak nucleate boiling heat fluxes with Kutateladze correlation (from Lyon[17]).

expected to decrease because as the critical point is approached there is no phase change associated with boiling and the latent heat vanishes. The theoretical expression indicates that the peak heat flux should go as the latent heat times the fourth root of the density difference. Both quantities vanish at the critical point. At low temperatures the latent heat approaches a constant and the temperature dependence is determined by the square root of the vapor density, which in turn decreases with temperature. Therefore, there should be a maximum in the peak heat flux. Analogous although somewhat less successful arguments can be used to describe the orientation dependence of q^*. Since the peak heat flux is proportional to the square root of the gravitational acceleration g, the buoyancy effects should decrease as the surface is turned from facing upward to facing downward. In fact, based on this simple argument, a surface facing downward should have $q^* = 0$. Experimentally, q^* obtains a minimum value a 180° orientation although its value is considerably greater than zero.

Measurements of the temperature and orientation dependence of q^* have been conducted most comprehensively by Lyon.[17] The orientation dependence of the maximum nucleate boiling heat flux is shown in Fig. 6.10. Note that for these experiments $q^* = 0.8\ \mathrm{W/cm^2}$ at 4.2 K which is quite close to the theoretical prediction. Furthermore, for each orientation there is a maximum value in the temperature dependence of q^*. Experimentally, this maximum occurs around 3.6 K. Finally, the orientation dependence is in qualitative agreement with theory. A minimum in q^* does occur at $\theta = 180°$, that is facing downward. However, the value of q^* at $\theta = 180°$ is still quite sizable, being about 25% of q^* for $\theta = 0°$.

The above correlations do not apply when the helium is subcooled to a state substantially off the saturated vapor pressure curve. In this case there can be no coexisting vapor. Some work[18–20] on the effect of subcool-

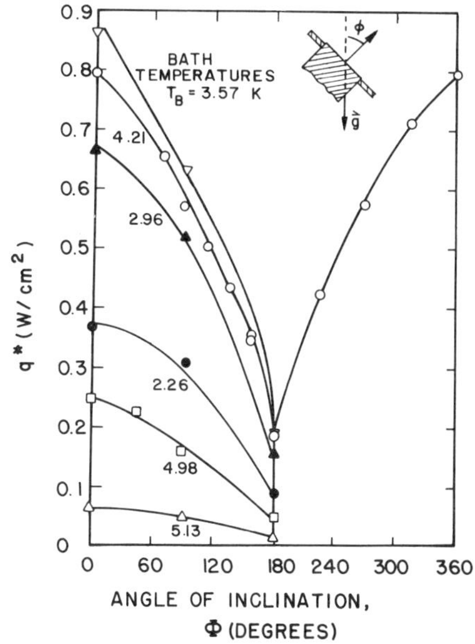

Fig. 6.10. Orientation dependence to the nucleate boiling heat flux (from Lyon[17]).

ing on heat transfer has attempted to treat the peak heat flux correlation in terms of a subcooled temperature difference ΔT_{sub}. The subcooled temperature difference is defined as the difference between the bath temperature T_b and the temperature corresponding to saturation T_{sat}. As heat is applied from the heat transfer surface, the temperature rises. However, in subcooled He I a larger temperature difference is required than in saturated He I. This is due to the need to bring the local environment first to saturation followed by the amount of superheat needed to cause nucleate boiling. Therefore, the absolute temperature needed to cause boiling at a given pressure should be independent of bath temperature. Such a hypothesis leads to a correlation for the enhancement of peak heat flux q^* with subcooling

$$\frac{q^*_{sub}}{q^*_{sat}} = 1 + \frac{aC_p \Delta T_{sub}}{\lambda} \tag{6.25}$$

where a is an empirical parameter found to be close to 1.75. Comparison of experimental results with the correlation given by (6.25) are shown in Fig. 6.11. Agreement is reasonable although the theory has received only limited application.

In addition to heat transfer in saturated and subcooled pool boiling helium, there have also been several investigations of free-convection heat transfer to helium near the critical point and in the supercritical

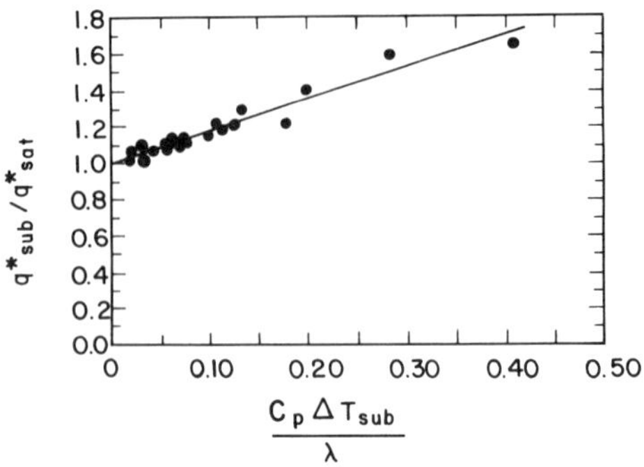

Fig. 6.11. Variation of the peak nucleate boiling heat flux with subcooling (from Ibrahim et al.[18]).

regime.[21–24] Near the critical point, the heat transfer is seen to be enhanced considerably over that at normal conditions. For example, near 2.24 bars, heat transfer coefficients as high as $10\ \mathrm{W/cm^2 \cdot K}$ have been observed. Such results correlate with the maxima in the thermodynamic properties near the critical point. Away from the critical point, the results are correlated best as a function of the Rayleigh number. A reasonable fit to much of the helium data in this regime can be obtained from the expression[25]

$$\mathrm{Nu} = 0.615\ \mathrm{Ra}^{0.258} \tag{6.26}$$

which is close to the form expected for laminar free-convection heat transfer. A compilation of convective heat transfer results for supercritical helium is shown in Fig. 6.12 along with several correlations. There is a characteristic trend to the data; however, the agreement between various experiments is variable.

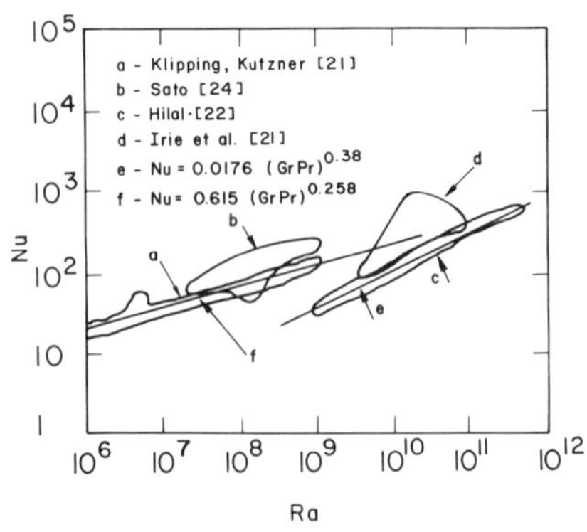

Fig. 6.12. Comparison of data and correlations on free convection heat transfer to supercritical helium (from Hilal et al.[22]).

6.4. FILM BOILING

Once the film boiling condition has been established, normally by exceeding q^* under steady-state conditions, as wholly different heat transfer process takes place. In the vicinity of the heat transfer interface the helium takes on a stable two-phase condition with a thin vapor layer blanketing the surface from the He I bath. For a surface facing upward, this condition is gravitationally unstable due to the density differential between the liquid and vapor. Experimentally, it is found that return to the nucleate boiling state requires a decrease in heat flux to q_R or q_{mfb} which can be substantially less than q^* for most configurations. It follows that there are two principal issues that should be addressed when evaluating the film boiling state. First, given the conditions of film boiling, how is the recovery process explained? In particular, is it possible to predict q_R? This process has to do with the stability of the He I–vapor interface. The second question pertains to the need to correlate the heat transfer coefficient in the film boiling condition. This process normally relies on dimensionless group analysis developed for other liquids.

6.4.1. Minimum Film Boiling Heat Flux

The stability of the vapor film blanketing the heat transfer surface can be evaluated in terms of a hydrodynamic condition referred to as the Taylor instability.[14] This interpretation is a standard approach to treating the interface between two dissimilar fluids. Imagine the condition shown in Fig. 6.13 which is an idealized film boiling heat transfer process. The liquid helium is heavier than the vapor so it would prefer to rewet the surface; however, it is being maintained in the present condition by the high-temperature vapor film. The stability of the liquid–vapor interface is controlled by the behavior of surface wave oscillations. The wave can be assumed to have an amplitude η_0 and a wavelength λ. Surface waves must be damped for the interface to be stable, otherwise the amplitude would grow beyond the vapor film thickness and rewet the surface. It is therefore reasonable to construct a model based on the assumption that the stability of these waves controls the recovery process.

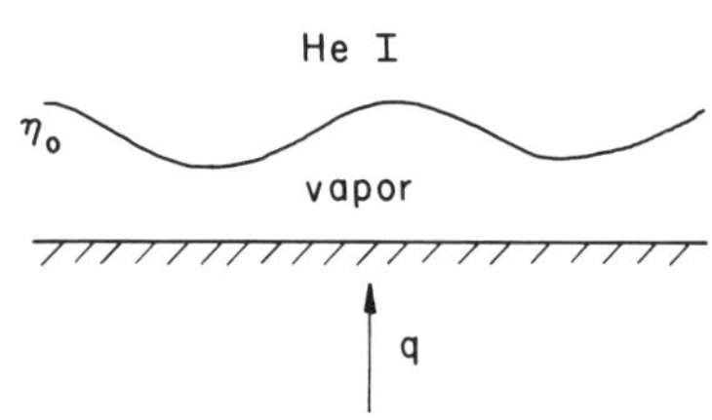

Fig. 6.13. Idealized film boiling heat transfer process.

To be more quantitative about the above argument, assume that the stability of a surface wave is assured if the energy associated with surface tension exceeds the combination of the kinetic and potential energies in the wave. Both these terms are related to the amplitude of the surface oscillation η_0, as well as physical parameters such as densities and surface tension. The kinetic and potential energies can be related as

$$\left.\frac{E}{\underset{\sim}{\lambda}}\right|_{\mathrm{KP}} = \frac{g(\rho_l + \rho_v)\,\eta_0^2}{2} \tag{6.27}$$

where $\underset{\sim}{\lambda}$ is the wavelength of the surface oscillation and g is the gravitational acceleration. Similarly, the surface term can be written

$$\left.\frac{E}{\underset{\sim}{\lambda}}\right|_{\sigma} = \frac{1}{\underset{\sim}{\lambda}} \int_0^{\underset{\sim}{\lambda}} \Delta p\, \eta\, dx \tag{6.28}$$

where Δp is taken to be the pressure difference and η is a sinusoidal varying wave, that is $\eta = \eta_0 \sin 2\pi x/\underset{\sim}{\lambda}$. By integrating over the wavelength of the oscillation and using the approximation that the surface wave oscillation is small compared to $\underset{\sim}{\lambda}$, we can evaluate the integral (6.28):

$$\left.\frac{E}{\underset{\sim}{\lambda}}\right|_{\sigma} = \frac{4\pi\sigma\eta_0^2}{\underset{\sim}{\lambda}^2} \tag{6.29}$$

As already stated, the condition for stability demands that the surface energy exceed the dyamic energy. This leads to a condition on $\underset{\sim}{\lambda}$ by demanding that (6.29) not exceed (6.27), that is,

$$\underset{\sim}{\lambda} < 2\pi \left(\frac{\sigma}{g(\rho_l - \rho_v)^{1/2}} \right) \tag{6.30}$$

For liquid helium near 4.2 K, (6.30) is obeyed for wavelengths less than about 2 mm, a dimension that must be comparable to a characteristic distance in the heat transfer probem, for example, the typical bubble dimension.

The minimum film boiling heat flux q_R can be understood by application of the Taylor instability theory to the film boiling heat transfer condition. Lienhard and Wong[26] and Zuber[15] used this analysis, identifying the breakdown of film boiling with the amplitude of the surface wave. The general relationship has one empirical constant and an explicit diameter dependence. For the special case of a flat plane, the correlation is simplified considerably:

$$q_R = 0.16\, \lambda \rho_v \left(\frac{g\sigma(\rho_l - \rho_v)}{(\rho_l + \rho_v)^2} \right)^{1/4} \tag{6.31}$$

It is interesting to compare (6.31) with (6.24), which predicts the peak heat flux q^*. Taking a ratio of these two expressions, the only important parameters turn out to be the relative densities of the liquid and vapor,

$$\frac{q_R}{q^*}=\left(\frac{\rho_v}{\rho_l+\rho_v}\right)^{1/2} \tag{6.32}$$

For example, considering liquid helium at 4.2 K, we find that the ratio described by (6.32) has a value of 0.35 at atmospheric pressure. Therefore, assuming $q^* = 0.85\ \text{W/cm}^2$ as measured by Lyon,[17] we find that (6.32) predicts a minimum film boiling heat flux of about $0.3\ \text{W/cm}^2$ in close agreement with experimental measurements.

The pressure dependence of the recovery hea flux is worth considering in light of the noted behavior of $q^*(p)$. As was observed in the previous section, subcooling increases q^* by about 50% per degree of ΔT_{sub} near atmospheric pressure. The subcooling effect on q_R is greater, as can be seen in Fig. 6.14. Note that q_R increases by about 90% per degree of ΔT_{sub} near atmospheric pressure.[18] The correlating relationships for the minimum film boiling heat flux predict this effect. Subcooling increases the ratio of the vapor to liquid density at saturation, which would result in an increase in q_R/q^*, as observed by experiment.

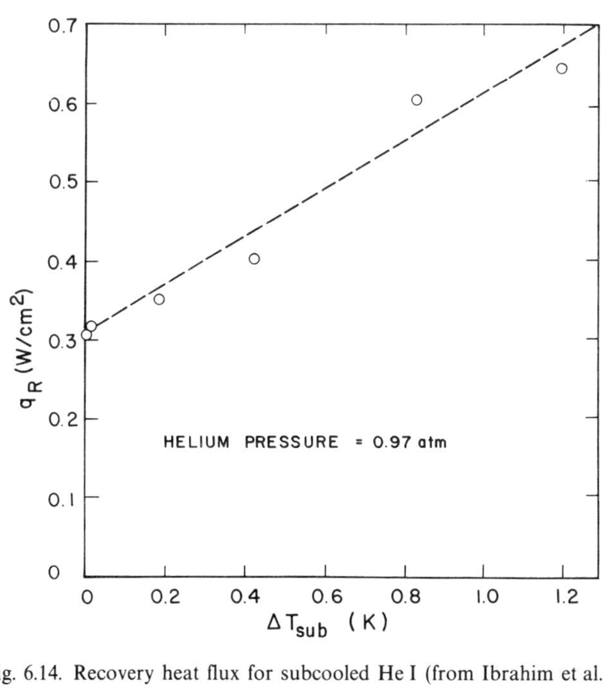

Fig. 6.14. Recovery heat flux for subcooled He I (from Ibrahim et al.[18]).

6.4.2. Heat Transfer Correlations

Once stable film boiling is established, it is of interest to be able to predict the magnitude of the film boiling heat transfer coefficient h_{fb} or the rate of heat flux q_{fb} for a given ΔT. Experimentally, h_{fb} has values that range over more than an order of magnitude between about 0.03 $W/cm^2 \cdot K$ and nearly 1 $W/cm^2 \cdot K$, with the latter being achieved for fine wires with diameters of $\sim 10\ \mu m$. The film boiling heat transfer also depends on fluid properties, being a function of the vapor and liquid densities, latent heat, and surface tension σ.

A number of semiempirical correlations exist for prediction of heat transfer on the film boiling condition. The best known and perhaps most accepted of those is due to Breen and Westwater.[27] As in the case of the minimum film boiling heat flux q_R, these authors base their theory on the Taylor instability. They consider the wavelength condition given by (6.30) as the minimum required to release vapor bubbles into the bulk fluid from the boiling film. The correlation depends on the thermal properties of the fluid including specific heat and viscosity in addition to geometrical conditions such as the diameter of the heat transfer sample. The correlation relates the film boiling heat transfer coefficient to these quantities and a number of numerical constants:

$$h_{fb}\left(\frac{\sigma}{g(\rho_l-\rho_v)}\right)^{1/8}\left(\frac{\eta_v(T_s-T_b)}{k_v^3\rho_v(\rho_l-\rho_v)\,g\lambda'}\right)^{1/4} = 0.37+0.28\left(\frac{\sigma}{gD^2(\rho_l-\rho_v)}\right)^{1/2} \tag{6.33a}$$

where

$$\lambda' = \frac{[\lambda+0.34\,C_{p_v}(T_s-T_b)]^2}{\lambda} \tag{6.33b}$$

and represents an effective latent heat. T_s and T_b are the surface and bath temperatures, respectively. D is the diameter of the heater surface, which for film boiling has an effect on the heat transfer coefficient. For the special case where the heater diameter is greater than a few millimeters, the second term on the right-hand side is small and (6.33) may be approximated by the relationship

$$h_{fb} = 0.37\left(\frac{g(\rho_l-\rho_v)}{\sigma}\right)^{1/8}\left(\frac{k_v^3\rho_v(\rho_l-\rho_v)\,g\lambda'}{\eta_v(T_s-T_b)}\right)^{1/4} \tag{6.34}$$

Furthermore, in helium for moderate temperature differences, that is $\Delta T \gtrsim 5$ K, the second term in (6.33b) is dominant. Under these conditions, it can be shown easily that film boiling heat transfer $q \approx (T_s-T_b)^{3/4}$.

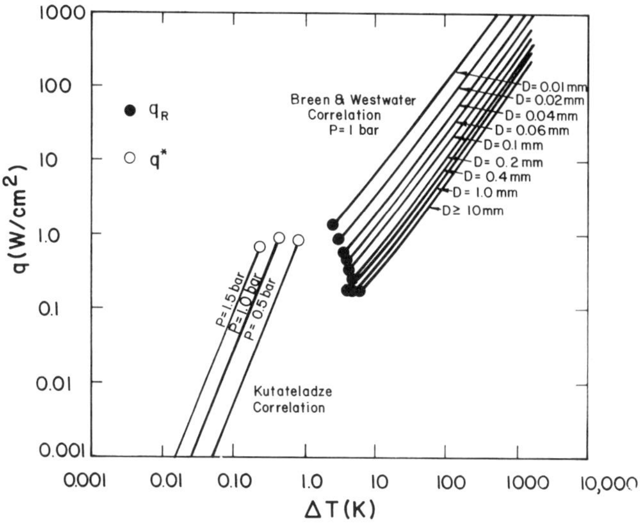

Fig. 6.15. A summary of pool boiling heat transfer correlations for He I.

In summary, Fig. 6.15 is a plot of the various predictive relationships for nucleate and film boiling heat transfer.[28] A Comparison between data for nucleate boiling and minimum film boiling has been made already. For the case of film boiling heat transfer, the Breen–Westwater correlation is only moderately successful a predicting experimental data.[16] In general, film boiling heat transfer coefficients measured on fine wires has yielded consistently higher heat transfer coefficients than calculated from the above correlation.

6.5. CHANNEL HEAT TRANSFER

As a very interesting and technically significant special case of pool boiling heat transfer, consider the channel heat transfer problem described schematically in Fig. 6.16. A channel of width w is formed between a heat transfer surface and a second adjacent surface. The channel may be oriented at any angle between vertical and horizontal, as shown in Fig. 6.16. This particular problem is significant because it is an attempt to model a cooling channel in a technical device such as a superconducting magnet. When heat is applied to the surface, the liquid can circulate owing to the thermosyphon effect where bubbles are transported under the influence of buoyancy forces. As the heat flux from the surface is increased, film boiling initiates at the top of the channel where the vapor quality is greatest.

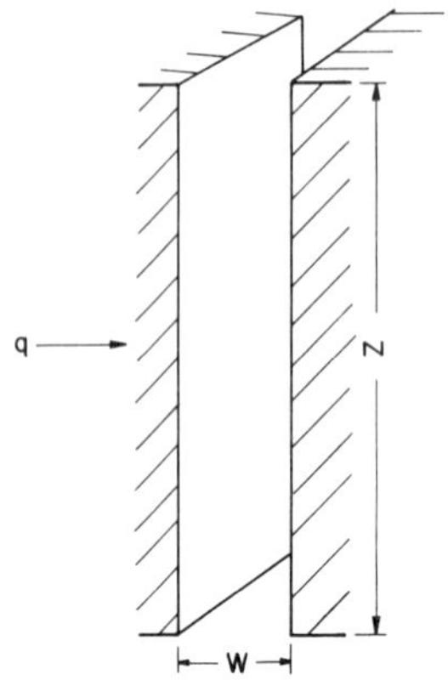

Fig. 6.16. Schematic of a heat transfer channel.

Because of the induced flow, the peak heat flux at the channel bottom can be quite large. The position dependence of the peak heat flux in one set of experiments is shown in Fig. 6.17. For vertically oriented surfaces in an open bath at 4.2 K, the peak heat flux is usually in the neighborhood of 0.6 W/cm^2. Therefore, the bottom of the channel has q^* quite close to that observed in an open bath. As the helium is vaporized and transported up the channel, the local peak heat flux is depressed. In the present example, the peak heat flux near the top approaches 0.34 W/cm^2.

Wilson[29] conducted one study of channel heat transfer in an experiment having variable height and width. The following set of observations were made. First, for small width w, the peak heat flux averaged over the entire surface area was directly proportional to the heat flux in the helium channel and therefore proportional to w. This suggests that the peak heat flux is governed by the bulk fluid flow. Second, for constant w, the peak heat flux was found to be inversely proportional to the square root of the channel height z. Based on the results as they depend on w, this indicates that the bulk heat flow varies as $z^{-1/2}$. Finally, it was found that for $w/z > 0.1$, the channel behaved effectively as an open bath with q^* approaching that of a vertically oriented surface.

Sydoriak and Roberts[30] derived a general relationship for frictionless homogeneous flow of a fluid in a evaporator, which Wilson applied to this

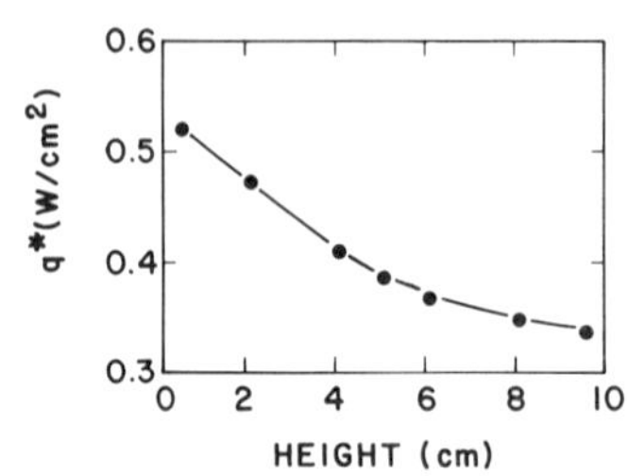

Fig. 6.17. Position dependence of the peak heat flux in a He I cooled channel (from Schmidt[34]).

channel heat transfer problem. The prediction for the critical power per unit area of heated surface, Q^*/A_s, is given by

$$\frac{Q^*}{A_s}=\frac{w}{\sqrt{z}}\frac{\lambda\rho_l}{2}\left[\frac{g\tilde{q}}{\beta-1}\left(1-\frac{\ln[1+\tilde{q}(\beta-1)]}{\tilde{q}(\beta-1)}\right)\right]^{1/2} \tag{6.35}$$

where $\beta=\rho_l/\rho_v$ and $\tilde{q}$ is the "critical quality"—the mass ratio of vapor to total (liquid plus vapor) at the channel top when film boiling initiates. This last quantity must be determined empirically although in general it should be a scalable function for different fluids. A plot of the critical power versus channel width w for different channel lengths z should give a linear plot from which the proportionality function $Q^*z^{1/2}/wA_s$ can be determined. Such a plot is displayed in Fig. 6.18 for four different channel heights. The linear plot obtained yields a critical quality $\tilde{q}$ ranging from 0.33 to 0.26 in the case of the largest channel. Thus, for calculational purposes, (6.35) should be a reasonable approximation for Q^*/A_s, assuming a constant value for $\tilde{q}\approx 0.3$.

One difficulty with the above described analysis is that it does not naturally lead into the open bath limit for $w/z>0.1$. In an effort to develop a more general equation for channel heat transfer, Lehangre *et al.*[31] suggested a correlation based on a series of experiments of different configuration:

$$\frac{Q^*}{A_s}=\frac{1}{1.7+0.125(z/D_h)^{0.88}} \tag{6.36}$$

where Q^*/A_s is in W/cm^2. The quantity D_h is the hydraulic diameter which is equal to four times the ratio of the flow cross section to the heated perimeter. Equation (6.36) has the appropriate limit for w/z large; however, for channels other than in the vertical orientation it is not applicable.

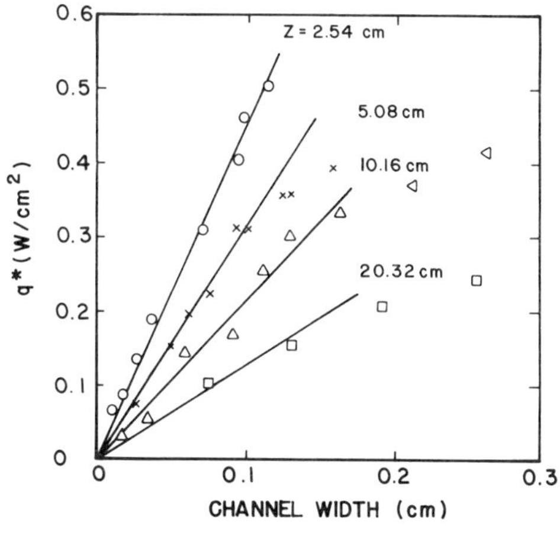

Fig. 6.18. Critical power versus channel width for He I heat transfer (from Wilson[29]).

In a recent set of experiments on the angular variation of channel heat transfer, Chen and Van Sciver[32] noted that for wide channels the maximum heat flux q^* should correlate with the open bath pool boiling results by Lyon[17] (see Fig. 6.10). The results of these experiments are shown in Fig. 6.19, where the angle θ is measured from the vertical orientation. There are two physical processes that lead to the observed angular dependence of q^*. The first process is nucleate boiling associated with movement of vapor bubbles normal to the heated surface. This effect, which is maximum at an angle of 90°, can be assumed to control the heat transfer process for large w. The other process is associated with the natural circulation of the heated fluid, where the movement is parallel to the heated surface. The natural circulation process is a maximum at $\theta = 0°$ and should dominate the heat transfer in the limit of small w. The combination of these two processes is necessary to describe the variation of q^* with θ and w. In particular, one would expect the maximum value of q^* to vary continuously between 0° and 90° as the channel width is increased.

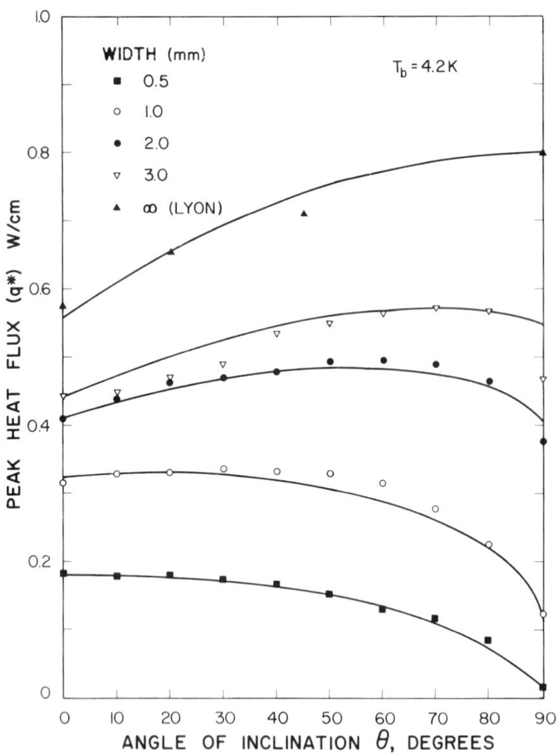

Fig. 6.19. Variation of the peak heat flux with channel orientation and width at 4.2 K. Channel is 127 mm in length.

A general correlation used to describe the angle and width dependencies of q^* is of the form

$$q^* = b \sin\left(\frac{\theta + 90^\circ}{2}\right) + c(\cos\theta)^{1/2} \tag{6.37}$$

where b and c are adjustable parameters, which should be functions of w only. The width dependence of these parameters is shown in Fig. 6.20. The basis for (6.37) is purely empirical evidence. The first term is used to describe the nucleate boiling heat transfer process. The angular dependence is in reasonable agreement with Lyon's pool boiling data. The second term represents the natural circulation process. The $(\cos\theta)^{1/2}$ angular dependence not only fits our experimental data for small w but is consistent with the Wilson's correlation which predicts $q^* \propto g^{1/2}$.

As a final comment on channel heat transfer, the steady-state conditions as described above generally take a considerable time to become established. This is because the natural circulation requires a substantial temperature rise and vapor production before it is fully established. Naturally, the time to reach steady state depends on the magnitude of Q^*/A_s, but for significant heat fluxes, of the order of 0.1 W/cm^2, this characteristic time can be as much as a second. The existence of a charac-

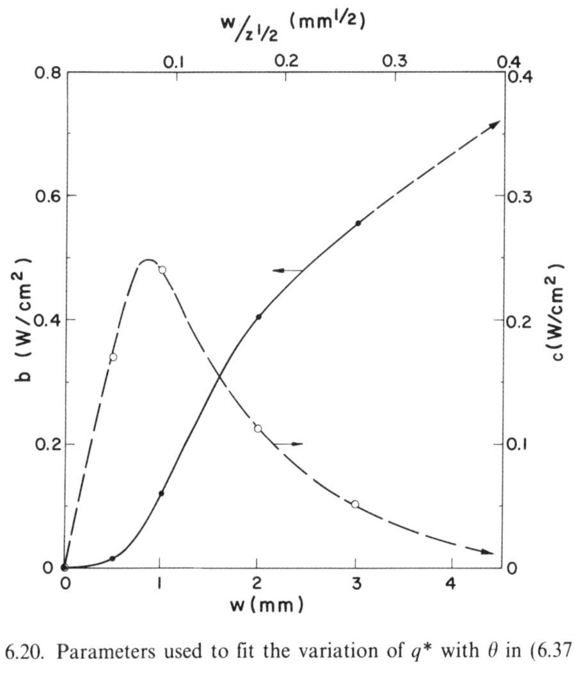

Fig. 6.20. Parameters used to fit the variation of q^* with θ in (6.37).

teristic time for the development of steady-state heat transfer has strong impact on technical applications. Since many systems experience transient heat transfer processes, which are on the millisecond time scale, it is important to appreciate that these heat transfer processes are far from steady state.

6.6. TRANSIENT HEAT TRANSFER

In the previous sections, it was assumed that the heat transfer process had been underway for sufficient time that steady-state conditions existed. The characteristic time required for the steady state to be achieved is equivalent to the time required for convection to become fully established. For nucleate boiling, enough heat must be transferred to vaporize bubbles and allow them to detach. In film boiling the characteristic time is associated with sufficient energy flux to vaporize a layer of helium. This can be represented approximately by

$$\Delta t \approx \frac{\rho \lambda \delta}{q} \tag{6.38}$$

where δ is the vapor film thickness. For an order-of-magnitude analysis assume that $\delta \approx 10\ \mu\text{m}$. Using the physical quantities for He I at 4.2 K, 1 bar, (6.38) predicts $\Delta t \approx 3/q$ ms, where q is in W/cm^2. For a heat flux of a few W/cm^2, the steady-state conditions are not established until well in excess of a few milliseconds. For times shorter than this value, the heat transfer processes are governed by nonconvective mechanisms such as conduction and radiation.

It is of considerable importance to be able to understand transient heat transfer in liquid helium. Transient heat transfer is fundamental to the analysis of a number of problems including the stability of superconducting magnets. There are several aspects to this problem which are worth noting at the beginning. First, for very short times, that is, $\Delta t < 1$ ms, heat transfer processes turn out to be controlled by physical mechanisms similar to those developed for the case of He II, that is, Kapitza conductance. Second, the transition between the region of heat transfer space governed by conduction and that which resembles the steady-state process is important. This transition occurs on time scales consistent with the rough calculation in (6.38). There are several parameters that are associated with transient heat transfer: (1) the peak heat flux $q^*(t)$ or critical energy $\Delta E^* = q^* \Delta t^*$; (2) the interfacial temperature difference ΔT_s; and (3) the effective heat transfer coefficient h. These parameters are governed primarily by the physical properties of helium and the rate of heat transfer.

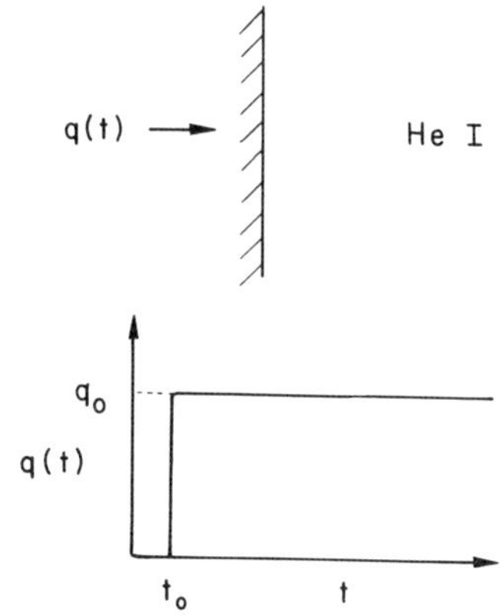

Fig. 6.21. Schematic representation of the transient heat transfer process.

The various regimes of transient heat transfer can be described best in terms of an idealized experiment, which in fact is not very different from actual experiments performed to investigate the problem. Imagine a solid heat transfer surface as shown in Fig. 6.21, which in this particular case is oriented vertically. Recall that the peak steady-state heat flux for this configuration is in the neighborhood of 0.6 W/cm^2. In this experimental configuration, a step function heat flux q_0 beginning at t_0 is applied to the sample. With suitable thermometry, which must have a response time faster than a millisecond, the surface temperature is recorded as it varies with time following the initiation of heat transfer.

As an example of the kind of data collected from this type of experiment, measurements by Steward[33] are shown in Fig. 6.22. In this particular case, the heat transfer surface is a thin carbon film which is both heater and thermometer. Note that for very short times, the surface temperature rise is not large, although it is an increasing function of heat flux. This short time regime, which is generally much less than 1 ms, is controlled by transient conduction mechanisms. As time progresses, the nucleation of convective mechanisms begins. For low heat fluxes, the heat transfer process eventually transforms to steady-state nucleate boiling. At high heat fluxes, that is, $q > 1$ W/cm^2, there is a transition to the film boiling state. This latter transition has associated with it an increase in ΔT_s of about an order of magnitude, consistent with the steady-state film boiling heat transfer characteristics. In the steady-state regime, the temperature differences presumably can be described in terms of the correlations introduced in previous sections of this chapter. There are, however, two effects in these data which need further discussion: (1) the size of the heat transfer coefficient in the transient conduction regime and (2) the time associated with the transition to steady-state heat transfer mechanisms.

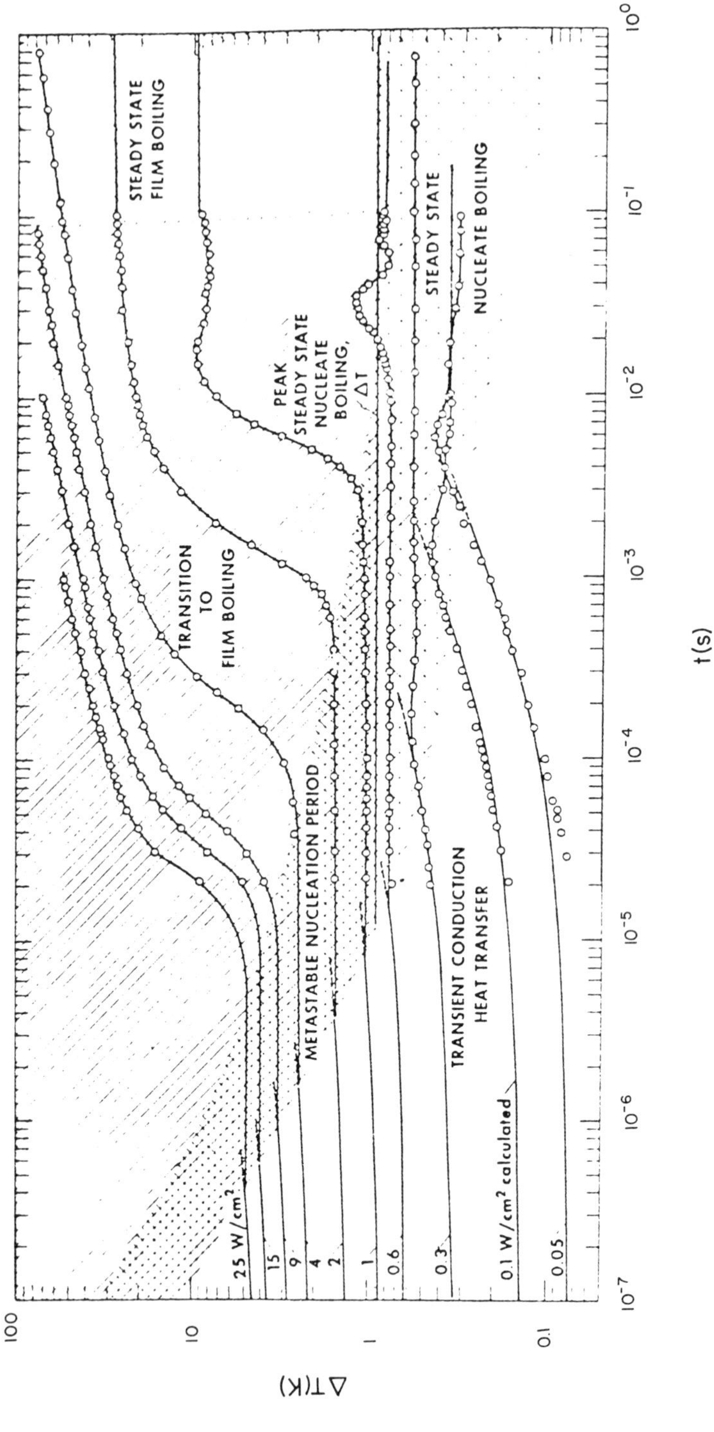

Fig. 6.22. Transient heat transfer to saturated liquid He I from a vertically oriented heater (from Steward[33]).

6.6.1. Surface Temperature Difference

Consider first the problem of the surface temperature difference during transient conduction. Since this mechanism is conduction dominated, it should be possible to understand in terms of diffusion theory. Returning to Fig. 6.21, imagine a steady heat flux through the interface. There are two potential contributors to the associated temperature difference. First, there is a thin fluid layer of thickness δ into which the heat diffuses. The temperature difference across the layer, defined as ΔT_f, should be determined exclusively by heat diffusion in the bulk fluid. Second, there is an interfacial temperature difference that is due to the mismatch of phonon heat transport between the two media. This mechanism, referred to as Kapitza conductance, is discussed in Chapter 5. It is a truly interfacial process occurring within a few atomic layers of the solid–helium boundary. The temperature difference due to Kapitza conductance is given the designation ΔT_K.

To evaluate the heat diffusion temperature difference, it is necessary to solve the heat diffusion equation with the proper set of boundary conditions. In one dimension this equation can be written

$$\frac{\partial^2 \Delta T_f}{\partial x^2} = \frac{1}{D} \frac{\partial \Delta T_f}{\partial t} \tag{6.39}$$

given a number of simplifying assumptions. In particular, it is assumed that the thermal diffusivity, $D = k/\rho C$, is a constant independent of temperature or position. Of course, this is not a particularly good assumption for liquid helium; however, it simplifies the model considerably to do so. Furthermore, it is assumed that the solid does not play a major role in the heat diffusion. This approximation has a negligible effect if the heat transfer surface has small heat capacity. The boundary conditions which are applied to solve (6.39) include

$$\Delta T_f(x, 0) = 0 \tag{6.40a}$$

$$\Delta T_f(\infty, t) = 0 \tag{6.40b}$$

and

$$q = -k \left. \frac{\partial \Delta T_f}{\partial x} \right|_{x=0} \tag{6.40c}$$

The last condition, being consistent with the constant flux assumption at the interface, is true only if the heat capacity of the heat transfer surface is neglected. The general solution to (6.39) with the above set of boundary conditions leads to

$$\Delta T_f = \frac{q}{k} \left[2 \left(\frac{Dt}{\pi} \right)^{1/2} \exp\left(\frac{-x^2}{4\,Dt} \right) - x \operatorname{erfc}\left(\frac{x}{2(Dt)^{1/2}} \right) \right] \tag{6.41}$$

where x is the dimension measured into the fluid. At the solid–fluid interface, $x=0$, the relationship simplifies considerably to yield

$$\Delta T_f = \frac{2q}{\sqrt{\pi}} \left(\frac{t}{\rho k C} \right)^{1/2} \tag{6.42}$$

For example, consider liquid helium at 4.2 K subjected to a heat flux of 1 W/cm^2 for a time of 10 μs, which is in the transient conduction regime. These conditions lead to a computed value for ΔT_f of about 0.3 K, which is approximately equal to the superheat required to produce homogeneous nucleation computed in Section 6.3. It is also worth noting that $\Delta T_f \approx 0.3$ K corresponds to about 1/3 the measured ΔT in Fig. 6.22.

The second mechanism which can lead to an interfacial temperature difference during transient heat transfer is due to Kapitza conductance. As has been described in Chapter 5, the temperature dependence of the Kapitza conductance in understood but the absolute value of the heat transfer coefficient is not predictable. Assuming that the mechanisms are the same independent of whether the fluid is He I or He II, it is reasonable to write

$$\Delta T_K \simeq \frac{q}{h_K} \tag{6.43a}$$

where $h_K \simeq AT^3$, consistent with experimental measurements of Kapitza conductance. For a copper surface below T_λ, A has been measured to have values around 0.1 $W/cm^2 \cdot K^4$, but with considerable uncertainty. Using this value for $T_b = 4.2$ K in He I, the interfacial temperature difference due to Kapitza conductance should vary as $\Delta T_K \approx 0.13\, q$. Therefore, for copper, the Kapitza temperature difference would be the same order as but somewhat smaller than that due to transient conduction except at very short times, $t \lesssim 1\ \mu s$.

On a very short time scale the heat flux dependence of the heat transfer coefficient in He I generally obeys the processes described above. Plotted in Fig. 6.23 is the surface temperature of copper samples measured in He I compared to that for He II.[34] On the right-hand side, the figure shows a typical ΔT_s versus q dependence observed for He II. Initially, there is a linear region; however, as the heat flux increases a considerable deviation from linearity occurs because the actual heat transfer varies as $T_s^n - T_b^n$ where $n \approx 3$. A somewhat different effect is observed for the case of He I. Here the dependence of ΔT versus q for high heat fluxes is almost linear, indicating a constant heat transfer coefficient. However, at low heat fluxes the surface temperature does not appear to extrapolate to the origin, indicating some failure in the linear modeling over the regime for which the

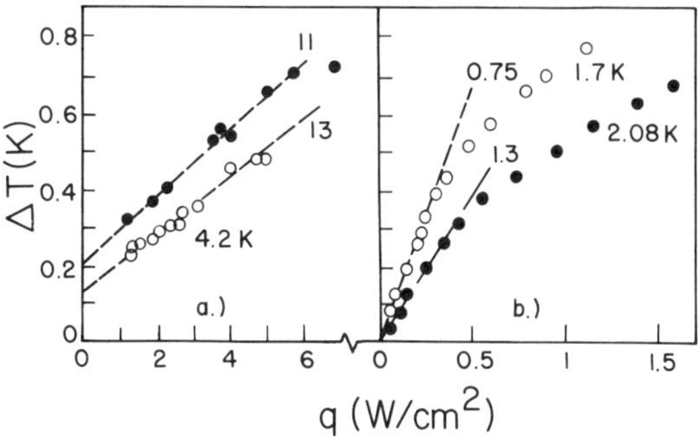

Fig. 6.23. Surface temperature of copper samples versus heat flux: (a) He I at 4.2 K and (b) He II at 1.8 K. Numbers on the straight lines refer to h values in $W/cm^2 \cdot K$ (from Schmidt[34]).

linear relationship holds. The Kapitza conductance appears to obey the cubic temperature dependence.

In principle, the temperature difference corresponding to a transient heat transfer event can be determined by a series summation of the two contributions described above. It follows that

$$\Delta T_s = \Delta T_K + \Delta T_f \tag{6.43b}$$

where, dependent on the choice of surface materials, one term can dominate the process. For comparison with theory, it should be possible to separate the two quantities in (6.43b) based on their different time dependencies.

6.6.2. Transition to Film Boiling

The other fundamental question related to transient heat transfer concerns the onset of film boiling. For a given heat flux q^*, the time required to reach the film boiling state, Δt^*, is of interest. This quantity establishes the limits to which enhanced heat transfer due to transient conduction can be credited in a heat transfer problem. The problem is understandable by means of a rather simple picture. Given that the transition is associated with the formation of a vapor film, a critical energy is defined as that required to vaporize the helium adjacent to the heat transfer surface and create the film.

A more quantitative model can be developed by assuming that ΔE^* is the energy needed to vaporize a layer of thickness δ, which corresponds to the diffusion length. Employing the transient diffusion model, we note that the diffusion length can be approximated by

$$\delta = \frac{\pi}{2} (Dt)^{1/2} \tag{6.44}$$

where again D is the thermal diffusivity. On a unit area basis, the critical energy is written

$$\Delta E^* = q^* \, \Delta t^* = \delta\lambda \tag{6.45}$$

where λ is the heat of vaporization of liquid helium at the existing temperature and pressure conditions. Substituting (6.44) into (6.45), we find that the heat flux needed to achieve film boiling is given by

$$q^* = \frac{\pi}{2}\lambda\left(\frac{k}{\rho C \, \Delta t^*}\right)^{1/2} \tag{6.46}$$

which for values associated with He I at 4.2 K, 1 bar, simplifies to

$$q^* \approx 0.07(\Delta t^*)^{-1/2} \tag{6.47}$$

where q^* is in W/cm^2 and Δt^* is in seconds. Note that this correlating equation does not contain any adjustable parameters. Plotted in Fig. 6.24 are numerous experimentally determined values for q^* along with the simple diffusion model fit given by (6.47). The correspondence is surprisingly good, which supports the basic physical ingredients to the critical energy analysis included in the simple model. A more precise empirical fit has been suggested[34]:

$$q^* = 0.127(\Delta t^*)^{-0.4} \tag{6.48}$$

This form is seen to agree with the data, particularly in the submillisecond regime.

As a final comment, the transient heat transfer analysis given above is oversimplified. Only two problems have been discussed, the heat transfer coefficient in the conduction regime and the critical energy for the onset of film boiling. In between these two regimes is the transition to film boiling including the dynamics of nucleation and the creation of the vapor film.

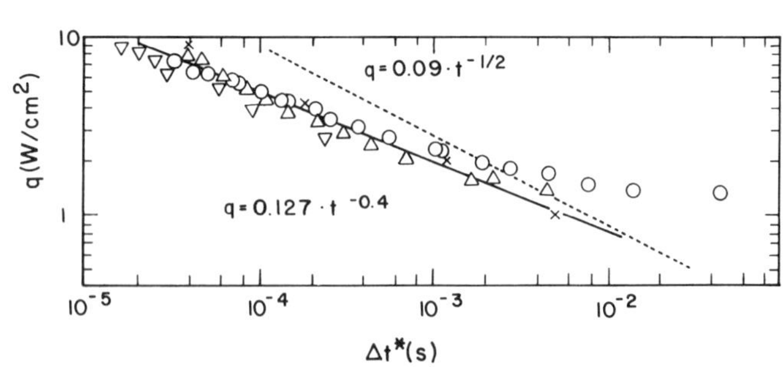

Fig. 6.24. Heat flux versus time for transition to film boiling (as compiled by Schmidt[34]).

These are considerably more complex phenomena than can be discussed here. This regime has been analyzed to some extent by Steward[33] by evaluating the orientation and pressure dependence to the type of data displayed in Fig. 6.22. The observations gleaned from this work are more qualitative but suggest that the transition regime is at least in part controlled by conventional heat transfer phenomena. Further work is required to grasp more fully the various aspects of the transient heat transfer problem.

6.7. SURFACE EFFECTS

For the most part, heat transfer analysis for He I takes little account of the character of the surface. In steady-state heat transfer, the surface is discussed only qualitatively in terms of activated nucleation sites. Heat transfer correlations used to describe nucleate boiling and both critical heat fluxes make no attempt to include the surface character in their treatment. This clearly is a weakness in the theory for there are considerable surface-induced changes in these values. For transient heat transfer, a greater effort is put forth to include the surface physics. As was discussed in Section 6.6 Kapitza conductance, which is a solid-state interfacial result, must be included when attempting to understand the transient conduction heat transfer. Since surface characteristics are not generally included in engineering correlations, it is of interest to consider how variations in surface character affect experimental results.

There has not been a great deal of research conducted on surface-dependent heat transfer in He I.[35–37] In the case of steady-state investigations, surface roughness, as measured by the coarseness of the surface abrasive, strongly affects the nucleate boiling regime. This effect can be understood qualitatively by considering arguments of activated nucleation sites. The smoother the surface, the fewer active nucleation sites. Since activated sites induce convection, a polished surface should have a larger surface temperature difference ΔT_s for the same heat flux. This result has been observed by Boissin et al.[35] Chemical treatment also has shown to affect the nucleate boiling regime.[37] These coatings combined with gross surface roughness have been employed to enhance cooling of composite conductors for large superconducting magnets.

Surface coatings also have been shown to affect the values of peak and recovery heat fluxes, q^* and q_R. The correlations used to describe these events in heat transfer do not include any of the surface characteristics. Cummings and Smith[36] have shown a clear increase in both the peak heat flux q^* and recovery value q_R with increased surface coatings. In their results shown in Fig. 6.25, the coatings were obtained by condensing H_2O

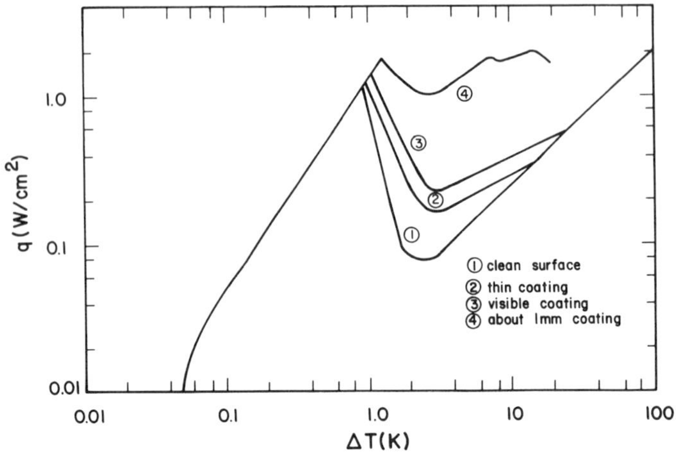

Fig. 6.25. Influence of H_2O coatings on pool boiling heat transfer (from Cummings and Smith[36]).

crystals on the surface. Similar behavior was observed by Ogata and Nakayama[37] on chemically treated surfaces. These results are not understood in terms of heat transfer models, but they represent interesting and technically significant improvements.

PROBLEMS

6.1. Calculate the critical radius of a vapor nucleus in He I at 4.2 K and 2.5 K under saturated vapor pressure. Assume a reaction rate of 1 $cm^{-3} \cdot s^{-1}$. Compare this value with the radius determined from (6.14). Estimate the reaction rate consistent with the radius determined from (6.14).

6.2. Determine the temperature at which the maximum in the peak nucleate boiling heat flux q^* in saturated He I occurs. What would be the change in this value if the system we pressurized to $p = 200$ kPa (2 bars)?

6.3. A vertically oriented, circular cross section channel containing He I at 4.2 K is exposed to a heat flux which varies linearly with height:

$$Q(z) = Q_m \left(\frac{z}{H}\right)$$

where H is the total channel height. Determine as a function of Q_m the position where the peak heat flux is first exceeded.

6.4. The Schmidt model for the transition to film boiling during transient heat transfer, (6.44), implies that the transition occurs when the surface reaches a temperature that is a constant for a given helium bath. Find an expression for the critical surface temperature in the Schmidt model and calculate its value for a saturated He I bath at 4.2 K. Compare your answer with the homogeneous nucleation temperature for the same conditions.

6.5. Consider film boiling heat transfer from a flat plate in He I. Calculate the plate surface temperature at the minimum film boiling heat flux q_R.

REFERENCES

1. E. R. G. Eckert and R. M. Drake, *Analysis of Heat and Mass Transfer*, McGraw–Hill, New York, 1972.
2. F. P. Incropera and D. P. Dewitt, *Fundamentals of Heat Transfer*, Wiley, New York, 1981.
3. J. A. Whitehead, Survey of Hydrodynamic Instabilities, in *Fluctuations, Instabilities and Phase Transitions*, Tormad Riste (Ed.), pp. 153–180, Plenum Press, New York, 1975.
4. R. P. Behringer and G. Ahlers, Heat Transport and Temporal Evolution of Fluid Flow Near the Rayleigh–Bénard Instability in Cylindrical Containers, *J. Fluid Mech.* **125**, 219 (1982).
5. P. G. J. Lucas, J. M. Pfotenhauer, and R. J. Donnelly, Stability and Heat Transfer of Rotating Cryogens. Part 1. Influence of Rotation on the Onset of Convection in Liquid ^{4}He, *J. Fluid Mech.* **129**, 251 (1983).
6. D. N. Sinha, J. S. Semura, and L. C. Brodie, Homogeneous Nucleation in ^{4}He: A Corresponding States Analysis, *Phys. Rev. A* **26**, 1048 (1982).
7. J. Frenkel, *Kinetic Theory of Liquids*, Chap. 7, Dover, New York, 1955.
8. E. Flint and J. Van Cleve, Heat Transport to He I from Polished Silicon Surface, *Adv. Cryog. Eng.* **27**, 283 (1982).
9. Y. Y. Hsu and R. W. Graham, An Analytical and Experimental Study of the Thermal Boundary Layer and the Ebullition Cycle in Nucleate Boiling, NASA TN D-594 (1961).
10. W. B. Bald, Cryogenic Heat Transfer Research at Oxford, Part I, Nucleate Pool Boiling, *Cryogenics* **13**, 457 (1973).
11. S. S. Kutateladze, Statistical Science & Technical Publications of Literature on Machinery, Atomic Energy Commission Translation 3770, Technical Information Services, Oak Ridge, TN (1949), 1952.
12. W. B. Bald and T. V. Wang, The Nucleate Pool Boiling Dilemma, *Cryogenics* **16**, 314 (1976).
13. C. Schmidt, Review of Steady State and Transient Heat Transfer in Pool Boiling Helium I, in *Stability of Superconductors*, pp. 17–32, International Institute of Refrigeration Commission A 1/2, Saclay, France, 1981.
14. L. S. Tong, *Boiling Heat Transfer and Two Phase Flow*, Chap. 2, Wiley, New York, 1965.
15. N. Zuber, M. Tribus, and J. W. Westwater, The Hydrodynamic Crisis in Pool Boiling of Saturated and Subcooled Liquids, in *International Development in Heat Transfer, Part.* II, pp. 230–234, ASME, New York, 1961.
16. R. V. Smith, Review of Heat Transfer to Helium I, *Cryogenics* **9**, 11 (1969).
17. D. N. Lyon, Boiling Heat Transfer and Peak Nucleate Boiling Fluxes in Saturated Liquid Helium Between λ and Critical Temperatures, *Adv. Cryog. Eng.* **10**, 371 (1965).
18. E. Ibrahim, R. W. Boom, and G. E. McIntosh, Heat Transfer to Subcooled Liquid Helium, *Adv. Cryog. Eng.* **23**, 333 (1978).
19. S. Capsi, Heat Transfer to Subcooled He I, *Adv. Cryog. Eng.* **29**, 281 (1984).
20. Yu. A. Kirichenko, K. V. Rusanov, and E. G. Tyurina, Heat Transfer in Subcooled Liquid Cryogens, *Cryogenics* **23**, 209 (1983).
21. F. Irie, G. Kippling, K. Lüders, T. Matsushita, U. Ruppert, and M. Takeo, Heat Transfer to Helium in the Near Critical Region, *Adv. Cryog. Eng.* **23**, 326 (1978).
22. M. A. Hilal, R. W. Boom, and M. M. El-Wakil, An Experimental Investigation of Free Convection Heat Transfer in Supercritical Helium, *Int. J. Heat Mass Transfer* **23**, 697 (1980).
23. G. Kippling and K. Kutzner, in *Pure and Applied Cryogenics*, Vol. 6, pp. 97–107, Pergamon Press, Oxford, 1967.
24. S. Sato, *Cryog. Eng. (Jp.)* **9**, 2 (1974).
25. M. A. Hilal, Analytical Study of Laminar Free Convection Heat Transfer to Supercritical Helium, *Cryogenics* **18**, 545 (1978).

26. J. H. Lienhard and P. T. Y. Wong, The Dominant Unstable Wavelength and Minimum Heat Flux During Film Boiling on a Horizontal Cylinder, *J. Heat Transfer* **86**, 220 (1964).
27. B. P. Breen and J. W. Westwater, Effects of Diameter of Horizontal Tubes on Film Boiling Heat Transfer, *Chem. Eng. Prog.* **58**, 67 (1962).
28. E. G. Brentari, P. J. Giarratano, and R. V. Smith, *Boiling Heat Transfer for Oxygen, Nitrogen, Hydrogen, and Helium*, NBS Technical Note 317, U.S. Government Printing Office, Washington, DC, Sept. 20, 1965.
29. M. N. Wilson, Heat Transfer to Boiling Liquid Helium in Narrow Vertical Channels, *Liquid Helium Technology, Proceedings of the International Institute of Refrigeration*, Commission I, Boulder, CO, Pergamon Press, Oxford, 1966, pp. 109–114.
30. S. G. Sydoriak and T. R. Roberts, Study of Boiling in Short Narrow Channels and Its Application to Design of Magnets Cooled by Liquid H_2 and N_2, *J. Appl. Phys.* **28**, 143 (1956).
31. S. Lehangre, J. C. Boissin, C. Johannes, and A. de La Harpe, Critical Nucleate Boiling of Liquid Helium in Narrow Tubes and Annuli, *Proceedings of the 2nd International Cryogenics Engineering Conference*, p. 274, Hiffe Science and Technology, Brighton, 1968.
32. Z. Chen and S. W. Van Sciver, Channel Heat Transfer in He I—Steady State Orientation Dependence, *Adv. Cryog. Eng.* **31**, (1986).
33. W. G. Steward, Transient Helium Heat Transfer Phase I—Static Coolant, *Int. J. Heat Mass Transfer* **21**, 863 (1978).
34. C. Schmidt, Transient Heat Transfer and Recovery Behavior of Superconductors, *IEEE Trans. Magnet.* **Mag-17**, 738 (1981).
35. J. C. Boissin, J. J. Thibault, J. Roussel, and E. Faddi, Boiling Heat Transfer and Peak Nucleate Boiling Flux in Liquid Helium, *Adv. Cryog. Eng.* **13**, 607 (1967).
36. R. D. Cummings and J. L. Smith, Boiling Heat Transfer to Liquid Helium, *Liquid Helium Technology, Proceedings of the International Institute of Refrigeration*, Commission I, Boulder, CO, Pergamon Press, Oxford, 1966, pp. 85–96.
37. H. Ogata and W. Nakayama, Boiling Heat Transfer to Helium from Machined and Chemically Treated Copper Surfaces, *Adv. Cryog. Eng.* **27**, 309 (1982).

Chapter 7

Forced Convection Normal Helium

In many applications of cryogenics, cooling is achieved best by confining the fluid to a tube or duct and circulating it through the system in a closed loop. Several aspects of this method distinguish it from immersion cooling in a pool of liquid. First, the fluid characteristics are more variable because the system is not restricted to operation in the liquid state. Subcritical fluids are obtained relatively easily in closed systems. Second, apart from the fluid properties, the mass flow or fluid velocity may be varied externally to obtain optimum flow or heat transfer characteristics. Finally, because forced flow systems are predominantly one dimensional, there is a greater possibility to interpret and predict the fluid behavior including pressure drop and heat transfer.

Considering low-temperature helium, we find a number of good examples where forced flow is the desired cooling scheme. These include but are not restricted to:

1. The flow of helium in refrigeration piping and heat exchangers. Refrigeration systems that use helium include conventional ^{4}He cycles as well as special devices such as ^{3}He–^{4}He dilution refrigerators for very-low-temperature research. In the former case the helium is usually well above the critical temperature and therefore single phase. The major consideration in this application is the appropriate optimization of the trade-off between heat transfer and pressure drop within the flow circuit.
2. The flow of liquid helium in transfer piping is perhaps the most often encountered helium flow application. It occurs whenever liquid is transferred from one storage system to another. Since helium transfer usually occurs near saturated vapor pressure, the process often involves two-phase flow. In this case, heat transfer is not the primary concern.

3. The use of helium as a coolant for superconducting magnet systems that employ an internally cooled conductor concept. The general approach here is to force the helium through a duct containing the composite superconductor.[1-4] Internally cooled conductors have considerably different behavior from that of traditional pool cooling, because the helium volume is limited to that contained within the tube. Most often, supercritical helium is suggested for this application, although two-phase subcritical helium is considered occasionally.[4]

To understand the details of forced flow helium, it is necessary to be aware of the physical principles associated with heat transfer and fluid flow. These principles are common to a variety of fluid dynamics problems. Since space is limited and the scope of the present discussion quite specialized, a brief review of this subject is included here. The reader is referred to more extended treatments in books on convection heat transfer for additional detail.[5]

7.1. GENERAL CONSIDERATIONS OF INTERNAL FLOW

Internal flow of ordinary fluids has been studied extensively and is fairly well understood. This does not mean that the problems have all been solved because, in general, fluid dynamics is sufficiently complicated to be exactly soluble only in very special cases or by complicated computer codes. Rather, the mode of problem solving is based on a combination of theoretical concepts and empirical evidence leading to semiempirical correlating functions.

The general problem to be discussed concerns internal flow hydrodynamics within a pipe or duct as shown schematically in Fig. 7.1. At any point within the pipe, the state of the fluid is given by its thermodynamic relationship between the local pressure p and temperature T. However, the fluid is not in global equilibrium because it is being transported through the pipe at a mass flow rate $\dot{m}$. Assuming there are no sinks or sources of fluid, the mass flow rate can be related to the local fluid velocity through the expression

$$\dot{\mathbf{m}} = \rho A \mathbf{v} \tag{7.1}$$

where the density ρ and velocity $\mathbf{v}$ are averaged locally in the direction of flow, and the pipe cross section is given by A.

The fluid is transposed through the duct under the influence of externally imposed forces. One such force might be gravity, which is the most

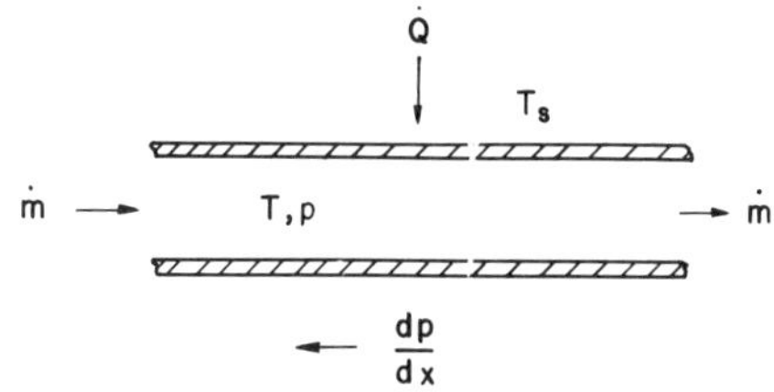

Fig. 7.1. Schematic representation of internal flow with heat transfer.

important when considering natural convection cooling loops. In most circumstances, the helium within the tube is forced under the influence of an externally imposed pressure drop. These external forces are needed to induce flow since the existence of viscosity results in a pressure drop for finite mass flow rates. Furthermore, since heat transport is of primary concern in forced flow helium, any heat flow in or out of the system must be determined. This heat transfer rate may result from joule heating within the pipe such as for superconducting cables in the normal state or heat transfer from an adjacent system such as occurs during heat exchanger operation. Heat transfer causes the surface temperatures T_s to be elevated above the local fluid temperature by an amount dependent on the local transfer coefficient.

In general, there are three sets of equations which must be solved to determine the behavior of a Newtonian fluid such as gaseous or liquid helium. A complete development of these equations is available in numerous text[6] on transport phenomena. The first is the continuity equation which is written

$$\frac{\partial \rho}{\partial t} + \nabla \cdot (\rho \mathbf{v}) = 0 \tag{7.2}$$

where $\mathbf{v}$ is a vector velocity and ρ is the local density. The continuity equation is derived from conservation of mass in a fluid element. The second equation is derived from conservation of momentum. It is actually a tensor equation. However, for the special case of constant viscosity, the momentum equation reduces to a fairly simple form,

$$\rho \frac{d\mathbf{v}}{dt} = -\nabla \mathbf{p} + \eta \nabla^2 \mathbf{v} + \rho \mathbf{F}_b \tag{7.3}$$

where $\mathbf{F}_b$ is the body force, that is, the force per unit mass of liquid. The total derivative of the velocity is given by

$$\frac{d\mathbf{v}}{dt} = \frac{\partial \mathbf{v}}{\partial t} + \mathbf{v} \cdot \nabla \mathbf{v} \tag{7.4}$$

The combination of (7.2) and (7.3) is generally referred to as the Navier–Stokes equations. The final conservation equation used to describe Newtonian fluids is the energy equation, which simply states that a fluid system must conserve energy. This energy equation may be written

$$\frac{d}{dt}\left(\frac{1}{2}\mathbf{v}^2+e\right)=-\frac{1}{\rho}\nabla\cdot(\rho\mathbf{v})+\dot{q}+\mathbf{F}_b\cdot\mathbf{v} \tag{7.5}$$

where e is the specific internal energy and $\dot{q}$ is the specific rate of heat input. The term $\mathbf{F}_b\cdot\mathbf{v}$ is the work done by an external force. It follows that complete solution to problems of Newtonian fluid mechanics requires simultaneous analysis of (7.2), (7.3), and (7.5). With the fluid properties varying with temperature, this process usually requires a numerical approach.

One case, which is treated extensively in elementary fluid mechanics texts, is that of a fluid within a tube or duct. It follows that the velocity can be described by two principal components—u in the direction of the duct axis and v normal to the duct axis and away from the surface. It is assumed further that the fluid is in a fully developed steady-state flow pattern and that it is incompressible, that is, $\rho=$ constant. This latter assumption is rather poor when dealing with helium. One problem that can be solved exactly for these conditions is laminar flow in the fully developed region. Laminar flow in the fully developed region implies that the velocity must obey the conditions

$$v=0 \quad \text{and} \quad \frac{\partial u}{\partial x}=0 \tag{7.6}$$

The solution to the fluid equations in the laminar flow regime predicts a parabolic velocity profile proportional to the pressure gradient divided by the viscosity. For a circular cross section channel the solution takes the form

$$u(r)=-\frac{1}{4\eta}\left(\frac{dp}{dx}\right)r_0^2\left[1-\left(\frac{r}{r_0}\right)^2\right] \tag{7.7}$$

where r_0 is the radius of the tube. A standard approach to developing approximate solutions to flow problems is to invoke the boundary layer approximation. This approach is based on the following set of conditions for the velocities:

$$u \gg v \quad \text{and} \quad \frac{\partial u}{\partial y} \gg \frac{\partial u}{\partial x}, \frac{\partial v}{\partial y}, \frac{\partial v}{\partial x}$$

In the boundary layer approximation the momentum equation (7.3) reduces to

$$u\frac{\partial u}{\partial x}+v\frac{\partial u}{\partial y}=-\frac{1}{\rho}\frac{\partial p}{\partial x}+\frac{\eta}{\rho}\frac{\partial^2 u}{\partial y^2} \tag{7.8}$$

For simplicity it is useful to define a friction factor f given in terms of the relevant quantities†

$$f\equiv\frac{-(dp/dx)\,D_h}{2\rho\bar{u}^2} \tag{7.9}$$

where $\bar{u}$ is the average velocity of the fluid along the duct axis. For the case of laminar flow the average velocity can be calculated by integration of (7.7) over the duct cross section,

$$\bar{u}=\frac{r_0^2}{8\eta}\left|\frac{dp}{dx}\right| \tag{7.10}$$

The hydraulic diameter is given in terms of the duct geometry,

$$D_h=\frac{4A}{P} \tag{7.11}$$

for which P is the perimeter of the channel and A the cross-sectional area. This quantity, referred to as the Fanning friction factor, is literally the ratio of the wall shear stress, $\tau\simeq\eta\,\partial u/\partial y$, divided by the kinetic energy density. In the case of laminar flow, the Fanning friction factor has a closed-form solution,

$$f=\frac{16}{\mathrm{Re}_D} \tag{7.12}$$

where $\mathrm{Re_D}=\rho\bar{u}D/\eta$. For turbulent flow which occurs for Reynolds numbers in excess of about 2000, no exact solution exists. Rather, empirical values such as those displayed in the Moody plot, Fig. 7.2., must be applied. These plots give the friction factor as a function of Reynolds number for different surface roughness conditions.

In a similar fashion the conservation of energy equation can be used to treat convective transfer. In the thermal boundary layer approximation,

$$\frac{\partial T}{\partial y}\gg\frac{\partial T}{\partial x}$$

† Note that some heat transfer books define the friction factor as four times the value given by (7.9).[7]

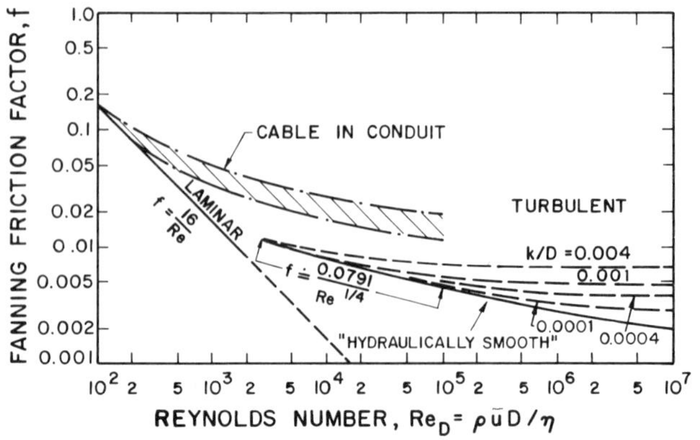

Fig. 7.2. Fanning friction factor for smooth and rough tubing compared to that of cable-in-conduit configuration (from Lue et al.[8]).

which is analogous to that for the velocity profile. Assuming for simplicity that the fluid possesses constant properties ρ, C_p, k, and η, we obtain the corresponding thermal boundary layer equation,

$$u\frac{\partial T}{\partial x}+v\frac{\partial T}{\partial y}=\alpha\frac{\partial^2 T}{\partial y^2}+\frac{\eta}{\rho C_p}\left(\frac{\partial u}{\partial y}\right)^2 \tag{7.13}$$

where $\alpha = k/\rho C_p$ is the thermal diffusivity. The first term on the right-hand side of (7.13) represents thermal diffusion. For most fluids, except some liquid metals, this term is not large and can be neglected. By suitable normalization, (7.13) can be shown to lead to the requirement that the average Nusselt number is a general function of Reynolds number and Prandtl number, that is,

$$\overline{\mathrm{Nu}} = f(\mathrm{Re}_D, \mathrm{Pr}) \tag{7.14}$$

The average Nusselt number is defined in terms of average heat transfer properties:

$$\overline{\mathrm{Nu}} = \frac{\bar{h}D_h}{\bar{k}} \tag{7.15}$$

where $\bar{k}$ is a suitably averaged fluid thermal conductivity. In fact, (7.14) is general and not dependent on the boundary layer approximation. Most empirical heat transfer correlations are constructed in a form consistent with (7.14).

A final general point concerns the enthalpy conservation within the channel. Explicitly, enthalpy conservation demands that the change in rate of enthalpy flow of a liquid, $\dot{m}h$, be equal to the difference between the total rate of heat transfer, $\dot{Q}$, and rate of work done on the fluid, $\dot{W}$; that is,

$$\dot{m}\frac{dh}{dx}=\frac{d}{dx}(\dot{W}-\dot{Q}) \tag{7.16}$$

If no work is done on the fluid, which is the case for simple fluid flow, then the rate of heat transfer to the surroundings is equal to the loss of fluid enthalpy.

To place the above concepts in a practical context, consider the flow of liquid helium through a tube of 30 m length and 10 mm diameter at an average velocity of 1 m/s. This example may represent a cooling loop in a superconducting magnet or some other indirectly cooled device. The first quantity to determine is the Reynolds number, $\mathrm{Re}_D=\rho uD/\eta=4.5\times10^5$, where the physical properties of liquid helium around 4 K are used in this calculation. Referring to the Moody diagram, Fig. 7.2, for this Re_D a smooth tube should have a friction factor $f=0.0034$. The total pressure drop is determined by integration of (7.10) over the length of the tube, which leads to the relationship

$$\Delta p=-\frac{2f\rho u^2L}{D_h} \tag{7.17}$$

For the case at hand, the pressure drop works out to be $\Delta p=2.5$ kPa (19 torr). We assume, for example, that the fluid is slightly subcooled, entering at 4.0 K, 100 kPa (1 atm); it will exit at a lower pressure of 97.5 kPa, which corresponds to a saturation temperature of 4.20 K.

It is necessary to consider one further point in this example. Any fluid undergoing a pressure drop is doing friction work on the tube. This friction work can be written in terms of the pressure drop,

$$\dot{W}_f=\frac{\dot{m}\,\Delta p}{\rho}=uA\,\Delta p \tag{7.18}$$

which for the example at hand gives 0.25 W. The internal heat generation rate results in a temperature increase in the fluid. Since it is subcooled no boiling occurs until the temperature reaches $T_b(p)=4.20$ K. For this example, the temperature increase would be only about 6 mK, which is insufficient to result in boiling. However, friction work can make a significant contribution to the temperature rise in helium cooling systems.

7.2. SUPERCRITICAL HELIUM

The discussion in Section 7.1 was intended to be a general introduction to the factors affecting forced flow cooling systems. The description was kept simple intentionally so as to introduce the important physical quantities such as friction factor and the origin of heat transfer correlations. This model is not particularly appropriate for low-temperature helium mainly because of the incompressibility assumption. Therefore, it is important to develop a more general fluid mechanics model. In developing this description emphasis is placed on the special case of supercritical helium ($p > 0.23$ MPa) where the fluid is single phase although quite compressible. It is assumed further that the temperature is above T_λ so that the fluid properties are describable by classical hydrodynamics.

7.2.1. Fluid Mechanics

To model the fluid mechanics of supercritical helium, the Navier–Stokes equations including the effect of finite compressibility are used.[9] Compressibility factors can have profound impact on the observed behavior of supercritical helium. For the steady one-dimensional case discussed in Section 7.1, the pressure change in the direction of flow can be derived from Eq. (7.2) and (7.3),

$$dp = -\frac{2f\rho u^2}{D}\,dx + u^2\,d\rho \tag{7.19}$$

where the first term is identical to that introduced for incompressible fluids while the second term represents the fluid acceleration due to density changes. It is desirable to replace $d\rho$ by a function of the extrinsic variables p and T. This can be achieved if the equation of state is known. However, a more general form can be developed through thermodynamic manipulations. Since the density is a function of p and T it can be expanded in terms of these variables:

$$d\rho = \left(\frac{\partial\rho}{\partial p}\right)_T dp + \left(\frac{\partial\rho}{\partial T}\right)_p dT \tag{7.20}$$

According to the thermodynamic definitions the bulk compressibility is

$$\kappa = \frac{1}{\rho}\left(\frac{\partial\rho}{\partial p}\right)_T$$

and the bulk expansivity is

$$\beta = \frac{1}{\rho}\left(\frac{\partial \rho}{\partial T}\right)_p$$

These quantities can be inserted into (7.20) to give the relationship

$$d\rho = \rho\kappa\, dp - \rho\beta\, dT \tag{7.21}$$

Due to density changes along the flow channel, the velocity of the compressible fluid is not constant. However, in the steady state the mass flow is constant. It is convenient to define a specific mass flow rate $G \equiv \dot{m}/A = \rho u$, where A is the channel cross section. Substituting this quantity and collecting terms between (7.19) and (7.21), we obtain a coupled relationship between the temperature and pressure variation along the the tube:

$$\left(1 - \frac{\kappa G^2}{\rho}\right)\frac{dp}{dx} = -\frac{2f}{\rho D} G^2 + \frac{\beta G^2}{\rho}\frac{dT}{dx} \tag{7.22}$$

Note that for small κ and β, this expression reduces to the incompressible form given by (7.10).

In addition to the fluid flow equation, it is necessary to consider enthalpy conservation in the fluid when it is subjected to a net heat flux q. This is achieved by evaluating the energy equation (7.5). The total heat input per unit length of the tube is related directly to the time rate of change in the stagnation enthalpy (enthalpy + kinetic energy). For a circular cross section with $A = \pi D^2/4$, this expression is

$$q = \frac{GD}{4}\frac{\partial}{\partial p}\left(h + \frac{u^2}{2}\right)\frac{dp}{dx} + \frac{GD}{4}\frac{\partial}{\partial T}\left(h + \frac{u^2}{2}\right)\frac{dT}{dx} \tag{7.23}$$

Using the compressibility and expansivity factors defined above, and noting that the Joule–Thomson coefficient, which governs isenthalpic expansion, is defined by the relationship

$$\mu_j = -\frac{1}{C_p}\left(\frac{\partial h}{\partial p}\right)_T \tag{7.24}$$

we can simplify (7.23) to provide a second coupled relationship between the variation of p and T with position along the tube:

$$q = -\frac{GD}{4}(\mu_j C_p + u^2\kappa)\frac{dp}{dx} + \frac{GD}{4}(C_p - u^2\beta)\frac{dT}{dx} \tag{7.25}$$

Given the two relationships for the gradient T and p, it is a straightforward calculation to eliminate variables by simultaneous solution, arriving at separate equations for each gradient; that is,

$$\frac{dp}{dx}=\frac{-2fG^2/\rho D+4qG\beta/\rho D(C_p-u^2\beta)}{1-(G^2/\rho)(\kappa+\beta\phi)} \tag{7.26}$$

and

$$\frac{dT}{dx}=\frac{4q/GD(C_p-u^2\beta)-2fG^2\phi/\rho D}{1-(G^2/\rho)(\kappa+\beta\phi)} \tag{7.27}$$

where an additional parameter has been defined,

$$\phi=\frac{\mu_j C_p+u^2\kappa}{C_p-u^2\beta} \tag{7.28}$$

Needless to say the above expressions (7.26), (7.27), and (7.28) are difficult to apply unless the physical properties have relatively simple behavior. Arp[9] has shown that by approximating the fluid by a nearly ideal gas, with $\kappa\approx 1/p$ and $\beta\approx 1/T$, the above expressions are reduced in complexity.

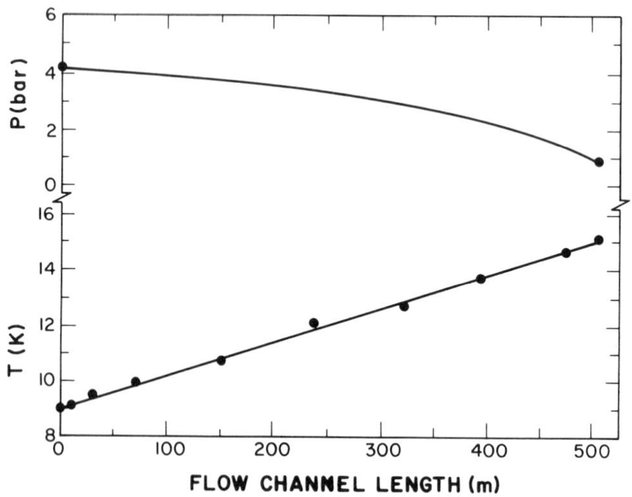

Fig. 7.3. Pressure and temperature profiles with 3.2 bar pressure loss. Computer data $m=0.928$ g/s, $D=4.8$ mm, $Q=0.074$ W/m, $p_{\text{in}}=4.2$ bars, $T_{\text{in}}=9.0$ K, $f=0.007$, $\text{Re}_D=1\times10^5$. Measured flow = 0.98 g/s. Solid line is computed value (from Dean et al.[10]).

Furthermore, if the velocities are fairly low, $u \lesssim 10\, m/s$ (7.26) and (7.27) take on more manageable forms:

$$\frac{dp}{dx} = -\frac{2G^2 f}{\rho D} + \frac{4qG\beta}{\rho D C_p} \tag{7.29}$$

and

$$\frac{dT}{dx} = \frac{4q}{GDC_p} - \frac{2fG^2}{\rho D}\left(\mu_j + u^2 \frac{\kappa}{C_p}\right) \tag{7.30}$$

Note that the first two terms in (7.29) and (7.30) are identical to that obtained for incompressible fluids. It follows that the second term in (7.29) becomes important at low velocities and high heat fluxes while in (7.30) the second term is significant at high velocities and low heat fluxes.

The above analysis has been tested experimentally by Dean *et al.*[10] in a 500 m long tube with an $L/D \approx 10^5$ operating in supercritical helium. Two sets of results from their investigation are shown in Figs. 7.3 and 7.4. The two cases show significantly different behavior. The data displayed in Fig. 7.3 represent the regime where the heat flux is large and the flow rate is small. As a result, the temperature profile is monotonically increasing in response to the dominance of the enthalpy flow term, $4q/GDC_p$. At the same time, the pressure profile is linear owing to the considerable impact of

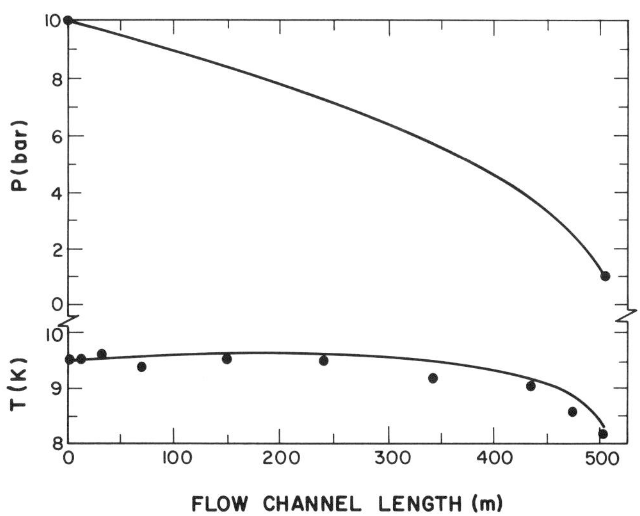

Fig. 7.4. Pressure and temperature profiles with 9 bar pressure loss. Computer data $m = 3.15$ g/s, $D = 4.8$ mm, $Q = 0.062$ W/m, $p_{\text{in}} = 10$ bars, and $T_{\text{in}} = 9.5$ K, $f = 0.005$, $\text{Re}_D = 3\text{–}4 \times 10^5$. Measured flow = 3.0 g/s. Solid line is computed value (from Dean et al.[10]).

the incompressibility term in (7.29). The second case, shown in Fig. 7.4, is an example of the regime where the heat flux is not large but the mass flow is high. The pressure profile is only slightly nonlinear since it is dominated by the friction-induced pressure drop. Perhaps more interesting is the temperature profile which actually decreases owing to the Joule–Thomson expansion effect. This process is well known and is applied in refrigeration systems to achieve cooling. Isenthalpic expansion is discussed more extensively in Chapter 8. An analytical comparison of the above experiments with the compressible fluid theory has been conducted and is shown by the solid curves in Figs. 7.3 and 7.4. One adjustable parameter, the friction factor f was used to fit both the pressure drop and temperature profile data. In both cases, the friction factor required to fit the data was considerably above that expected for turbulent flow at the Reynolds numbers corresponding to the experimental system.

Supercritical helium represents a fairly good example of a simple compressible fluid. The hydrodynamic equations appropriate for compressible flow adequately describe the pressure drop and heat transport character. However, the flow characteristics do not represent the entire problem. In particular, it is also of interest to be able to predict the heat transfer coefficient between the tube or duct and the helium stream. The conventional approach to this problem demands the development of engineering correlations appropriate to the particular configuration of interest.

7.2.2. Heat Transfer Correlations

Investigations of heat transfer to forced supercritical helium have shown that traditional engineering correlations are best at describing the data. Results of analysis of the thermal boundary layer indicate that the average Nusselt number should be couched in a form consistent with (7.14). As an aside, be aware that the local heat transfer coefficient can be a considerable function of temperature and therefore varies along the length of the tube. The local heat transfer coefficient is defined in terms of the local mean fluid temperature,

$$h = \frac{q}{T_s - T_m} \tag{7.31}$$

where T_m is obtained by taking an appropriate average across the channel.

There are numerous single-phase fluid heat transfer correlations for internal flow. Several important factors must be considered when selecting a correlation to apply to a particular system. First, determine whether the fluid is in the laminar or turbulent flow regime. The critical Reynolds number for single-phase internal flow is around 2000. Second, determine whether the entry region has significant impact on either the hydrodynamics or temperature development. This requirement demands fairly large L/D ratios. Helium has a Prandtl number of the order of unity, so it is expected that these developments will occur almost simultaneously. Finally, once the conditions of flow are established, it is necessary to select among several possible correlations dependent on whether the range of parameters is appropriate for the particular empirical fit.

Apart from the very simplest cases of fully developed laminar flow with constant surface or constant heat flux boundary conditions, all engineering correlations are of the form of (7.14) where the average Nusselt number is a function of the fluid Reynolds and Prandtl numbers. By far the most work on the heat transfer correlations appropriate for forced convection supercritical helium are for the turbulent regime. This fact is in part pragmatic because helium has such a low viscosity that the turbulent transition occurs at velocities in the range of a few centimeters per second, far lower than is generally of engineering interest. In the turbulent regime, a number of good correlations exist for single-phase turbulent heat transfer. The Dittus–Boelter expression,[11] perhaps the most common heat transfer correlation, is written in the form

$$\overline{\mathrm{Nu}} = 0.023\ \mathrm{Re}_D^{4/5}\ \mathrm{Pr}^{2/5} \tag{7.32}$$

where $\overline{\mathrm{Nu}}$ represents the average Nusselt number over the tube length. Correct application of this expression demands consideration of the temperature dependence of the fluid properties. The appropriate properties must be evaluated at the film temperature T_f, defined by

$$T_f = \frac{T_s + T_m}{2} \tag{7.33}$$

which is a simple average between the surface and mean fluid temperatures. The Dittus–Boelter correlation has been used quite effectively for cryogenic fluids.

In the case of supercritical helium, Giarratano et al.[12] have suggested that a prefactor of 0.022 gives a better fit to their data. A standard deviation of 14.8% was obtained between experimental results and the Dittus–Boelter correlation. An improved fit to experimental data was achieved by use of a correlation of the form[13]

$$\mathrm{Nu} = 0.0259\ \mathrm{Re}_D^{4/5}\ \mathrm{Pr}^{2/5} \left(\frac{T_s}{T_m}\right)^{-0.716} \tag{7.34}$$

where now explicit temperature variation of the properties is taken into account by the last factor. The relationship given by (7.34) has been shown to correlate to a standard deviation of 8.3% with several sources of experimental data. A normalized form of this comparison is shown in Fig. 7.5, where the heat transfer coefficient is plotted against reduced temperature T_b/T_c.

In general, fully developed forced flow supercritical helium can be assumed to have a well-established engineering basis. Since the fluid is single phase, its hydrodynamics can be evaluated in terms of the Navier–Stokes equation of motion including compressibility factors. This problem is quite difficult owing to the variability of physical properties with pressure and temperature. Consequently, its solution requires numerical integration of complicated nonlinear equations. Be aware that this particular problem represents only one special case of forced flow helium. Other problems concerning two-phase flow and transient effects, subjects of subsequent sections in this chapter, are more complex in physical nature.

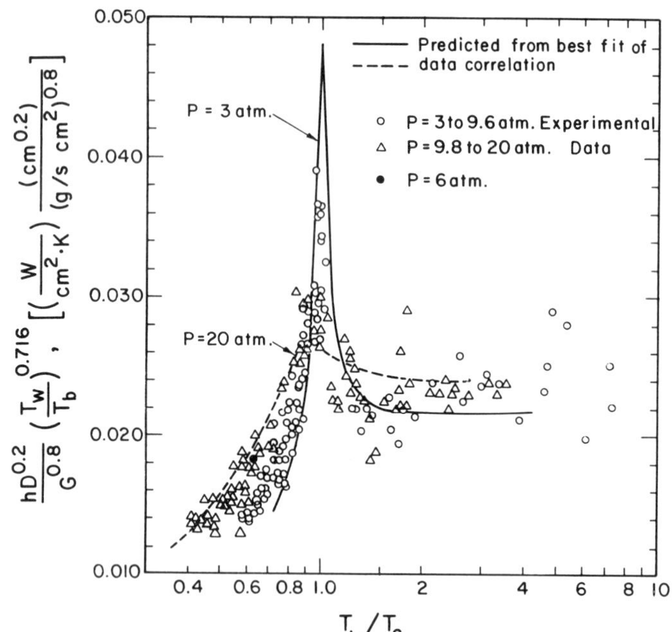

Fig. 7.5. Experimental and predicted heat transfer results for supercritical helium using (7.34) (from Giarratano et al.[12]).

7.3. TWO-PHASE FLOW AND HEAT TRANSFER

Whenever liquid helium is allowed to flow at pressures less than the critical pressure, $p_c = 0.23$ MPa, it is common to expect two-phase phenomena to occur. This occurrence often is associated with boiling and heat transfer; however, adiabatic systems can also experience two-phase flow as a result of pressure drop and friction loss factors. Two-phase phenomena have also been discussed in Chapter 6 in connection with nucleate pool boiling; however, for internal flow it is not possible to separate the heat transfer characteristics from the fluid dynamics. Two-phase internal flow systems are very common and important in helium cryogenic systems. Most often they are encountered in helium transfer systems where the liquid helium is siphoned from one vessel to another through a vacuum insulated line. Obvious factors of concern are the possibilities of excessive pressure drop or flow oscillation, both of which contribute inefficiencies to the helium transfer process.

Liquid helium transfer as well as other applications have made it essential to be aware of the physical phenomena that commonly occur in flowing subcritical liquid helium. Unfortunately, this subject has received very little attention from the research community and not a great deal is known. There are several reasons for the incomplete understanding of two-phase flowing helium. First, a full understanding of this problem is an awesome task. It is a multivalued problem dependent on such factors as temperature, pressure, mass flow rate, system configuration, and relationship between the two phases present (liquid and vapor). Consequently, it would take a substantial quantity of empirical data to describe the behavior even in a small number of special cases. Second, two-phase flow is very specialized because it only concerns itself with a region of the phase diagram where coexistence is possible. Finally, there has not been a concerted effort to achieve this understanding because the problem has limited application. On the other hand, two-phase flow in water systems has received considerable attention because of numerous engineering applications.[14]

7.3.1. Flow Regimes and Transitions

Consider the general character of two-phase flowing liquid and vapor helium. Based on knowledge obtained from experiments with two-phase flows of noncryogenic fluids, some general statements can be made about flow characteristics. There are approximately seven different types of flow pattern that can be obtained in a two-phase mixture.[15] These different patterns are shown schematically in Fig. 7.6. The factors that determine which flow pattern is stable for a particular set of conditions include mass flow

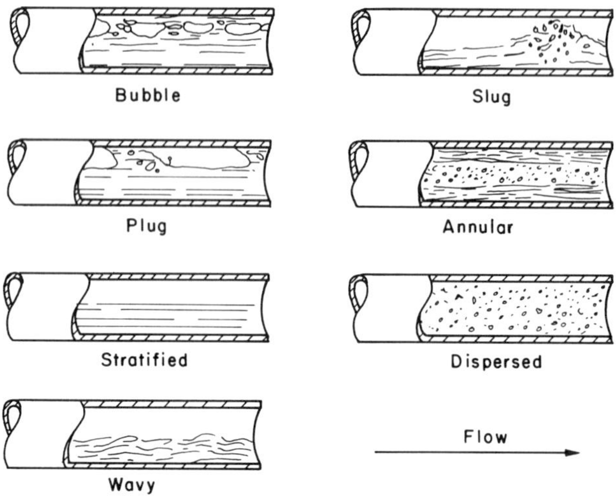

Fig. 7.6. Sketches of flow patterns present in horizontal tubes (from Baker[15]).

rate, densities of each of the components, viscosty, latent heat of transformation, and surface tension. The qualitative description of each of the flow regimes is as follows[15]:

1. *Bubble flow*: Gas bubbles flow along with the liquid at approximately the same velocity.
2. *Plug flow*: Gas bubbles coalesce to form plugs in the channel. This condition usually occurs for higher mass flow rates and vapor volumes than in bubble flow.
3. *Stratified flow*: Complete separation of liquid and gas occurs owing to density differences. The vapor and liquid occupy unchanging fractions of the cross-sectional area. This flow pattern occurs most commonly in large-diameter channels at low mass flow rate.
4. *Wavy flow*: Similar to stratified flow but due to higher mass flow rate, the interface between liquid and vapor experiences an oscillatory motion.
5. *Slug flow*: The wave amplitude increases until it touches the tube wall resulting in a pattern of vapor slugs separated by totally liquid regions.
6. *Annular flow*: At fairly high vapor fractions and velocities, the flow pattern will overcome the gravitationally induced stratification causing the vapor to flow through the center and the liquid to adhere to the walls. In this flow pattern, the vapor velocity is much greater than that of the liquid.

7. *Dispersed flow*: The liquid film becomes unstable and breaks up into small droplets which are carried along with the vapor stream. This flow regime occurs at the highest mass flow rates.

The above list of flow patterns are at least approximately ordered in terms of increasing flow rate and vapor fraction. Each of these flow patterns are potentially present in liquid–vapor helium systems. Similar behavior is expected to occur in vertically oriented tubes although with slightly different regimes of occurrence. Based on empirical evidence from room temperature fluids, there have been several attempts to correlate two-phase flow into a map which is dependent on system parameters.[15,16] Unfortunately, due to the limited existence of data on helium systems, very few attempts have been made to compare the behavior of helium with other common fluids.[17–19]

To make a more quatitative discussion of the subject of two-phase flow, it is necessary to define several physical quantities. The three main definitions pertaining to two-phase flow are:

1. *Void fraction* α: Defined as the ratio of the local vapor volume to the total flow volume. It is normally given in terms of the fraction of cross-sectional area that is vapor, that is,

$$\alpha = \frac{A_v}{A_l + A_v} \tag{7.35}$$

2. *Flow quality* χ: The ratio of vapor mass flow rate to total mass flow of both vapor and liquid:

$$\chi = \frac{\dot{m}_v}{\dot{m}_l + \dot{m}_v} \tag{7.36}$$

   By definition, χ is always less than or equal to unity.

3. *Slip ratio* S: Defined as the ratio of the vapor to liquid velocity,

$$S = \frac{u_v}{u_l} \tag{7.37}$$

   In two-phase flow the slip ratio is almost always greater than or equal to unity since the less-dense vapor travels more rapidly than the bulk liquid.

These quantities can be shown to relate to each other through the condition,

$$S = \left(\frac{\chi}{1-\chi}\right)\left(\frac{1-\alpha}{\alpha}\right)\left(\frac{\rho_l}{\rho_v}\right) \tag{7.38}$$

The most important two-phase flow problem for any liquid–gas system is to be able to measure and model the behavior of all three quantities as a function of external parameters.

Although there is no complete description of two-phase flow in helium, several flow transitions have been noted by Zust and Bald.[17] The first such transition is the bubble flow–slug flow transition which was theoretically investigated by Radovcick and Moissis.[20] The prediction is for a transition which is a continuous function of the mass flow and void fraction. The void fraction, consistent with the transition, is an inverse function of the mass flow rate. The critical void fraction is expected to vary between 0.1 and 0.3 for the smallest mass flow rates. The second transition represents the upper limit to slug flow. Wallis[21] has predicted a maximum flow quality χ based on the stability varying critical quality given by

$$\chi_{c_1} = \frac{\rho_v}{\rho_l - \rho_v} + \frac{0.105}{G}\left(\frac{gD\rho_v\rho_l^2}{\rho_l - \rho_v}\right)^{1/2} \tag{7.39}$$

where ρ_l and ρ_v are the two phase densities and g is the gravitational acceleration. The final transition is associated with the onset of pure annular flow. The criterion due to Wallis[21] leads to the correlation

$$\chi_{c_2} = \frac{0.9[gD\rho_v(\rho_l - \rho_v)]^{1/2}}{G} \tag{7.40}$$

These three correlations have not been tested extensively for helium, although experiments show only qualitative agreement with the theory.

7.3.2. Pressure Drop

The pressure drop in two-phase helium systems has been investigated in several experiments.[18,19] Across a tube of length L this quantity can be written as a summation of three terms,

$$\Delta p_T = \Delta p_{gr} + \Delta p_a + \Delta p_f \tag{7.41}$$

where Δp_{gr} is the gravitational pressure drop due to change in elevation of the fluid. Unlike single-phase incompressible flow, the pressure drop described by (7.41) must take into account the variation in the fluid density along the channel. This contribution is particularly important in flows involving heat transfer. The gravitational pressure drop for vertically oriented tubes is given by the relationship

$$\Delta p_{gr} = -g\int_0^L \langle\rho\rangle\, dx \tag{7.42}$$

where the average density $\langle \rho \rangle$ is defined in terms of the void fraction

$$\langle \rho \rangle = \alpha \rho_v + (1-\alpha)\, \rho_l \tag{7.43}$$

Therefore, it is necessary to obtain the void fraction before the gravitational pressure drop can be determined. The acceleration contribution to the pressure drop, Δp_a, also has an established form dependent on void fraction and flow quality. It is given by

$$\Delta p_a = G^2 \left(\frac{\chi^2}{\alpha \rho_v} + \frac{(1-\chi)^2 - 1}{(1-\alpha)\, \rho_l} \right) \tag{7.44}$$

where in this case it is necessary to have knowledge of both the void fraction and flow quality. The final pressure drop contribution, Δp_f, in two-phase flow is that due to frictional interaction between the tube and wall. This term is expected to be similar in form to that obtained for single-phase fluids with the additional complication associated with the two-fluid nature of the problem.

There are several theoretical developments which have been successful at interpreting certain aspects of the pressure drop in two-phase cooling systems. One such analysis is known as the Lockhart–Martinelli correlation.[22] The explicit assumptions of this model are as follows:

1. The static pressure drop in the vapor phase is equal to that in the liquid phase for all mass flow rates.
2. The two-phase fluid volume is made up of a linear combination of the vapor and liquid volumes.
3. Friction losses are assumed to dominate, and the effect of momentum change and hydrostatic head are ignored.

The Lockhart–Martinelli correlation for pressure drop introduces several parameters which are functions of measurable quantities. The first of these is the dimensionless two-phase pressure gradient,

$$\phi_l^2 = \frac{(dp/dx)_{TP}}{(dp/dx)_l} \tag{7.45}$$

where the subscript TP stands for two-phase. The reference pressure gradient is that of the liquid and corresponds to the traditional incompressible flow equation given by (7.17). The other parameter which is commonly used in relating data to the Lockhart–Martinelli correlation is a ratio of the pressure gradients associated with each phase,

$$\chi_{tt}^2 = \frac{(dp/dx)_v}{(dp/dx)_l} \tag{7.46}$$

where the subscript tt refers to both phases being turbulent, a situation that is usually the case for low-temperature helium systems. This parameter can be correlated against the properties of each of the two fluids in the form

$$\chi_{tt} = \left(\frac{\rho_v}{\rho_l}\right)^p \left(\frac{\eta_l}{\eta_v}\right)^q \left(\frac{1-\chi}{\chi}\right)^r \tag{7.47}$$

where the coefficients take on values that are typically, for example, $p \simeq 0.6$, $q \simeq 0.1$, and $r \simeq 1$. Usually, correlations are indicated by plots of ϕ_l versus χ_{tt}.[23]

Several simplified forms to the Lockhart–Martinelli correlation exist in the literature. Levy[24] produced reasonable agreement when correlating experiments to the form

$$\phi_l^2 = \frac{(1-\chi)^{1.75}}{(1-\alpha)^2} \tag{7.48}$$

The above expression is based on a combination of the Lockhart–Martinelli correlation and the relationship for void fraction.

Another rather direct model for pressure drop in two-phase systems is referred to as the homogeneous model, which is based on a similar set of assumptions as the Lockhart–Martinelli correlation. The homogeneous model describes a two-fluid system as having the following set of characteristics:

1. Equal liquid and vapor velocities.
2. Thermodynamic equilibrium between phases.
3. Applicability of the single-phase friction factor to the two-phase flow.

Based on these assumptions, the two-phase friction multiplier ϕ_l^2 can be expressed as

$$\phi_l^2 = \left[1 + \chi\left(\frac{\rho_l}{\rho_v} - 1\right)\right]\left[1 + \chi\left(\frac{\eta_l}{\eta_v} - 1\right)\right]^{-n} \tag{7.49}$$

where the coefficient n is taken to be approximately 0.25 in most analyses.[21] The above expression can be related to experimental data if the variation of physical quantities with flow quality is known.

There have been only limited experimental investigations of two-phase pressure drop.[17–19,25–31] Khalil et al.[18] studied two-phase flow in a vertically oriented tube with diameter 6.35 mm and pressures between 0.106 and 0.204 MPa. In this work it was possible to measure pressure drop Δp, overall mass flow $\dot{m}$, and local average void fraction $\langle\alpha\rangle$. These measurements allowed determination of both void fraction and slip ratio as

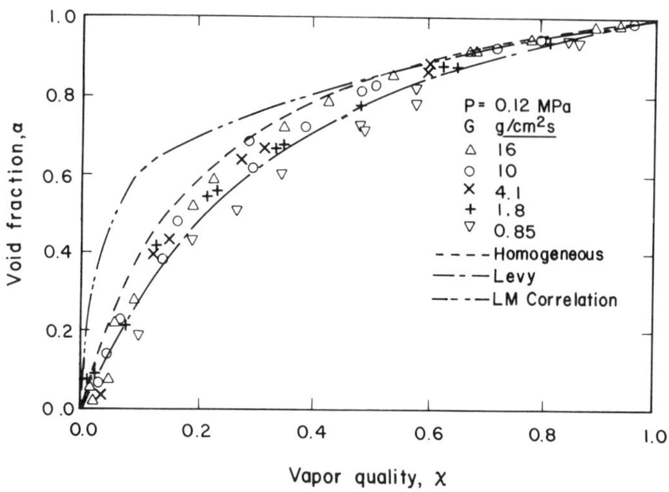

Fig. 7.7. Void fraction for helium flow as a function of vapor quality (from Khalil[18]).

they depend on vapor quality. The void fraction was compared with the available theoretical developments. Plotted in Fig. 7.7 is the variation of void fraction with vapor quality for different mass flow rates at 0.12 MPa. The theoretical developments do not predict a variation of α with G and no systemetic dependence is observed in the data. As can be seen in the figure, the Levy and homogeneous models do a better job of modeling behavior than the Lockhart–Martinelli correlation which predicts a higher void fraction at low vapor qualities. Experimental results of the slip ratio variation with vapor quality are shown in Fig. 7.8. The slip ratio is always greater

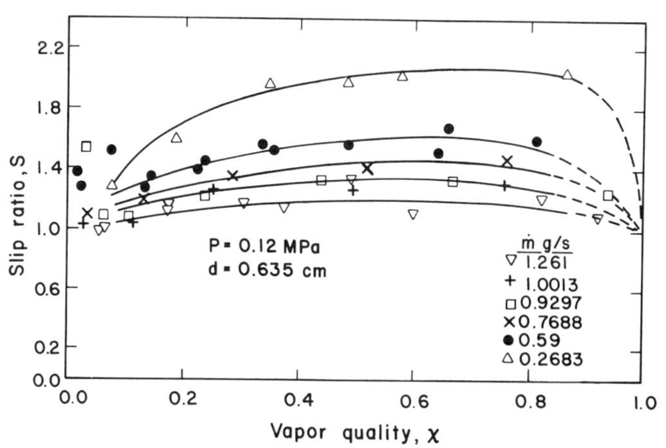

Fig. 7.8. Slip ratio as a function of vapor quality and mass flow rate at 0.12 MPa (from Khalil[18]).

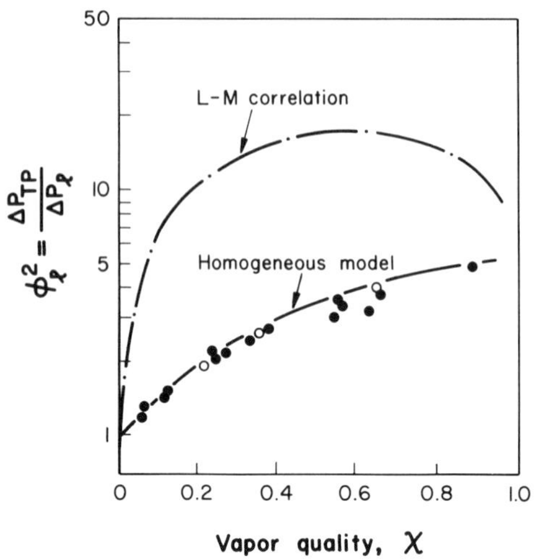

Fig. 7.9. Dimensionless pressure drop in two-phase helium flow (from Nakagawa et al.[31]). The solid circles (●) are for the test loop with 11.4 mm ID and the open circles (○) are for the test loop with 18.4 mm ID.

than unity for finite mass flow rates, with the biggest deviation occuring at the highest G.

Other measurements of two-phase pressure drop in helium confirm the analysis based on the homogeneous model. Nakagawa et al.[31] studied the variation of the dimensionless two-phase pressure drop (7.45) with vapor quality. This work was carried out on two test loops of diameter 11.4 and 18.4 mm. A summary of these experimental results is shown in Fig. 7.9. As with previous measurements of two-phase flow in helium systems, the authors observed that the homogeneous model was better than the Lockhart–Martinelli correlation for analyzing the data. It is surprising that the homogeneous model, which assumes $S = 1$, adequately describes the pressure drop, particularly in light of the slip ratio measurements displayed in Fig. 7.8. However, it is fortuitous since the homogeneous model is fairly straightforward in its application.

7.3.3. Heat Transfer Correlations

Two-phase flow in subcritical helium is a complex problem; however, the difficulty of understanding the processes involved increases significantly when heat transfer is included. In addition to mass flow rate, vapor quality, and void fraction determining the flow conditions, the effect of heat flux into the fluid also must be considered. In particular, heat transfer can lead to rapid variations of vapor quality along the tube section, as illustrated in Fig. 7.10. A low-quality fluid enters a heated tube section from below. Initially, nucleate boiling occurs at the fluid–tube interface. These bubbles are stripped from the wall and produce local bubbly flow. As the fluid con-

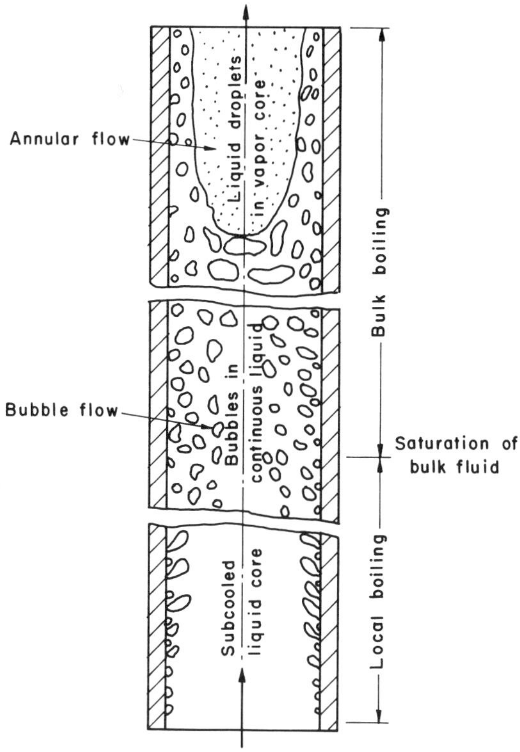

Fig. 7.10. Flow patterns in a vertical heated channel (from Tong[32]).

tinues through the tube more heat is transferred, increasing the vapor quality until slug flow and finally annular flow occur. Thus in a heated section the fluid cannot be described by a single-flow regime as is a reasonable approximation in adiabatic flow systems.

Experimental investigations of two-phase flow heat transfer have been carried out by de La Harpe et al.[27] and Johannes.[28] These studies consist principally of forced flow helium at 4.2 K confined to a tube of a few mm ID with temperature and pressure probes necessary to determine the heat transfer coefficient and critical heat flux for boiling. The latter quantity is strongly geometry dependent and cannot be generalized easily in other systems.

The heat transfer coefficient obtained for two-phase helium is discussed best in terms of classical correlating relationships. The approach is to determine first the Nusselt number corresponding to the Dittus–Boelter equation for the liquid flow only, that is,

$$\mathrm{Nu}_l = 0.023\,(\mathrm{Re}_l)^{0.8}(\mathrm{Pr}_l)^{0.4}(1-\chi)^{0.8} \qquad (7.50)$$

where the last multiplier is to indicate the contribution of only the liquid. Note that $\mathrm{Nu}_l \equiv hD_h/k_l$, where k_l is the liquid thermal conductivity. Measurements of the two-phase heat transfer coefficient have shown that the actual Nusselt number normalized to (7.50) can be correlated to the Lockhart–Martinelli parameter,

$$\frac{\mathrm{Nu}_{\mathrm{exp}}}{\mathrm{Nu}_l} = A\chi_{tt}^{-n} \tag{7.51}$$

where χ_{tt} is given by (7.47). The best fit to the data of Johannes[28] of the form of (7.51) are displayed in Fig. 7.11 for which the appropriate values are $A = 5.40$ and $n = 0.385$. Results of de La Harpe[27] are also displayed as the dashed line in the figure with agreement in form to the data of Johannes, although possessing substantially different values for the coefficient A. Although the above correlation seems appropriate for low to moderate values of the Lockhart–Martinelli parameter, in the limit of large χ_{tt}, which occurs for small vapor quality, the ratio given by (7.51) must approach unity. Therefore, to be universally applicable the correlation should reflect this fact.

Two-phase heat transfer as with two-phase flow is a complex process which is very difficult to understand fully. If faced with a problem in this area, the best approach is to apply one of the accepted correlations. However, these calculations are only approximate and should be used only as a guide. If greater accuracy is required, experimental modeling is the required approach.

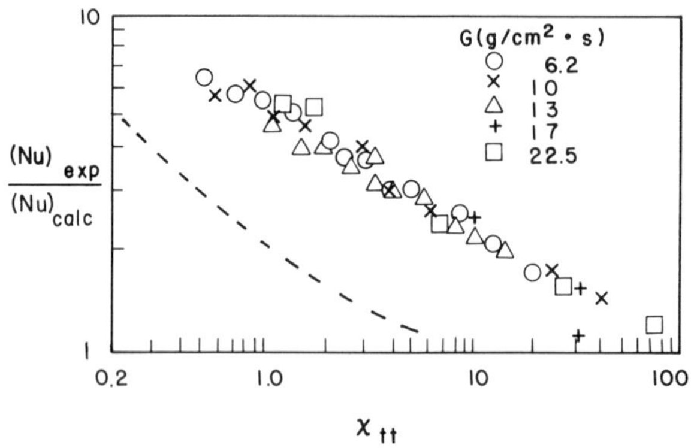

Fig. 7.11. Two-phase heat transfer correlation for helium (from Johannes[28]). The dashed line is a comparison to previous measurements of de La Harpe et al.[27]

7.4. TRANSIENT EFFECTS

Up until now the only processes considered have occurred in steady-state fully developed flows of helium within a tube. These of course represent a fraction of the problems occurring with forced flow helium. For example, the development of the temperature and velocity profile in entrance regions of the channel requires knowledge of the time variation of the thermal and momentum distributions. These are important effects. However, perhaps more important from the viewpoint of application is the behavior of forced flow helium when it experiences a transient thermal disturbance causing the temperature to rise locally. This mechanism in turn can induce local pressure changes due to bulk expansivity. Theoretical analysis of this problem requires solution of the fully time-dependent fluid dynamic equations.

At the outset it is useful to consider the technological significance of transient heat transfer in forced flow helium. An important system for this kind of heat transfer condition is in the internally cooled cabled superconductors (ICCS) for large magnet applications. To prevent such a device from entering the normal state or "quenching," it is necessary to provide sufficient thermal stability to guard against reasonable sized disturbances. Since disturbances are usually transient in nature and caused by local mechanical motion or magnetic field fluctuations, the transient behavior of the helium coolant plays an important role.

There are two principal issues concerning this transient behavior. First, since the helium within the channel is all that is available to absorb the thermal pulse and L/D is sufficiently large that the system is nearly adiabatic, there must be enough enthalpy available to absorb the disturbance and any other associated heat generation. Furthermore, there must be adequate heat transfer between the tube and helium so that the surface temperature is kept under control. Second, since helium and particularly supercritical helium is a very compressible fluid, it is necessary to evaluate the local transient pressure rise to determine whether it exceeds the allowable limits for the cooling system. These problems have their basis in compressible fluid dynamics.

To solve the transient heat transfer problem in general, the fundamental fluid hydrodynamics equation of Navier–Stokes—(7.2), (7.3), and (7.5)—must be applied. This requires numerical solution. However, there are essentially two classes of problem for which approximate solutions exist. These involve (1) local disturbances where the fluid within the conduit can be considered adiabatic and (2) global disturbances where the length of the conduit and the time evolution of flow are of tantamount importance. Problems of class (1) are approximated by neglecting wall friction and solving for the fluid inertia in response to a local heat deposition.

Class (2) problems neglect fluid inertia but consider the interaction between the fluid and wall.

For the general case of one-dimensional flow with wall friction, the time variation of compressible fluids obeys a simplified set of equations. In particular, the continuity equation may be written

$$\frac{\partial \rho}{\partial t}+\rho\frac{\partial u}{\partial x}+u\frac{\partial \rho}{\partial x}=0 \tag{7.52}$$

and for the case where the body force is associated with wall friction, the momentum equation takes the form

$$\frac{\partial u}{\partial t}+u\frac{\partial u}{\partial x}=-\frac{1}{p}\frac{\partial p}{\partial x}-\frac{2fu^2}{D_h} \tag{7.53}$$

where for large L/D the variation of velocity in the direction perpendicular to flow can be neglected. Furthermore, in the linear approximation which is valid for velocities much less than c, the speed of sound, and pressures $p \ll \rho c^2$, the energy equation becomes

$$\frac{d}{dt}\left(\frac{1}{2}u^2+e\right)=-\frac{1}{\rho}\frac{\partial(pu)}{\partial x}+\dot{q} \tag{7.54}$$

where e is the specific internal energy of the fluid and $\dot{q}$ is the rate of heat input per unit mass.

Dresner[33] has solved Eqs. (7.52), (7.53), and (7.54) for problems of class (1), that is, no wall friction. A detailed description of this analysis is beyond the scope of the present discussion. However, by considering the above three expressions several factors should be apparent. The addition of a heat pulse will induce an acceleration in the fluid. This fact can be seen in (7.54) where the time rate of change of the kinetic energy is a function of $\dot{q}$. It is not difficult to show by dimensional arguments that the induced velocity is approximately given by

$$u \approx \dot{q}l\left(\frac{\beta}{\rho C_p}\right) \tag{7.55}$$

where l is the characteristic length of the heated zone and β is the bulk thermal expansivity of helium. Inserting some typical values into (7.55) such as $\dot{q}=10$ W/cm^3 for 10 ms and $l=1$ m, a velocity in excess of 2 m/s is calculated. This velocity obviously is symmetric about the point at which the heat is applied. Its contribution is additive to the imposed external velocity and may in fact be the dominant local velocity under transient conditions.

Heat-induced fluid flow has significant impact on the heat transfer. To a first approximation the effect can be seen by referring to steady-state heat transfer correlations. For example, the Dittus–Boelter correlation indicates that the heat transfer coefficient should be proportional to $u^{4/5}$, so the larger induced flow produces better heat transfer. However, there are additional heat transfer mechanisms that must be considered. For short-duration, small heat pulses, the induced flow is not present and the heat transfer is not governed by fully developed correlations but rather by transient conduction. This subject has been discussed in Section 6.6 on transient heat transfer to normal helium. The general result stated there was that heat transfer for short durations is dictated by Kapitza conductance phenomena until a critical energy is exceeded. Subsequent to exceeding this critical energy, the heat transfer transforms to ordinary convection processes. This hypothesis has yet to be proved to occur in forced flow systems; however, it can be inferred from stability experiments on superconductors. A qualitative representation[34,35] of the transient heat transfer behavior in forced flow systems is shown in Fig. 7.12. For small energy pulses, that is, short times, the heat transfer is via conduction mechanisms which provides a rather high effective heat transfer coefficient. This behavior is represented by region CD of the figure. Above the critical energy the heat transfer deteriorates along path DA. As discussed in Chapter 6, the energy required to cause this transformation is proportional to the square root of the pulse duration. For higher-energy pulses, the heat transfer is controlled by near-steady-state convection mechanisms as indicated by AB. Thus, the heat transfer coefficient has a minimum value at the point where convection takes control over conduction.

Two problems of class (2) have been solved, namely, (i) the transient pressure rise and (ii) the thermal expulsion of helium resulting from energy deposition over its entire conduit length. These problems have been

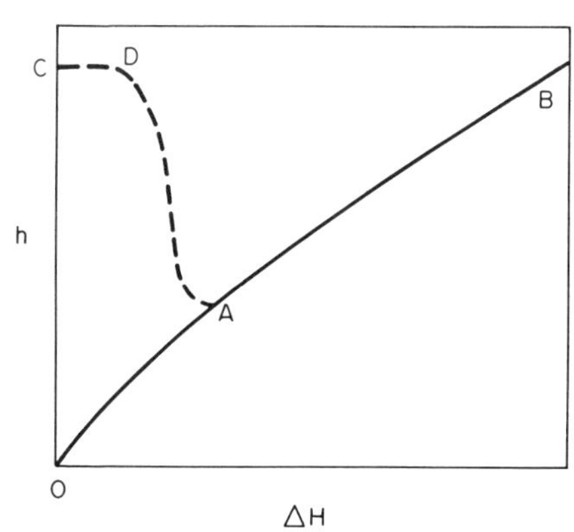

Fig. 7.12. The dependence of the induced heat transfer coefficient h on the size of the heat pulse ΔH, assuming a fixed pulse duration (figure courtesy of Oak Ridge National Laboratory).

studied[36–38] from the viewpoint of the nonlinear flow equations with strong wall friction. The approach is to note that the momentum equation (7.53) relates the local pressure to gradients in the velocity which are in turn caused by the heat pulse. Miller et al.[36] developed an approximate method for calculating the transient pressure rise and expulsion velocity which avoids simultaneous solution of the three fluid equations (7.52), (7.53), and (7.54). The complete approach to this problem obviously requires numerical methods. For a tube with a large L/D ratio, work done by the pressure gradient goes mostly into overcoming the wall friction rather than accelerating the fluid. By neglecting the inertial term in the momentum equation, an approximate solution is obtained for the local pressure rise at the center of the tube, that is,

$$\Delta p = \frac{2f}{D_h} \int_0^L \rho u^2 \, dx \tag{7.56}$$

where the local density is determined in part by application of the heat pulse $\dot{q}$. To produce a tractable solution it is necessary to make some assumptions about the velocity profile. If the velocity varies linearly with position, the pressure rise and density may be written simply as

$$\Delta p = \frac{2f}{D_h} \rho u_m^2 \frac{L}{3} \tag{7.57a}$$

and

$$\frac{1}{\rho} \frac{d\rho}{dt} = -\frac{u_m}{L} \tag{7.57b}$$

where L is the half length of the tube and u_m is the mean velocity. This form for the velocity profile cannot be proved independently but represents a good assumption based on comparison with experiments.

As with the steady-state problem considered in Section 7.2, a thermodynamic expression for the density in terms of the pressure and entropy can be used to reduce the unknowns in the equation. The expression can then be placed in a useful form by inserting (7.57a) and (7.57b), the result of which is a differential equation coupling the time variation of the density, pressure, and heat:

$$\frac{d}{dt}(\Delta p) = \frac{\beta c^2}{C_p}(\rho \dot{q}) - \frac{\rho c^2}{L}\left(\frac{3D_h \, \Delta p}{2f\rho L}\right)^{1/2} \tag{7.58}$$

where c is the speed of sound in the helium, and Δp is the local pressure rise. In (7.58) the first term on the right-hand side is the adiabatic heating term while the second term is the decrease in pressure associated with the acceleration of the fluid. It is therefore a fairly straightforward numerical integration to determine the local pressure rise at any point in the tube.

The peak pressure in the system, $\Delta p_{\max}$, represents a special case for (7.58). It follows that at $\Delta p_{\max}$, $d(\Delta p)/dt = 0$, which leads to the simplified solution

$$\Delta p_{\max} = \frac{2}{3}\left(\frac{\beta^2}{\rho C_p^2}\right)\frac{fL^3}{D_h}(\rho\dot{q})^2 \tag{7.59}$$

This expression would be only a slight improvement over (7.58) except that the thermodynamic parameter group $(\beta^2/\rho C_p^2)$ can be approximated by a single function of pressure, as can be seen in Fig. 7.13. Over much of the range of interest, the group is only weakly dependent on pressure and can be approximated by the relationship

$$\left(\frac{\beta^2}{\rho C_p^2}\right) = 0.45p^{-1.8} \qquad \text{(SI units)} \tag{7.60}$$

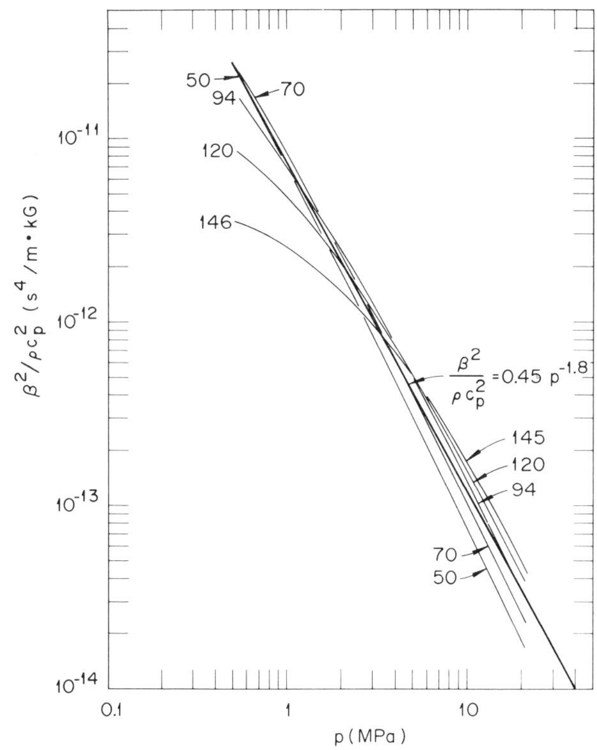

Fig. 7.13. Plot of the thermodynamic group $\beta^2/\rho C_p^2$ versus pressure. Numerical values on the curve refer to the fluid density in g/cm^3 (figure courtesy of Oak Ridge National Laboratory).

Substitution of (7.59) into (7.58) results in the following form for the peak pressure:

$$p_{\max} = 0.65 f^{0.36} \left(\frac{\rho \dot{q}^2 L^3}{D_h}\right)^{0.36} \left(1 - \frac{p_0}{p_{\max}}\right)^{-0.36} \tag{7.61}$$

where p_0 is the ambient pressure in the tube. Assuming a typical Reynolds number, $\mathrm{Re}_D = 10^5$, gives a smooth tube friction factor of $f = 0.0044$. Inserting this value into (7.61) leads to a fairly simple correlation between the maximum pressure and the rate of heat input:

$$p_{\max} = 0.10 \left(\frac{(\rho \dot{q})^2 L^3}{D_h}\right)^{0.36} \tag{7.62}$$

This expression is plotted in Fig. 7.14 along with experimental data obtained for two different channel lengths and two heat pulse types. Also show in the figure is a calculation by Krafft and coworkers[39,40] which is a numerical integration of the complete fluid equations.

In addition to a determination of the peak pressure, the solution to the class (2) problem yields an approximate value for the expulsion velocity, that is, the velocity at the conduit ends during a global disturbance. This problem has been evaluated by similar methods as discussed above.[38] The derivation is beyond the present treatment but the result is a fairly simple expression,

$$u = 0.952 \left(\frac{\beta c \dot{q}}{\rho C_p}\right)^{2/3} \left(\frac{D_h t}{f}\right)^{1/3} \tag{7.63}$$

where t is the characteristic time and c is the speed of sound in the helium. This expression can be shown to predict velocities in excess of 10 m/s for

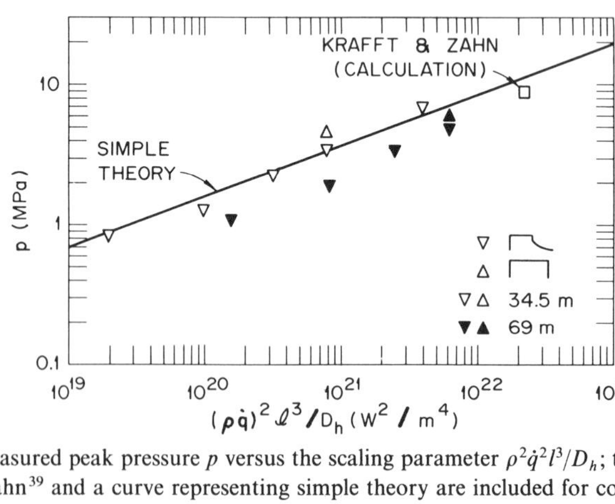

Fig. 7.14. Measured peak pressure p versus the scaling parameter $\rho^2\dot{q}^2 l^3/D_h$; the calculation of Krafft and Zahn[39] and a curve representing simple theory are included for comparison (figure courtesy of Oak Ridge National Laboratory).

reasonable operating parameters. This velocity is considerably larger than that associated with the imposed flow circuit and therefore may induce transient flow damage if the conduit is part of a closed cryogenic system.

The time variation of the temperature profile also can be obtained from the above analysis. Combining the thermodynamic relationships for the entropy change with temperature and pressure, we obtain a form for the time variation of temperature:

$$\frac{dT}{dt}=\frac{(\rho\dot{q})}{\rho C_p}+\frac{\beta T}{\rho C_p}\frac{d(\Delta p)}{dt} \tag{7.64}$$

As before, the first term on the right-hand side represents the adiabatic heating while the second is the expansion work done by the fluid. Unfortunately, a simple solution to (7.64) does not exist and numerical integration is required. This analysis has been carried out by Miller et al.[36] in comparison to their experimental results on transient heat transfer to forced flow supercritical helium. The experiment and calculation were shown to agree in form although the absolute magnitude of temperature is not reproduced accurately.

The dynamic behavior of supercritical helium seems to be well described by the Navier–Stokes equations for a single-component compressible fluid. This fact can be seen by the accuracy to which the theory is able to correlate data on pressure and temperature within a long tube with internal heat generation. Prediction of the fluid dynamics unfortunately says nothing about the heat transfer coefficient h. No direct measurements exist for the transient heat transfer coefficient. However, evidence inferred from stability experiments supports the type of heat transfers shown in Fig 7.12. Until actual heat transfer experiments are performed, however, this can be only a qualitative description. For problems that are more complex than the special cases described above, it is necessary to solve the full hydrodynamic equations. This is a difficult problem owing to the complexity of the equation and the variation of the helium properties. Analysis of these problems requires sophisticated numerical methods.[39,40]

Problems

7.1. Derive the separate equations for the one-dimensional pressure and temperature gradients in a compressible single-phase fluid. Show that for low velocities and nearly ideal gas behavior these expressions reduce to the forms given by (7.29) and (7.30). Determine a relationship for the pressure gradient under the condition where $dT/dx = 0$.

7.2. A heat exchanger cools supercritical helium at 10 bars from 6 to 4.5 K. The design consists of a tube immersed in the saturated bath of He I at 4.2 K. Determine the length and diameter of the tube consistent with the following specification: $\dot{m} = 1$ g/s and $\Delta p = 0.1$ bar. For simplicity assume isothermal conditions on the external surface of the tube.

7.3. Make a plot of the two-phase friction multiplier ϕ_l^2 as a function of void fraction χ at 4.2 and 2.5 K. On the same graph plot the normalized Nusselt number.

7.4. Derive Eq. (7.60) for the time rate of change of the pressure rise in a one-dimensional channel subjected to a transient heat pulse $\dot{q}$. Estimate the heat transfer rate $\dot{q}$ required to make a significant change in the pressure within a 100 m long, 10 mm diameter channel.

REFERENCES

1. M. O. Hoenig, Y. Iwasa, D. B. Montgomery, and A. Bejan, Cryostabilized Single Phase Helium Cooled Conductors for Large High Field Superconducting Magnets, in *Proceedings of the 6th Symposium on Engineering Problems of Fusion Research*, pp. 586–592; IEEE Nuclear and Plasma Sciences Society, Pub. No. 75CH1097-5-NPS, 1976.
2. G. Krafft, P. Komarek, and A. Hofmann, Forced Flow Cooling in Single Phase Helium, in *Stability of Superconductors*, pp. 33–42, International Institute of Refrigeration Commision A1/2, Saclay, France, 1981.
3. V. E. Keilin, E. Ju. Klimenko, and B. N. Samoilov, Forced-Cooled Superconducting Systems, *Cryogenics* **10**, 224 (1970).
4. L. Dresner, Stability of Internally Cooled Superconductors: A Review, *Cryogenics* **20**, 558 (1980).
5. See, for example, E. R. G. Eckert and R. M. Drake, Jr., *Analysis of Heat and Mass Transfer*, Mc-Graw-Hill, New York, 1973.
6. See, for example, R. B. Bird, W. E. Steward, and E. N. Lightfoot, *Transport Phenomena*, Wiley, New York, 1966.
7. F. P. Incropera and D. P. Dewitt, *Fundamentals of Heat Transfer*, Chap. 8, Wiley, New York, 1981.
8. J. W. Lue, J. R. Miller, and J. C. Lottin, Pressure Drop Measurement on Forced Flow Cable Conductors, *IEEE Trans. Magnet.* **Mag-15**, 53 (1979).
9. V. Arp, Forced Flow, Single Phase Helium Cooling Systems, *Adv. Cryog. Eng.* **17**, 342 (1972).
10. J. W. Dean, W. Stewart, and J. K. Hoffer, Temperature Profiles in a Long Gaseous Helium Cooled Tube, *Adv. Cryog. Eng.* **23**, 250 (1978).
11. F. W. Dittus and L. M. K. Boelter, University of California, Berkeley, Publ. Eng. Vol. 2, p. 443, 1930.
12. P. J. Giarratano, V. D. Arp, and R. V. Smith, Forced Convection Heat Transfer to Supercritical Helium, *Cryogenics* **11**, 385 (1971).
13. M. F. Taylor, NASA-TND-4332 (1968).
14. D. S. Scott, Properties of Concurrent Gas–Liquid Flow, *Advances in Chemical Engineering*, Vol. 4, pp. 199–273, Academic Press, New York, 1963.
15. J. L. Baker, *Flow Regime Transitions at Elevated Pressures in Vertical Two-Phase*, ANL Report 7093, Argonne National Laboratory, Argonne, IL, 1965.
16. P. Griffith and G. B. Wallis, Two Phase Slug Flow, *Trans. ASME* **83**, 307 (1961).
17. H. K. Zust and W. B. Bald, Experimental Observations of Flow Boiling of Liquid Helium I in Vertical Channels, *Cryogenics* **21**, 657 (1981).
18. A. Khalil, "Experimental Measurements of Void Fraction in Cryogenic Two-Phase Upward Flow," Ph.D. Thesis, University of Wisconsin–Madison, 1978; *Cryogenics*, 411 (1981).
19. H. Haraguchi, S. Kimiya, S. Nakagawa, A. Iwata, M. Yoshiwa, T. Kato, O.Takahashi, E. Tada, and S. Shinamoto, Flow Characteristics of Liquid Helium in a Tube. I. The Frictional Pressure Drop of Helium Two-Phase Flow, Japanese Cryogenic Engineering Conference, May 16–18, 1983.

20. N. A. Radovcick and R. Moissis, MIT Report No. 7-7673-22, Massachusetts Institute of Technology, Cambridge, MA, 1962.
21. G. B. Wallis, *One Dimensional Two Phase Flow*, McGraw-Hill, New York, 1969.
22. R. W. Lockhart and R. C. Martinelli, Proposed Correlation of Data for Isothermal Two Phase Two Component Flow in Pipes, *Chem. Eng. Prog.* **45**, 39 (1949).
23. R. V. Smith, Fluid Dynamics, in *Cryogenic Fundamentals*, G. G. Haselden (Ed.), Chap. 5, Academic Press, New York, 1971.
24. S. Levy, Stream Slip—Theoretical Prediction from Momentum Model, *Trans. ASME, J. Heat Transfer* **82**, 113 (1960).
25. G. Hildebrandt, Heat Transfer to Boiling Helium I Under Forced Flow in a Vertical Tube, in *Proceedings of the 4th International Cryogenics Engineering Conference*, pp. 295–300, IPC Science and Technology Press, London, 1972.
26. V. Keilin, E. Yu. Klimenko, and I. A. Kovalev, Device for Measuring Pressure Drop and Heat Transfer in Two-Phase Helium Flow, *Cryogenics* **15**, 141 (1975).
27. A. de La Harpe, S. Lehongre, J. Mollard, and C. Johannes, Boiling Heat Transfer and Pressure Drop of Liquid Helium I Under Forced Circulation in a Helically Coiled Tube, *Adv. Cryog. Eng.* **14**, 170 (1963).
28. C. Johannes, Studies of Forced Convection Heat Transfer to Helium I, *Adv. Cryog. Eng.* **17**, 352 (1972).
29. V. Keilin, A. Kovalev, V. V. Likov, and M. M. Pozvonkov, Forced Convection Heat Transfer to Liquid Helium I in the Nucleate Boiling Regime, *Cryogenics* **15**, 141 (1975).
30. H. Ogata and S. Sato, Forced Convection Heat Transfer to Boiling Helium in a Tube, *Cryogenics* **14**, 375 (1974).
31. S. Nakagawa et al., Pressure Drop and Heat Transfer in Helium Two Phase Flow, in *Proceedings of the 10th International Cryogenics Engineering Conference*, pp. 570–573, IPC Science and Technology Press, London, 1984.
32. L. S. Tong, *Boiling Heat Transfer and Two-Phase Flow*, Chap. 3, Kriger, New York, 1975.
33. L. Dresner, Heat Induced Flows in Cable-in-Conduit Conductors, *Cryogenics* **19**, 653 (1979).
34. L. Dresner, Parametric Study of the Stability Margin of Cable-in-Conduit Superconductors: Theory, *IEEE Trans. Magnet.* **Mag 17**, 753 (1981).
35. J. W. Lue, J. R. Miller, and L. Dresner, Stability of Cable-in-Conduit Superconductors, J. Appl. Phys. **51**, 772 (1980).
36. J. R. Miller, L. Dresner, J. W. Lue, S. S. Shen, and H. T. Yeh, Pressure Rise During Quench of a Superconducting Magnet Using Internally Cooled Conductors, in *Proceedings of the 8th International Cryogenic Engineering Conference*, C. Rizzuto (Ed.), pp. 321–329, IPC Science and Technology Press, London, 1980.
37. L. Dresner, Thermal Expulsion of Helium From a Quenching Cable in Conduit Conductor, in *Proceedings of the 9th Symposium on Engineering Problems of Fusion Research*, IEEE Pub. No. 81CH1715-2-NPS, pp.618–621, IEEE, New York, 1981.
38. J. W. Lue, J. R. Miller, L. Dresner, and S. S. Shen, Simulation of the Quenching of an Internally Cooled Superconducting Magnet, *Proceedings of the 9th International Cryogenics Engineering Conference*, K. Yasukochi (Ed.), pp. 814–818, IPC Science and Technology Press, London, 1982.
39. G. Krafft and G. Zahn, Experimental and Theoretical Investigations of Heat Induced Transients in Forced Flow Helium Cooling Systems, in *Proceedings of the 8th Symposium on Engineering Problems of Fusion Research*, C. K. McGregor and T. H. Batzer (Eds.), IEEE publication No. 79CH1441-5 NPS, pp. 1724–1728, IEEE, New York, 1979.
40. J. Benkowitsch and G. Krafft, Numerical Analysis of Heat Induced Transients in Forced Flow Helium Cooling Systems, *Cryogenics* **20**, 209, 1980.

Chapter 8

Liquefaction and Refrigeration Systems

Although the emphasis of this book is on the development of physical understanding of helium as a cryogenic fluid, the discussion would be incomplete without reviewing the various methods of obtaining low-temperature helium. Such methods are based solidly in engineering thermodynamics and rely primarily on a combination of processes which make up a cycle. A cycle consists of a closed circuit where the working fluid, for example helium, is compressed, expanded, and heat exchanged in such a way as to achieve cooling. The simplest and most thermodynamically ideal cycle is the Carnot cycle which consists of a combination of isothermal and isentropic processes. The Carnot cycle is difficult to achieve in a practical system. However, real refrigeration and liquefaction systems are made up of similar processes which are usually compared in their performance to that of the ideal Carnot cycle. In all gas refrigeration systems, the cooling is achieved by an expansion process. There are essentially two types of expansion which are applied. *Isentropic expansion* is that where the fluid does work and expands keeping its entropy constant ($\Delta S = 0$). This is the best method of expansion because it produces the largest temperature change over a given pressure change. The other expansion process is *isenthalpic expansion*, where the fluid undergoes a pressure change without heat transfer ($\Delta Q = 0$). This method is in common application in practical refrigeration systems for its ease of use. However, it is of lower thermodynamic efficiency because it is an irreversible process leading to nonidealities.

The present chapter begins with an ideal liquefaction process as a point of comparison. A number of practical cycles developed for cryogenic liquefaction and refrigeration are then discussed with some aspects of component nonideality included. Examples involve the liquefaction and refrigeration of most common cryogenic fluids.

8.1. IDEAL LIQUEFACTION

The thermodynamic limitations of the liquefaction process can be demonstrated best by consideration of an ideal system, that is, a refrigeration cycle that performs at Carnot efficiency and consequently consists of reversible isothermal and isentropic processes. Consider a Carnot refrigerator producing cooling in a low-temperature reservoir. Furthermore, unlike the discussion in the previous section, assume that the refrigerator consists of a large number of cycles operating over small temperature differences. This process can be thought to occur physically by a large number of reservoirs operating at intermediate temperatures T_i, as in Fig. 8.1.

The ideal Carnot refrigerator achieves a certain amount of cooling at T_C by performing a continuous process. The work done to carry out this process can be approximated by an integral expression,

$$w = -\int_{T_H}^{T_C} \frac{\delta Q}{T} (T_H - T) \tag{8.1}$$

where each refrigerator operates between T_H and T. To produce liquefaction in a liquid–gas system, there are two contributions to the work which must be included:

1. w_g is the work required to cool a unit mass of gas from ambient temperature to its boiling point T_{bp}.
2. w_l is the work required to convert a unit mass of gas at T_{bp} to a unit mass of liquid.

The first of these terms can be derived by application of (8.1), keeping in mind that $\delta Q = C_p \delta T$. The work needed to bring a unit mass of gas to T_{bp} in a liquid–gas system is written

$$w_g = -\int_{T_H}^{T_{bp}} C_p\, dT \left(\frac{T_H}{T} - 1\right) \tag{8.2}$$

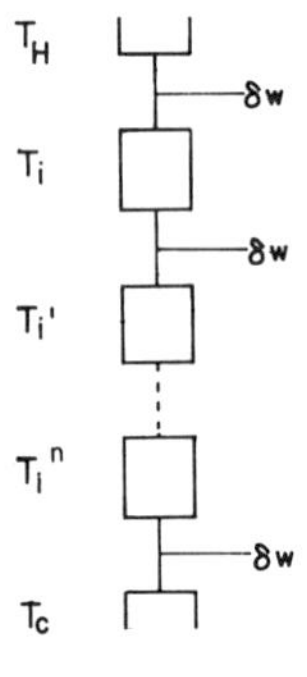

Fig. 8.1. Infinitesimal Carnot refrigerators.

where the sign convention has been set to permit the work to be a positive quantity.

The second term, which is the liquefaction work, always occurs between T_{bp} and T_H. For this term $\delta Q = \lambda$, the latent heat per unit mass of liquefied gas. The integral in (8.1) then collapses to its value at the low-temperature end T_{bp}, yielding the second contribution to the work,

$$w_l = \lambda\left(\frac{T_H}{T_{bp}} - 1\right) \tag{8.3}$$

It is quite apparent in (8.3) that the term in brackets simply represents the coefficient of performance of a Carnot refrigerator operating between T_{bp} and T_H.

The total work w_T to liquefy a unit mass of gas is simply a sum of the above two terms. Rearranging w_T in a form that is easier to evaluate on physical grounds, we obtain

$$w_T = T_H\left(\int_{T_{bp}}^{T_H} \frac{C_p}{T}\, dT + \frac{\lambda}{T_{bp}}\right) - \left(\int_{T_{bp}}^{T_H} C_p\, dT + \lambda\right) \tag{8.4}$$

Cast in this way, the total work can be seen to consist of two quantities. The first represents the entropy difference between the gas at room temperature and the liquid at T_{bp}. The second term is the heat or enthalpy difference between the gas at room temperature and the liquid at T_{bp}. Note that this quantity enters as a negative because it is thermodynamically recoverable.

The above definitions can be placed on clearer footing by consideration of a *T-S* diagram for a real fluid. An idealized example of such a diagram is shown in Fig. 8.2. Achievement of liquefaction as described

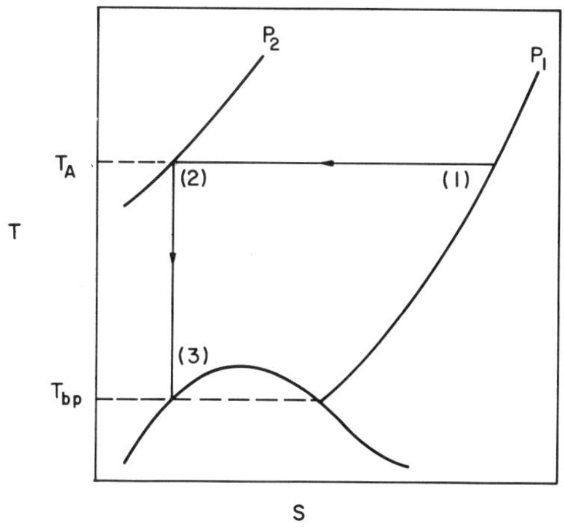

Fig. 8.2. *T-S* diagram showing ideal liquefaction process 1 → 3.

above can be identified on this T-S diagram by the path 1–2–3. Here, step 1–2 represents an isothermal process followed by step 2–3 which is an isentropic expansion of the fluid to the normal boiling point T_{bp}. In practical terms, the flow circuit used to achieve this process is shown in Fig. 8.3. Here there are two principal components. An isothermal compressor brings the fluid to high pressure, p_2, at which the entropy is the same as the liquid at the normal boiling point. The isentropic expansion engine then reduces the fluid temperature until liquefaction occurs. The total work per unit mass of fluid (8.4) is therefore given by the difference in the state functions evaluated at the end points of the ideal liquefaction path,

$$w_t = T_H(s_1 - s_3) - (h_1 - h_3) \tag{8.5}$$

Typically, for real fluids, these quantities are tabulated, permitting the above calculation to be performed trivially. However, for the case concerning helium, which is nearly an ideal gas, these functions can be calculated with reasonable accuracy from (8.4). It is instructive to perform this calculation as an example to appreciate better the numerical values associated with the liquefaction process.

The two contributions to the liquefaction work of helium are the enthalpy and entropy terms. The enthalpy term is defined by

$$\Delta h = \int_{T_{bp}}^{T_H} C_p \, dT + \lambda \tag{8.6}$$

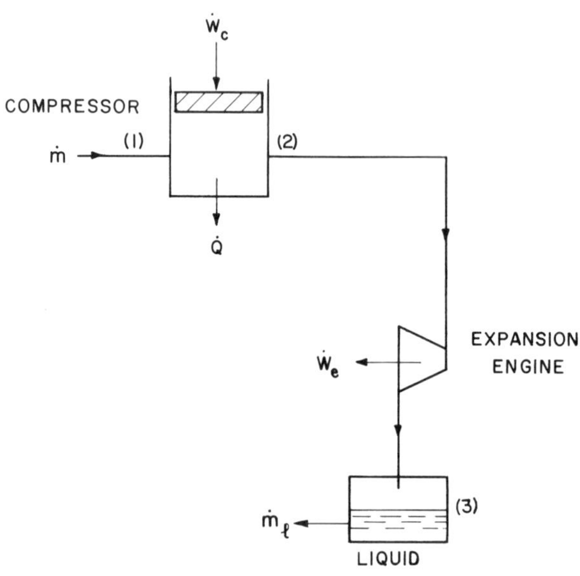

Fig. 8.3. Flow circuit for ideal expansion liquefier.

Since helium approximates an ideal monatomic gas, $C_p \simeq \frac{5}{2}R =$ 20.8 J/mol · K. The latent heat of helium is a temperature-dependent quantity but at its normal boiling point of 4.2 K, $\lambda = 84$ J/mol. Combining these two contributions in (8.6) the result is

$$\Delta h = 20.8(300 - 4.2) + 84 = 6237 \text{ J/mol}$$

where the high temperature is set arbitrarily to 300 K. It is important to note that helium has a very small latent heat, a fact due to weak intermolecular binding which is discussed further in Chapter 3, and the principal amount of work must be done when cooling the gas.

The entropy contribution to liquefaction of helium can be calculated similarly,

$$\Delta s = \int_{T_{bp}}^{T_H} \frac{C_p}{T} dT + \frac{\lambda}{T_{bp}} \tag{8.7}$$

which for helium, as a nearly ideal gas, can be evaluated directly between ambient temperature and the normal boiling point:

$$\Delta s \simeq 20.8 \ln\left(\frac{300}{4.2}\right) + \frac{84}{4.2} = 109 \text{ J/mol} \cdot \text{K}$$

Here it is useful to note that the latent heat contribution is significant, representing about 20% of the entropy term. Combining the enthalpy and entropy terms, we calculate that the ideal work to liquefy a unit of helium is 26,463 J/mol or 6615 J/g. At saturated vapor pressure densities this work per unit volume is 0.229 kW · h/L. Thus, it appears that for an ideal liquefier the cost of liquefaction in terms of electricity is not a major factor. It should be emphasized, however, that real liquefiers of helium are far from ideal, actually operating at between 5 and 20% Carnot efficiency.

As has already been indicated, the above calculation was only for demonstration purposes. Actual procedure uses tabulated values for entropy and enthalpy which can then be calculated trivially by means of (8.5). Charts that give these values for helium and other common cryogenic fluids are given in Appendix 3. Also, there are tabulated data available in the literature.[4–6] To use the *T-S* diagrams directly to calculate ideal liquefaction work, points 1 and 3 of the liquefaction path must be located. Point 1 represents helium at a temperature and pressure of 300 K and 1 bar. At this condition the entropy and enthalpy are

$$s_1 = 31.4 \text{ J/g} \cdot \text{K} \qquad \text{and} \qquad h_1 = 1573 \text{ J/g}$$

Table 8.1. Ideal Liquefaction Work of Cryogenic Fluids

Fluid	T_{bp} (K)	W_T (J/g)	W_T (kW · h/L)
O_2	90	634	0.202
Ar	87	476	0.188
Air	80	722	0.176
N_2	77	761	0.171
H_2	20.4	11,890	0.231
He	4.2	6,807	0.236

At point 3, the saturated liquid state at 4.2 K,

$$s_3 = 3.5 \text{ J/g} \cdot \text{K} \qquad \text{and} \qquad h_3 = 9.7 \text{ J/g}$$

Insertion of these values into (8.5) yields $w_T = 6807$ J/g, quite close to that calculated using the ideal gas law. This result is a demonstration of the near ideality of helium gas.

In summary, similar data are tabulated for other cryogenic fluids and charts for these fluids can be found in the literature. Comparison can be made between these different fluids according to their ideal work to liquefy a unit volume. Listed in Table 8.1 are most of the common cryogenic fluids including helium with their ideal work to liquefy a unit volume. It is interesting to note that they all have w_T within 20% of 0.2 kW · h/L, a quantity has no fundamental significance but is useful to keep in mind for calculation purposes.

8.2. FIRST LAW OF STEADY FLOWS

In advance of discussing the more practical aspects of helium liquefaction and cooling, it is necessary to introduce the concept of the first law of steady flows. This concept is useful because most actual liquefaction processes occur in the steady state and not equilibrium, so that the behavior of a thermodynamic system under flow conditions is of tantamount importance. Furthermore, the process of interest usually involves the refrigeration power or liquefaction rate of a particular cryogenic

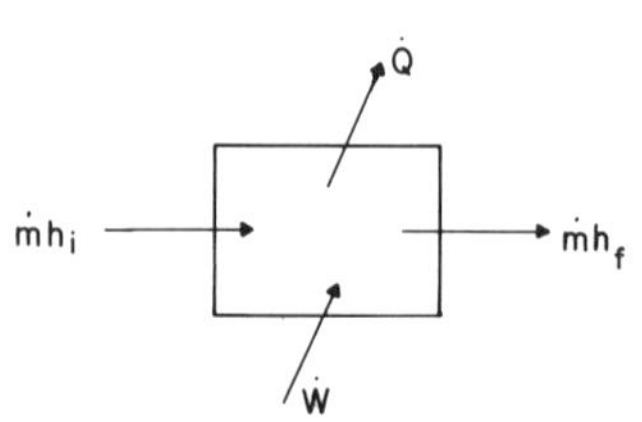

Fig. 8.4. Unit volume illustrating the first law of steady flows.

system. The first law of steady flows can be understood best by reference to Fig. 8.4. Here a unit volume of helium is considered where the symbols are identified as follows:

$\dot{Q}$ = rate of heat removed from unit volume
$\dot{W}$ = rate of work done on fluid in volume
$\dot{m}$ = mass flow rate
h = specific enthalpy of the liquid
s = specific entropy of the fluid

A simplification that has been applied to this process is to ignore the effect of the potential and kinetic energies. For helium these terms are generally quite small and their inclusion would serve only to complicate the derivations at hand. It is reassuring to note that introduction of the above concept does not modify the thermodynamic relationship already introduced for the work necessary to liquefy a unit mass of fluid (8.5). The only variation is to redefine the work on a unit mass flow basis such that (8.5) is rewritten in the form

$$\frac{\dot{W}}{\dot{m}} = T_H \Delta s - \Delta h \tag{8.8}$$

where the entropy and enthalpy differences are per unit mass.

To understand better the discussion of practical refrigeration systems, it is helpful to define several quantities related to continuous flow refrigeration:

1. Work per unit mass compressed, $\dot{W}/\dot{m}$.
2. Work per unit mass liquefied, $\dot{W}/\dot{m}_l$.
3. Fraction of mass flow liquefied, $y = \dot{m}_l/\dot{m}$.

An optimized practical design endeavors to minimize $\dot{W}/\dot{m}_l$ and maximize the yield y. In addition, there are several terms useful in the identification of the thermodynamic efficiency of a particular refrigeration cycle. It follows from (1.3) that the coefficient of performance (COP) in a flow system is the ratio of the heat removed at low temperature to the work at high temperature:

$$\mathrm{COP} = \frac{\dot{Q}_c}{\dot{W}} \tag{8.9}$$

Obviously, for a Carnot cycle, $\mathrm{COP} = T_C/(T_H - T_C)$. As a measure of the deviation from Carnot efficiency, a quantity referred to as the figure of merit (FOM) is defined as the ratio of the ideal Carnot work ($\dot{W}_i$) and actual work ($\dot{W}$) such that

$$\text{FOM} = \frac{\dot{W}_i}{\dot{W}} \tag{8.10}$$

Proper design will attempt to maximize FOM as is reasonable. Typically, the highest efficiency helium refrigerators have FOM $\approx 25\%$.

8.3. ISENTHALPIC EXPANSION

Up until now refrigeration or liquefaction has been considered only in terms of ideal processes. Furthermore, the methods by which this refrigeration occurs have not been identified. In an effort to approach reality, the present section considers one of the most common liquefaction techniques, that of isenthalpic expansion which is commonly referred to as the Joule–Thomson effect.

8.3.1. Joule–Thomson Effect

The Joule–Thomson effect was one of the first methods used in liquefaction of cryogenic fluids. The method consists of performing an isenthalpic expansion of the fluid through a valve capable of sustaining a relatively large pressure drop. By definition there is no work done by the fluid and the process is adiabatic but irreversible. In practice, there are several methods by which the Joule–Thomson effect is achieved. A simple and direct method is to place a porous plug within an insulated channel such that the pressure drop through the media is about right for the desired flow conditions as illustrated in Fig. 8.5a. This approach is close to ideal

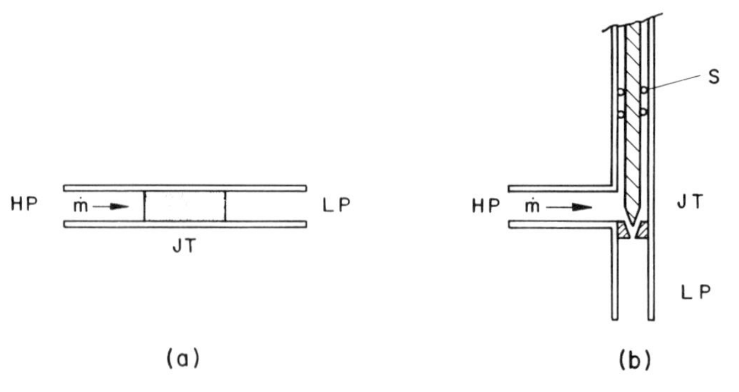

Fig. 8.5. Two types of Joule–Thomson valve: (a) porous plug and (b) controllable needle valve.

because the channel can be insulated, giving adiabatic conditions ($\dot{Q}=0$). However, porous plugs are by definition unadjustable so care must be exercised in choosing the proper physical dimensions. An alternative to the porous plug J-T valve is to use a needle valve which restricts the pressure drop to occur over a small region (Fig. 8.5b). For the needle valve, thermodynamic ideality is more difficult to achieve because the system is not totally insulated ($\dot{Q}\neq 0$). However, this trade-off must be weighted against adjustability and ease of operation.

In either case, the J-T valve has one purpose and that is to allow isenthalpic expansion of the fluid. Isenthalpic expansion means that the specific enthalpy of the fluid entering the valve must equal that of the existing fluid, which is at lower pressure. It follows that cooling can only be achieved if the constant enthalpy curve in p-T space has a positive slope. This requirement can be explained in more common terms by introduction of what is called the Joule–Thomson coefficient μ_j, which is simply the slope a constant enthalpy curve in p-T space:

$$\mu_j \equiv \left(\frac{\partial T}{\partial p}\right)_h \tag{8.11}$$

To achieve cooling by the Joule–Thomson effect, this quantity must be positive. However, for real gases, μ_j can be either positive or negative depending on the state of the fluid. Note that the Joule–Thomson coefficient was shown to be an important parameter in the description of forced flow single-phase helium in Chapter 7.

The form of μ_j as given in (8.11) is not particularly useful for identifying the physical processes involved; however, it can be cast in more meaningful context by some standard thermodynamic manipulations. Using a theorem of partial differentiation, we can write the Joule–Thomson coefficient as

$$\mu_j = -\left(\frac{\partial T}{\partial h}\right)_p \left(\frac{\partial h}{\partial p}\right)_T \tag{8.12}$$

The enthalpy h is a state function which is expandable in terms of p and T such that

$$dh = \left(\frac{\partial h}{\partial T}\right)_p dT + \left(\frac{\partial h}{\partial p}\right)_T dp \tag{8.13}$$

But the enthalpy also has a familiar differential form,

$$dh = T\,ds + v\,dp \tag{8.14}$$

The first term on the right-hand side of (8.14) can be rearranged by use of a $T\,ds$ equation,

$$T\,ds = C_p\,dT - T\left(\frac{\partial v}{\partial T}\right)_p dp \tag{8.15}$$

where

$$C_p = T\left(\frac{\partial s}{\partial T}\right)_p$$

and the second term was modified by the use of the Maxwell relation,

$$\left(\frac{\partial v}{\partial T}\right)_p = -\left(\frac{\partial s}{\partial p}\right)_T$$

Equating common terms in (8.13), (8.14), and (8.15), we can do a straightforward calculation to produce the relationship for μ_j in terms of useful thermodynamic functions:

$$\mu_j = \frac{1}{C_p}\left[T\left(\frac{\partial v}{\partial T}\right)_p - v\right] \tag{8.16}$$

Based on (8.16), the only additional information required to complete the form for μ_j are the equation of state and the constant pressure specific heat C_p. In actual fact, the most important issue is to determine the locus of points in p-T space below which the Joule–Thomson coefficient is positive. The curve that describes these points is referred to as the inversion curve and is defined by the location where $\mu_j = 0$.

The inversion curve can be derived for any gas with an established equation of state. The most simple of equations of state is that for an ideal gas, $pv = RT$, for which the Joule–Thomson coefficient is always zero. Therefore, there would be no cooling (or warming) of an ideal gas associated with isenthalpic expansion. In general, the Joule–Thomson effect is a second-order process brought on by nonideality of gases.

To understand physically how μ_j can be nonzero, return to the original thermodynamic relationship (8.12). Note that the specific enthalpy may be written

$$h = e + pv \tag{8.17}$$

where e is the specific interal energy of the fluid. Substituting (8.17) into (8.12), we obtain a different relationship for μ_j:

$$\mu_j = -\frac{1}{C_p}\left[\left(\frac{\partial e}{\partial p}\right)_T + \left(\frac{\partial pv}{\partial p}\right)_T\right] \tag{8.18}$$

The first term within the parentheses of (8.13), $(\partial e/\partial p)_T$, represents the deviation from ideal gas behavior of Joule's law: $e = C_v T$. For an ideal gas this term is zero, but for real gases where there are interactive potentials between molecules, $(\partial e/\partial p)_T$ is always negative. It is easy to understand why $(\partial e/\partial p)_T$ is less than zero. Consider a volume containing a unit mass of gas at constant temperature. If the volume is reduced, a result of increasing the external pressure, the average intermolecular spacing decreases. Joule's law is derived by assuming all the energy within the gas is in the form of random kinetic energy of the molecules. However, as the real gas is compressed, some of this kinetic energy is converted to potential energy, thus decreasing the internal energy of the system and making $(\partial e/\partial p)_T < 0$.

The second term in (8.18) can be either positive or negative. In simple terms, it measures the deviation from Boyle's law, which states that $pv = f(T)$ only, a behavior obviously obeyed by an ideal gas. This term can be understood best by taking limiting values. At high temperature and high pressure, the molecules in the gas are close together and are thus dominated by the repulsive term in their interaction. The temperature must be high so the attractive term in the potential is negligible. This should occur for temperatures much above the critical temperature T_c. The repulsive interaction makes the gas less compressible than an ideal gas and therefore

$$\left(\frac{\partial(pv)}{\partial p}\right)_T > 0$$

The behavior of this term at relatively low pressures and temperatures is considerably different. Specifically, this region is dominated by the attractive term of the potential. The attractive term of the potential tends to make the gas more compressible because the fluid is approaching the condensation temperature and

$$\left(\frac{\partial(pv)}{\partial p}\right)_T < 0$$

The behavior of both contributions to the pv term in (8.18) are summarized in Fig. 8.6.

8.3.2. Real Gases

The thermodynamic properties of a real gas are described best in terms of an empirical equation of state. In Chapter 3, several equations of state appropriate for helium gas have been introduced. These equations are somewhat difficult to apply in practical problems of refrigeration. Thus,

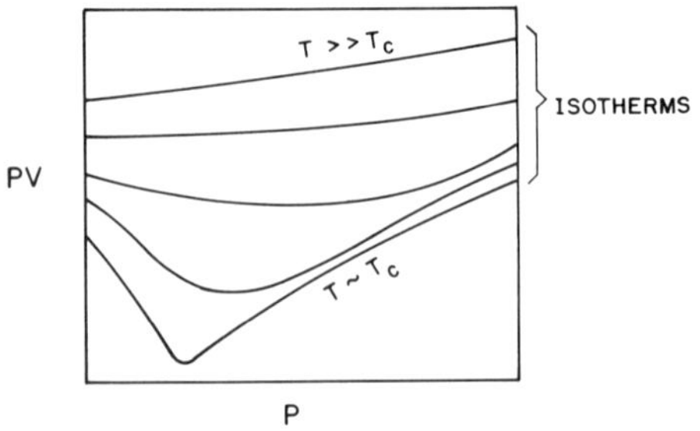

Fig. 8.6. Variation of pv as a function of absolute pressure for a real gas (from Barron[7]).

to achieve some physical understanding of the behavior of the Joule–Thomson coefficient, it is more advantageous to study its dependence on the classical equation of state.

The most common classical equation of state is for a van der Waals gas. This equation of state is given by

$$\left(p+\frac{a}{v^2}\right)(v-b)=RT \tag{8.19}$$

Here a and b are constants that depend on the particular gas being approximated. Physically, a represents the strength of the hard core potential and b is an effective reduction in volume due to intermolecular attraction. For helium the values of these two constants are $a = 0.0340\ \mathrm{L^2 \cdot atm/mol^2}$ and $b = 0.0235$ L/mol. For the van der Waals gas the form of μ_j can be computed by differentiation and rearrangement of (8.19). Specifically, the volume derivative can be shown to be

$$\left(\frac{\partial v}{\partial T}\right)_p=\frac{R}{RT/(v-b)-(2a/v^3)(v-b)} \tag{8.20}$$

Substituting into (8.16) yields the equation for the Joule–Thomson coefficient,

$$\mu_j=\frac{1}{C_p}\left[\frac{(2a/RT)(1-b/v)^2-b}{1-(2a/vRT)(1-b/v)^2}\right] \tag{8.21}$$

The above equation can be used to compute values of μ_j for a van der Waals gas as a function of temperature and specific volume, albeit a tedious process. As has been mentioned above, it is of most interest to determine the location of the inversion temperature, below which the Joule–Thomson effect produces cooling. Since the inversion temperature

T_{inv} is defined at $\mu_j = 0$, its form for a van der Waals gas can be determined by setting the numerator of (8.21) to zero. Thus,

$$T_{\text{inv}} = \frac{2a}{bR}\left(1 - \frac{b}{v}\right)^2 \tag{8.22}$$

which appears to be a relatively simple solution except for the fact that v is coupled to T through the equation of state. In p-T space the inversion curve of a van der Waals gas is a polynomial function.

The most straightforward value to compute from the inversion curve is that of the maximum inversion temperatures T^m_{inv}. This value can be calculated from (8.22) by taking the low-density limit, $p \to 0$ or $v \to \infty$. For a van der Waals gas the solution is in closed form in terms of the constants a and b:

$$T^m_{\text{inv}} = \frac{2a}{Rb} \tag{8.23}$$

Substituting the values for helium into (8.23) yields the value of $T^m_{\text{inv}} =$ 35.2 K. Plotted in Fig. 8.7 is the inversion curve for helium.[4] Because of the small slope to the high-temperature isenthalps, it is difficult to determine accurately T^m_{inv}; however, for helium $T^m_{\text{inv}} \approx 45$ K, a quantity quite close to that determined for the van der Waals gas model.

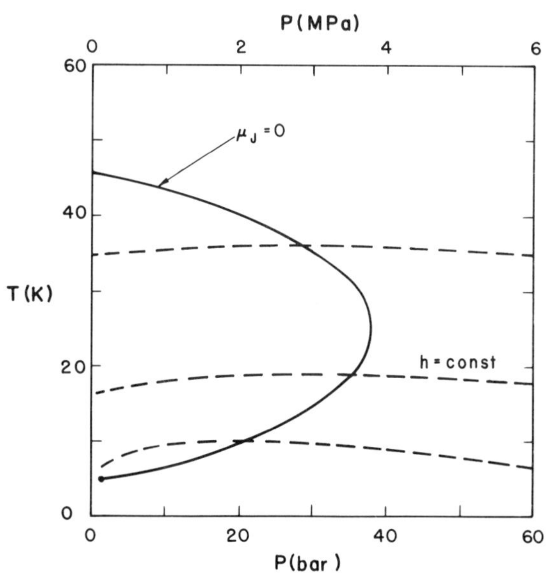

Fig. 8.7. Inversion curve for helium.

There is practical significance to the fact that the helium inversion curve lies below 45 K. To achieve cooling by isenthalpic expansion requires that the helium entering the Joule–Thomson valve be at a temperature substantially below this value. Hydrogen and neon are the only fluids with normal boiling points below 45 K, hydrogen being the most abundant of the two. Consequently, in the early development of helium liquefaction capability it was necessary first to liquefy hydrogen and then to use the hydrogen bath to precool the incoming helium flow. Subsequently, other methods were developed to allow liquefaction without LH_2 precooling. However, it still remains a challenge to precool the helium to below the inversion curve so that the Joule–Thomson effect can be used.

It is worth noting that the inversion curve for a van der Waals gas has a universal form dependent only on the strength of the intermolecular potentials as measured by the coefficients a and b. In general, the intermolecular potential strength can be scaled by the critical point. Therefore, a universal form for the inversion curve is possible if the equation of state can be scaled relative to the critical point. The law of corresponding states defines such a general equation of state in terms of the critical temperature T_c, pressure p_c, and volume v_c, that is,

$$\frac{p}{p_c} = \phi(T/T_c, v/v_c) \tag{8.24}$$

where ϕ is an undetermined function that relates these quantities.[8] The van der Waals gas is a special case of (8.24). For the law of corresponding states to apply, several assumptions are necessary. The law ignores quantum effect. As discussed in Chapter 3, helium has substantial quantum effects and would be expected to deviate from the law of corresponding states. A further assumption has to do with the symmetry of interaction. The law is established on the basis of symmetric molecular potentials. In fact, many molecules have very asymmetric bonding and configuration. A good example of an asymmetric potential is H_2O, but others include CO_2 and organic compounds. The law of corresponding states ignores these effects.

The law of corresponding states can be used to determine the coefficients in a particular equation of state. In the vicinity of the critical point, the isotherm have an inflection point; see, for example, Fig. 1.2. It follows that the equation of state must have a zero in its first and second derivative when evaluated at the critical point:

$$\left(\frac{\partial p}{\partial v}\right)_{T_c} = \left(\frac{\partial^2 p}{\partial v^2}\right)_{T_c} = 0 \tag{8.25}$$

This represents two equations which must be satisfied by the equation of state. For the van der Waals equation of state, there are two coefficients, a and b, that can be determined by the conditions set in (8.25). In terms of critical pressure and temperature, it is a straightforward calculation to obtain

$$a = \frac{27}{64}\frac{R^2 T_c^2}{p_c} \tag{8.26}$$

and

$$b = \frac{RT_c}{8p_c} \tag{8.27}$$

Listed in Table 8.2 are values of these coefficients appropriate for various gases of interest to cryogenics.

Since the Joule–Thomson coefficient is a function of the equation of state, the above arguments indicate that the inversion curve also should be a universal function provided it can be written in terms of reduced quantities, $\tau = T/T_c$ and $\pi = p/p_c$. For a van der Waals gas, the ratio of the maximum inversion temperature given by (8.23) to the critical temperature can be shown to have a constant value of 6.75. Furthermore, the entire curve can be shown to obey a reduced form displayed in Fig. 8.8.[9] This graph can be used to determine an approximate inversion curve for most common gases. It should be pointed out that the van der Waals equation of state is only one approximation to a real gas and others exist which give comparable shape to the inversion curve. Listed in Table 8.2 are the critical temperatures and maximum inversion temperatures for some cryogenic fluids. The reduced inversion temperature is approximately constant, with an increase for the case of helium where quantum effects would be expected to be largest.

Table 8.2. Critical Point Coefficients to the van der Waals Equation of State and Maximum Inversion Temperature for Various Cryogenic Fluids[a]

Fluid	T_c (K)	p_c (atm)	a ($L^2 \cdot$ atm/mol)	b (L/mol)	T_{inv}^m (K)	T_{inv}/T_c
He	5.2	2.275	0.0340	0.0235	43	8.3
H_2	33.2	12.98	0.243	0.0263	202	6.1
Ne	44.4	26.84	0.210	0.0170	260	5.9
Air	133.3	38.7	1.312	0.0354	603	4.5
N_2	126.0	33.5	1.354	0.0387	623	4.9
Ar	150.8	48.3	1.345	0.0321	723	4.8
O_2	154.6	50.1	1.363	0.0317	761	4.9

[a] See Refs. 7 and 10.

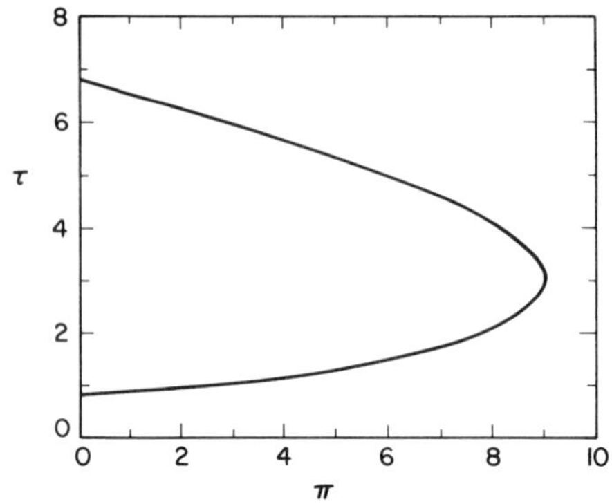

Fig. 8.8. Inversion curve for a van der Waals gas in terms of reduced pressure ($\pi = p/p_c$) and temperature ($\tau = T/T_c$).

Isenthalpic expansion for cooling of cryogenic fluids has a number of advantages and disadvantages. What makes the method so useful is its ease of application. No moving parts are required and the J-T value is a relatively small component. A further advantage is that there is no problem with expanding into the region of two-phase coexistence. Thus, the common application of a J-T value is at the final stage of a helium liquefaction system where the liquid actually is produced. It is not possible to achieve helium liquefaction with only the Joule–Thomson effect unless the incoming fluid is below about 40 K. Thus, additional methods are necessary to combine with the J-T expansion stage. These methods may include precooling with another cryogenic fluid or adiabatic expansion of the incoming fluid requiring it to do work. The latter method has definite advantages over the J-T expansion in that it approaches a constant entropy process and therefore produces a larger temperature difference for the same expansion pressure drop.

8.4. JOULE–THOMSON LIQUEFIER

The simplest practical method to achieve liquefaction of helium is by direct application of the Joule–Thomson effect. For this method to work, it is necessary to begin below the inversion curve, which for helium implies an initial temperature below about 40 K. This limitation will be overlooked for the time being. Future discussion will indicate how to bring the fluid below the inversion temperature initially. The Joule–Thomson liquefier is sometimes referred to as a simple Linde–Hampson system.[7] This system provides a simple inexpensive method for liquefaction of a gas whose inversion temperature is above ambient.

The principal components of the Joule–Thomson liquefier are shown in Fig. 8.9. In addition to the compressor stage, the system requires two

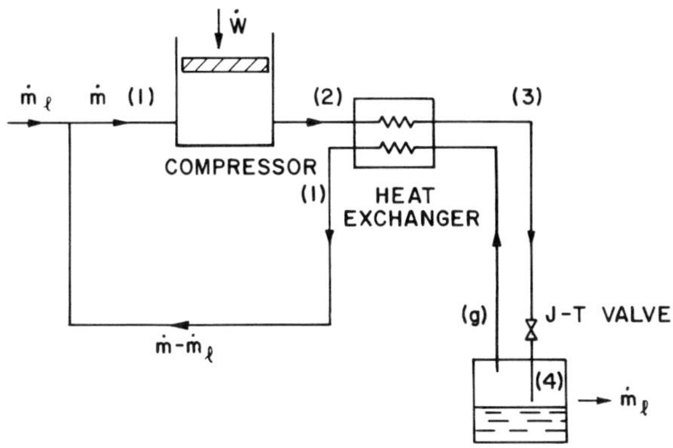

Fig. 8.9. Schematic diagram of Joule–Thomson liquefier.

components: a heat exchanger and a Joule–Thomson value. The heat exchanger, simplistically shown in Fig. 8.9, is actually a rather complex component of design. A good heat exchanger has a number of desirable traits, all of which require careful attention to detail:

1. Maximum surface area for heat transfer.
2. Minimum flow resistance to fluid (Δp small).
3. Minimum mass (for cooldown).

Of the three characteristics listed, 1 and 2 are at least moderately counteractive, and therefore good design involves trade-offs between these two characteristics. It is necessary to be familiar with heat transfer and fluid flow characteristics of helium before discussing heat exchanger design. Chapter 7 considers heat transfer and flow characteristics of helium in some detail. For the present discussion, it will be assumed that the heat exchanger is an ideal component, meaning that there is only heat transfer between the high- and low-pressure gas flows. Furthermore, friction pressure drop across the heat exchanger will be neglected. It should be noted that the above assumptions can be substantially different from isothermal conditions, where the temperatures of the two gas streams are the same. This distinction will become more clear as examples are considered. In the thermodynamic language of cryogenic systems, heat exchangers are assumed to conserve enthalpy.

The other major component in the simple liquefier shown in Fig. 8.9 is the Joule–Thomson valve. As described in the previous section, the J-T valve performs an isenthalpic expansion of the high-pressure stream. Provided the inlet temperature is below the inversion curve, the J-T expansion can produce a two-phase mixture of liquid and vapor helium. The yield of this expansion stage is determined by an enthalpy balance between the incoming high-pressure stream and the two coexisting phases at ambient pressure.

Another way of looking at the yield of a Joule–Thomson refrigerator is by reference to a *T-S* diagram; see Fig. 8.10. Here the points on the flow chart in Fig. 8.9 are indicated. The compressor stage involves step 1–2. A truly isothermal compression would follow a horizontal line on the *T-S* diagram. Also shown in the figure is a more realistic five-stage real compression process that actually consists of an isentropic compression followed by an isobaric heat transfer to the reservoir. The reasons for this multistage process are discussed later. The compression stage is followed by an isobaric process which brings the gas from 2 to 3, reducing its temperature and entropy. The cooling used to bring the gas from 2 to 3 is transferred from the return low-pressure stream through the heat exchanger where the exiting gas is brought back to ambient conditions. Point 3 represents the inlet to the J-T valve. Step 3–4 is then an isenthalpic expansion into the two-phase coexistence region. In principle, it is easy to determine the liquefaction yield y from this diagram by using the two-phase lever rule, where the y is given by the ratio of the length 4 to g divided by ℓ to g. The fraction of gas which is not liquefied follows the isobaric return stage g to 1, where its enthalpy is used to cool the incoming fluid through the heat exchanger. Finally, since the return gas has a smaller mass flow by the amount liquefied, in advance of recompression there is a make-up flow added to the stream to maintain $\dot{m}$ constant.

The above description is a qualitative indication of the type of analysis that is possible for the Joule–Thomson liquefier. To make a more quan-

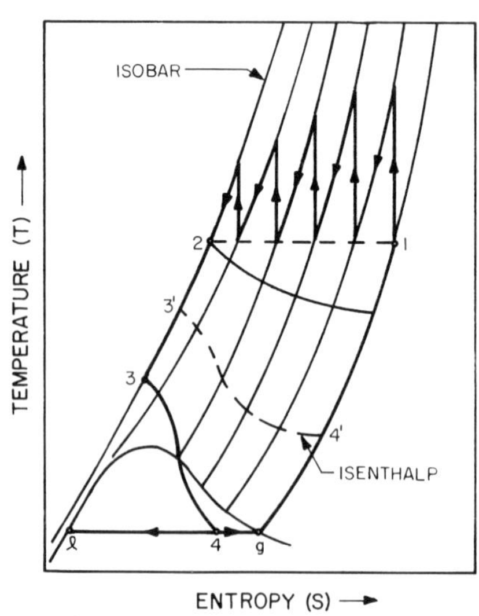

Fig. 8.10. *T-S* diagram showing isenthalpic expansion processes.

titative determination of the liquefaction fraction or yield of this system, it is necessary to invoke the first law of steady flows. For simplicity, assume no work or external heat is applied to the system and that the heat exchanger conserves enthalpy. With this set of assumptions, the analysis requires writing down enthalpy conservation for the combination of the heat exchanger, Joule–Thomson valve, and liquid yield, that is,

$$\dot{m}h_2 = (\dot{m} - \dot{m}_l)\, h_1 + \dot{m}_l h_l \tag{8.28}$$

where $\dot{m}_l$ is the liquid mass flow rate and h_l is the specific enthalpy of the liquid. Defining the yield or fraction liquefied as $y = \dot{m}_l/\dot{m}$, we can rearrange to give an expression for this value:

$$y = \frac{h_1 - h_2}{h_1 - h_l} \tag{8.29}$$

It is apparent from (8.29) that the working fluid must be below the inversion curve for the fluid to cool through the J-T valve. Unless the specific enthalpy of the high-pressure stream, h_2, is lower than that of the low-pressure side, h_1, the quantity described by the right-hand side of (8.29) will always be negative, indicating that no liquid yield is possible. Note that the yield is only a function of the fluid properties and the compression ratio p_2/p_1. For a given fluid and initial temperature, the pressure is the only quantity which may be varied in the optimization process. This is a standard procedure for determining the best operating conditions of a particular system.

The maximum yield for the simple Joule–Thomson refrigerator occurs when h_2 in (8.29) is a minimum. This is a result of h_1 and h_l being fixed by initial conditions. The minimum in h_2 is found by taking its derivative with respect to pressure and setting it equal to zero:

$$\left(\frac{\partial h_2}{\partial p}\right)_T = 0 \tag{8.30}$$

Referring to the definition of the Joule–Thomson coefficient, (8.11), note that the above derivative can be written in terms of μ_j:

$$\left(\frac{\partial h}{\partial p}\right)_T = -\mu_j C_p \tag{8.31}$$

Therefore, the best place to begin an isenthalpic expansion is on the inversion curve, where $\mu_j = 0$.

To attach numerical significance to the above description, consider a simple Joule–Thomson refrigeration system operating such that points 1 and 2 were fixed at 20 K. Although this seems like an unphysical system, it could be achieved by precooling the helium with a liquid hydrogen bath. The first question to ask is: What is the best operating pressure for p_2? The answer of course is derived by reference to the inversion curve, Fig. 8.7. Here at 20 K, the pressure corresponding to the inversion curve is about 3.8 MPa (38 bars), a not unreasonable value for a helium compressor. Given the initial temperature and pressure, it is a straightforward process to determine the yield by substituting the tabulated values for the specific enthalpies into (8.29):

$$h_1(20\text{ K},\ 1\text{ bar}) = 117.9\text{ J/g}$$

$$h_2(20\text{ K},\ 38\text{ bars}) = 106.8\text{ J/g}$$

$$h_l(4.2\text{ K},\ 1\text{ bar}) = 9.7\text{ J/g}$$

The result of that calculation gives $y = 0.10$.

The yield of the refrigeration system is not the only factor to be considered when evaluating the quality of design. In particular, it is important to have a thermodynamically efficient process in addition to a high yield. For an idealized system such as the one considered here, it is possible to evaluate the work required to produce a unit mass of liquid and compare this value with the thermodynamically ideal process described earlier. To perform this calculation, assume that the various thermodynamic processes are ideal. The work required for liquefaction is entirely that of the compressor which operates at constant temperature. The first law of steady flows, when applied to the compressor in the Joule–Thomson refrigerator, gives an expression for the compressor work $\dot{W}_c$,

$$\frac{\dot{W}_c}{\dot{m}} = T(s_1 - s_2) - (h_1 - h_2) \tag{8.32}$$

where $\dot{m}$ is the total mass flow. To cast (8.32) in a form that is better compared with an ideal system, substitute the definition of yield,

$$y = \frac{\dot{m}_l}{\dot{m}} = \frac{h_1 - h_2}{h_1 - h_l}$$

A general expression for the work per unit mass liquefied by this process is then

$$\frac{\dot{W}_c}{\dot{m}_l} = \left(\frac{h_1 - h_l}{h_1 - h_2}\right) [T(s_1 - s_2) - (h_1 - h_2)] \tag{8.33}$$

Rather than work out the value for (8.33) given a particular system, it may be more instructive to compare it with the ideal liquefaction system given by

$$\frac{\dot{W}_i}{\dot{m}_l} = T(s_1 - s_l) - (h_1 - h_l) \tag{8.34}$$

Then for the Joule–Thomson liquefier, the ratio of these two quantities, (8.33) and (8.34), gives the figure of merit,

$$\text{FOM} = \left(\frac{h_1 - h_2}{h_1 - h_l}\right)\left(\frac{T(s_1 - s_l) - (h_1 - h_l)}{T(s_1 - s_2) - (h_1 - h_2)}\right) \tag{8.35}$$

For the example considered above, where a helium Joule–Thomson liquefier is initially at 20 K, with the high-pressure side at 38 bars, the entropy of each state is known:

$$s_1(1\text{ bar, } 20\text{ K}) = 17.32\text{ J/g}\cdot\text{K}$$

$$s_2(38\text{ bars, } 20\text{ K}) = 9.18\text{ J/g}\cdot\text{K}$$

$$s_l(1\text{ bar, } 4.2\text{ K}) = 3.45\text{ J/g}\cdot\text{K}$$

Using these values fixes the figure of merit, FOM = 11.4%. It should be noted that this value only takes into consideration the liquefaction between 20 and 4.2 K. To consider the overall efficiency of the process, it would be necessary to include the stage operating between 300 and 20 K. The above proviso makes the calculation of the work per unit mass liquefied ($\dot{W}_c/\dot{m}_l$) from (8.33) incomplete as well.

A number of factors contribute to inefficiencies in the system, tending to reduce the FOM below what is calculated for an ideal system. With respect to the compression stage, it must be borne in mind that a truly isothermal compressor is not realizable. Real compressors actually operate in a two-step process, the first being an adiabatic compression followed by isobaric heat exchange with a reservoir. This process is shown schematically in step 1–2 of Fig. 8.10. The isothermal compression is approximated by a multistage two-step process. The necessity to provide a multistage process occurs because the adiabatic compression causes rather large temperature increases in the gas which can cause damage to the equipment. Furthermore, a multistage process more closely approaches an isothermal process, as can be seen in the *T-S* diagram, and consequently is more thermodynamically efficient.

The problem of temperature rise in the adiabatic compression stage is demonstrated most dramatically by reference to an example. Assume it were desirable to achieve a single-stage compression of helium between 1 and 10 bars at 300 K. This would consist of an adiabatic compression stage followed by an isobaric heat exchange. Equating the entropy before and after compression,

$$s_1(330\text{ K},\ 1\text{ bar}) = s_2(T_2,\ 10\text{ bars})$$

determines the maximum temperature T_2. The result gives a rather high temperature at the end of compression, $T_2 \approx 750$ K, which can cause a number of undesirable effects including physical damage to the compressor and its lubricants. For efficiency as well as hardware limitations, it is desirable to have multistage compression.

The above discussion assumes that the liquefaction system already has achieved the steady state and consequently all temperatures 1 through 4 are constant. In practice, the cooldown process is of equal concern. The partial cooldown of a simple Joule–Thomson liquefier may be represented on the *T-S* diagram by the cycle 1–2–3′–4′. During this stage of the operation, there would be no liquefaction and the mass flow would be constant throughout. The net effect of continuing the cooldown would be to reduce gradually 3′ to 3 and 4′ into the two-phase region to 4. Naturally, the rate at which the cooldown takes place depends on the thermal inertia of the system. This fact points out the desirability of minimizing the mass of the refrigeration components, which is trait 3 of the characteristics of good refrigerator and heat exchanger design.

8.5. ISENTROPIC EXPANSION

A thermodynamically more efficient method to achieve cooling of a liquid–gas system is to make the fluid do work against an external engine. This kind of expansion can be adiabatic and reversible, thereby approaching an isentropic process, $\Delta s = 0$. Since the Carnot cycle is comprised of isothermal and isentropic stages, it is clear that the expansion by the performance of work is more efficient than simple isenthalpic expansion.

To determine the effectiveness of an expansion process for the production of cooling it is useful to define a parameter similar to the Joule–Thomson coefficient. This parameter, referred to as the isentropic expansion coefficient μ_s, is written in terms of the slope of an isentrope in pressure–temperature space:

$$\mu_s = \left(\frac{\partial T}{\partial p}\right)_s \tag{8.36}$$

As before, this quantity is defined in terms of common thermodynamic variables. By using an identity of partial differential equations and a Maxwell relation, the expression for μ_s becomes

$$\mu_s = \frac{T}{C_p}\left(\frac{\partial v}{\partial T}\right)_p \tag{8.37}$$

where use has been made of the definition

$$C_p = T\left(\frac{\partial s}{\partial T}\right)_p$$

Comparison of (8.37) with (8.16) indicates that a simple relationship exists between the Joule–Thomson coefficient μ_j and the isentropic expansion coefficient μ_s:

$$\mu_j = \mu_s - \frac{v}{C_p} \tag{8.38}$$

This latter expression clearly shows that isentropic expansion is a more thermodynamically efficient process, since v/C_p is always a positive quantity. In a practical sense, (8.38) implies that to begin with a fluid in a given state and to performing either isentropic work or isenthalpic expansion over the same change in pressure, the resulting temperature decrease will be larger for the isentropic process.

By application of the ideal gas law to (8.37) or by recalling that $\mu_j = 0$ for an ideal gas, it can be shown easily that the isentropic expansion coefficient $\mu_s = v/C_p$ is always positive. Thus, the isentropic process is not dependent on the nonideality of gases to achieve cooling. For a real gas, μ_s can be either greater or less than the value computed for an ideal gas. In particular, for the case of a van der Waals gas, applying the equation of state (8.19) to (8.37) yields an expression for μ_s,

$$\mu_s = \frac{v}{C_p}\left[\frac{1 - b/v}{1 - (2a/vRT)(1 - b/v)^2}\right] \tag{8.39}$$

where consistent with the initial statement, the quantity in brackets can be either greater or less than unity.

8.6. CLAUDE LIQUEFIER

The present section summarizes how to employ the performance of work as part of a helium liquefaction or refrigeration system. Since the maximum inversion temperature for helium is about 40 K, the performance of work is an essential component of an operating helium refrigerator. Without this procedure it would be possible to liquefy helium only by initially precooling the gas with either liquid hydrogen ($T \approx 20$ K) or liquid neon ($T \approx 27$ K) temperatures. This method is cumbersome and probably not very efficient. Fortunately, by diverting some of the high-pressure gas flow through an expansion engine, near-isentropic processes can be used to precool the helium below T_{inv}.

It is most instructive to deal with this subject by way of a specific system. The Claude system as it was first developed provides a good working example.[3,7,10] A flow scheme for the Claude liquefaction system is shown in Fig. 8.11. The system consists of a compressor, three heat exchangers with an expansion engine operating on the second, and a Joule–Thomson valve. The purpose of the expansion engine is to divert a fraction of the incoming high-pressure gas stream performing work as the gas expands into the low-pressure side. Modern liquefiers and refrigerators are more complex than the Claude system but use similar combinations of expansion processes. Two coupled expansion circuits, such as the Claude system provides, have several advantages over the simple Joule–Thomson refrigeration scheme. First, it has the potential of being more efficient because part of the process is isentropic with inherently higher thermodynamic efficiency. Second, the efficiency can be improved by taking advantage of the work produced by the expansion circuit, an option that makes sense particularly for large systems. Finally, since there are now two or more coupled flow circuits, the main cooling circuit and that through the expander, it is possible to vary the fraction of the flow to achieve higher performance characteristics.

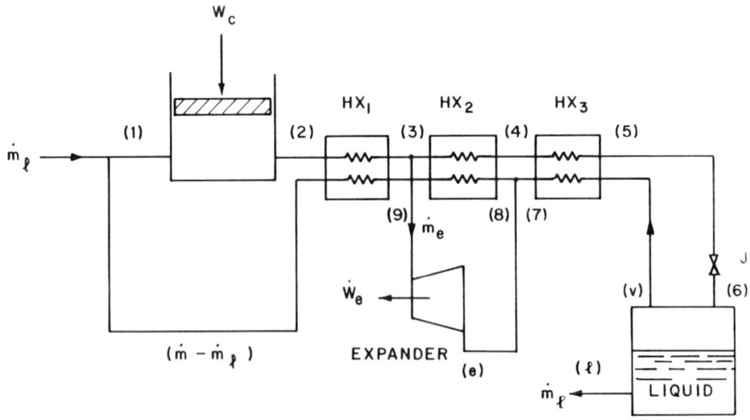

Fig. 8.11. Flow scheme for Claude liquefaction system.

The Claude cycle is displayed on a T-S diagram in Fig. 8.12. Note that the points as indicated on the flow chart are also shown in Fig. 8.12. Point (3) is particularly interesting because it indicates the position where the two circuits separate. The isentropic expansion reduces the temperature to point (e). The higher-pressure stream continues until point (5) where the J-T valve produces an isenthalpic expansion into the two-phase region. Thus, in the Claude system there are two free parameters to select, the high-pressure value, p_2, and the fraction of gas through the expansion engine circuit, $\dot{m}_e/\dot{m} = x$. To see how these parameters enter the calculations of yield and net work, it is necessary to consider the thermodynamics of both processes.

We apply the first law of steady flows to the entire system and assume that the expander work is recoverable; the energy balance through the main circuit then yields

$$-\frac{\dot{W}_e}{\dot{m}} = (1-y)\,h_1 + yh_l - h_2 \tag{8.40}$$

where as before $y = \dot{m}_l/\dot{m}$. The expander work $\dot{W}_e$ enters (8.40) in the negative sense because the gas is doing work against an external load. In addition to the above expression, there exists a separate independent equation for the expansion circuit only,

$$-\frac{\dot{W}_e}{\dot{m}} = x(h_e - h_3) \tag{8.41}$$

where the fraction of flow through the expansion circuit, $x = \dot{m}_e/\dot{m}$, is an adjustable parameter. The above two equations allow the determination of

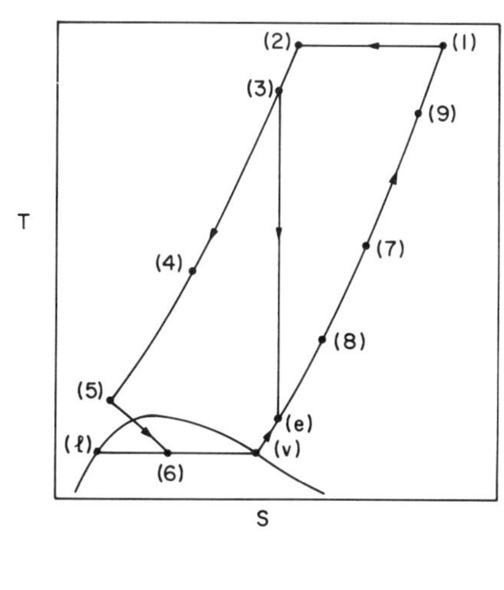

Fig. 8.12. T-S diagram for operating Claude system.

the net yield y and the thermodynamic efficiency or FOM of the Claude system.

The yield is independent of whether the expander work is used in the thermodynamic cycle. An expression for y can be established by simply equating (8.40) and (8.41) which gives

$$y = \left(\frac{h_1 - h_2}{h_1 - h_l}\right) + x\left(\frac{h_3 - h_e}{h_1 - h_l}\right) \tag{8.42}$$

The first term in (8.42) is that of the simple Joule–Thomson liquefier. As has already been dicussed, this quantity is only positive when $T < T_{\text{inv}}$. However, the second term is always positive and so it is possible, based on suitable choices of x and p_2, to have a positive yield at temperatures above the inversion curve. Two cautionary steps must be exercised when evaluating the yield from (8.42). First, conservation of mass is not a component of this analysis. Based only on (8.42), it appears that to maximize the yield x should take on its maximum value. Conservation of mass, however, requires that $y + x \leqslant 1$. Furthermore, it is unreasonable to set $y + x = 1$ because this would allow no mass flow on the low-pressure side of the last heat exchanger, a physically undesirable situation since the heat exchanged is cooled by this return flow. Subject to the above restrictions, the yield appears to increase monotonically with x.

The second step which must be taken when evaluating the yield of a system that includes an expander is to establish the absolute value of the enthalpy at either the inlet or outlet of the engine. Since the pressure drop is given, the absolute values of h_e or h_3 must be known to establish $h_3 - h_e$. This quantity should be maximized to achieve maximum yield. Obviously, h_e and/or h_3 are not free parameters and are estalished by absolute temperatures at (3) and (e). As can be seen in Fig. 8.12, T_3 is always less than T_2. However, the absolute value of T_3 cannot be established from what has been presented so far. Rather, it requires some complex analysis of heat transfer and fluid flow for a particular system, both of which are beyond the scope of the present discussion. The examples presented below require some assumptions for these values. However, a particular assumption can be checked against thermodynamic laws, which prevent an efficiency in excess of Carnot.

In addition to the yield of the Claude system, it is useful to determine the net work involved in liquefying a unit mass of fluid. Assuming that the net work is equal to the difference between the compressor work and expander work ($\dot{W} = \dot{W}_c - \dot{W}_e$), we obtain

$$\frac{\dot{W}}{\dot{m}} = [T(s_1 - s_2) - (h_1 - h_2)] - x(h_3 - h_e) \tag{8.43}$$

The first term is identical to that of the Joule–Thomson refrigerator, while the second term is the net gain achieved by using the expander work. As with the yield, to minimize the net work the second term should be optimized. Clearly, the net work for the Claude system must be greater than the ideal expansion system allowing the assumptions applied to (8.43) to be checked for correctness.

To evaluate the practical aspects of this discussion, consider the use of the Claude cycle for the liquefaction of helium. Here for the sake of clarity, assume all processes are thermodynamically ideal. The Claude system is as described above and displayed schematically in Fig. 8.11. The first parameter which must be selected is the value of the high-pressure side, p_2. Two practical considerations enter here. First, whether the hardware can withstand the pressure difference particularly in terms of compressor and expander pressure drop. Second, since the last step of the process is through a J-T valve, it probably makes little sense to have p_2 above the inversion curve. With these factors under consideration, a realistic value to choose in this example is $p_2 = 20$ bars. Therefore, the thermodynamic states at three points on the cycle are established:

(1) Low-pressure return side:

$$p_1 = 1 \text{ bar}$$
$$T_1 = 300 \text{ K}$$
$$s_1 = 31.4 \text{ J/g} \cdot \text{K}$$
$$h_1 = 1573 \text{ J/g}$$

(2) High-pressure inlet side:

$$p_2 = 20 \text{ bars}$$
$$T_2 = 300 \text{ K}$$
$$s_2 = 25.2 \text{ J/g} \cdot \text{K}$$
$$h_2 = 1597 \text{ J/g}$$

(l) Saturated liquid state:

$$p_l = 1 \text{ bar}$$
$$T_l = 4.2 \text{ K}$$
$$s_l = 3.5 \text{ J/g} \cdot \text{K}$$
$$h_l = 9.7 \text{ J/g}$$

As has been stated above, there are two values to select in determining the yield and net work, namely, x and T_3. Based on (8.42), a desirable choice is a maximum value for x; however, mass conservation sets an upper limit for $y = 1 - x$. For the present discussion assume $x = 0.33$ and $T_3 = 200$ K which establishes the thermodynamic state at (3):

(3) Inlet to expansion engine:

$$p_3 = 20 \text{ bars}$$
$$T_3 = 200 \text{ K}$$
$$s_3 = 23.1 \text{ J/g} \cdot \text{K}$$
$$h_3 = 1060 \text{ J/g}$$

The expansion engine performs an isentropic expansion on the gas, defining the thermodynamic state of the exiting gas ($s_e = s_3$):

(e) Exit from expansion engine:

$$p_e = 1 \text{ bar}$$
$$T_e = 60 \text{ K}$$
$$s_e = 23.1 \text{ J/g} \cdot \text{K}$$
$$h_e = 326 \text{ J/g}$$

With the above four thermodynamic states and the value of x fixed, it is a straightforward computation to determine y by substitution of numerical values into (8.42). This calculation results in a yield of $y = 0.15$. Note that once T_3 is selected, the yield is directly proportional to x. For this set of parameters $x + y = 0.48$.

It is also an easy calculation to determine the net work for the Claude system and to compare this value with a thermodynamically ideal system. Since (8.43) is written on a per unit mass compressed basis ($\dot{W}/\dot{m}$), it is necessary to divide by the yield from (8.42) to get the work per unit mass liquefied, which is more directly comparable with an ideal system. For the example above, the net work per unit mass of liquid can be determined:

$$\frac{\dot{W}}{\dot{m}_l} = \frac{1}{y}\left(\frac{\dot{W}}{\dot{m}}\right) = 10{,}747 \text{ J/g}$$

This value has little physical significance by itself but does when compared with the thermodynamically ideal case ($\dot{W}_i/\dot{m}$):

$$\frac{\dot{W}_i}{\dot{m}_l} = T(s_1 - s_l) - (h_1 - h_l) = 6865 \text{ J/g}$$

The ratio of these two values gives the figure of merit for this particular Claude system:

$$\text{FOM} = \frac{\dot{W}_i}{\dot{W}} = 64\,\%$$

Determination of the figure of merit represents a thermodynamic check on the correctness of the above assumptions. If the FOM computed above worked out to be greater than unity, it would be in violation of the second law of thermodynamics. Nothing in the analysis restricts this from happening. Clearly, a complete calculation which must be done numerically has built into it a check to prevent this unphysical process from occuring. The above example indicates the advantages of performance of work on an ideal liquefaction system. Although the Claude system is very simple in its introduction of this concept, it accrues advantages. In fact, modern refrigerators and liquefiers of helium use cycles employing the same basic principles. As the systems evolve into complexity, the ability to evaluate their performance analytically diminishes, leading to an increased number of numerical assumptions.

8.7. OTHER HELIUM LIQUEFACTION SYSTEMS

Numerous helium liquefaction systems have been developed in an attempt to improve thermodynamic efficiency or reliability. In general, these systems use the same concepts introduced for the simple Joule–Thomson liquefier and the Claude system. Since these systems contain considerably more complexity than those already discussed, only their general nature will be surveyed. Actual numerical analysis of these other systems is beyond the level of the present discussion. This section consists of a brief survey of the principles behind three helium liquefaction systems: cascade, Collins, and Stirling cycle.

8.7.1. Cascade System

The use of the intermediate-temperature fluids in the liquefaction of low-temperature liquids has been an established technique since the beginning of cryogenics. In fact the concept was used by Pictet in the first liquefaction of O_2. The cascade liquefaction system, which is a special case of this concept, consists of a series of closed-cycle systems each using the change of state from liquid to gas to achieve cooling.[3,7] Thus, a true cascade system employs working fluids with overlapping two-phase coexistence regions. The cascade system can have high thermodynamic efficiency but it is generally cumbersome because of the number of stages required. Furthermore, it is not possible to have a true cascade liquefier which produces liquid helium or even liquid hydrogen and neon because the critical points of these fluids are below the triple point of the next-higher-temperature liquid. Some aspects of the cascade system are applied in the liquefaction of helium. Precooling of the gas in a liquefier consisting of heat exchangers, expansion engines, and a Joule–Thomson valve, cannot be considered a true cascade system, but the benefits are similar. Helium liquefiers based on the Claude system or a modification of it have considerably higher liquefaction yield when liquid nitrogen precooling is included in the system.

The application of such a hybrid cascade system to the liquefaction of helium is displayed in Fig. 8.13. The flow diagram is for a helium liquefier using this method combined with a simple Joule–Thomson liquefier. The system consists of several circuits, each containing a different working fluid. All circuits, except the helium liquefier, operate in a closed-cycle mode, where the liquid simply provides the refrigeration for the next lower stage. Since this system uses the Joule–Thomson effect, it is only necessary for the working fluid to be below its inversion temperature before cooling can occur. Consequently, each working fluid can have a wider temperature span than with a conventional cascade system. The working fluids must have inversion temperatures above the boiling point of the next-higher working fluid in the system. In the case displayed in Fig. 8.13, the helium is precooled by a closed-cycle hydrogen liquefier. In turn, since the inversion temperature of hydrogen is 202 K, it can be precooled by liquid nitrogen or a similar fluid with $T_{bp} < 202$ K. In this circuit it is possible to produce the liquid necessary to precool the hydrogen circuit with a simple nitrogen

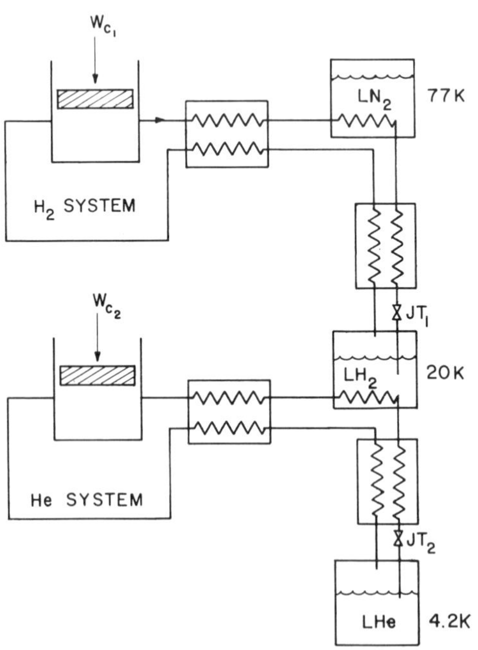

Fig. 8.13. Flow diagram for hybrid cascade system for liquefaction of helium.

cycle, since T_{inv} of nitrogen is above room temperature. In principle, the above described hybrid cascade system would be expected to produce liquid helium, although probably with low thermodynamic efficiency because of the compounding effect of numerous Joule–Thomson liquefiers, all of which operate at rather low figures of merit.

8.7.2. Collins Helium Liquefaction System

In the 1940s the Collins helium liquefier evolved out of an effort to produce a commercial system using expansion engines to precool several heat exchangers.[7,11] The Collins system is actually quite similar to the Claude system except that the former uses between two and five expansion engines rather than just one. A schematic diagram of a two-engine Collins system is shown in Fig. 8.14. In designing the Collins system, there are a greater number of degrees of freedom over that of the Claude system. Depending on the number of expansion circuits, i, there are an equal number of expansion circuit mass flows to select, $\dot{m}_{e_i}$. Furthermore, the

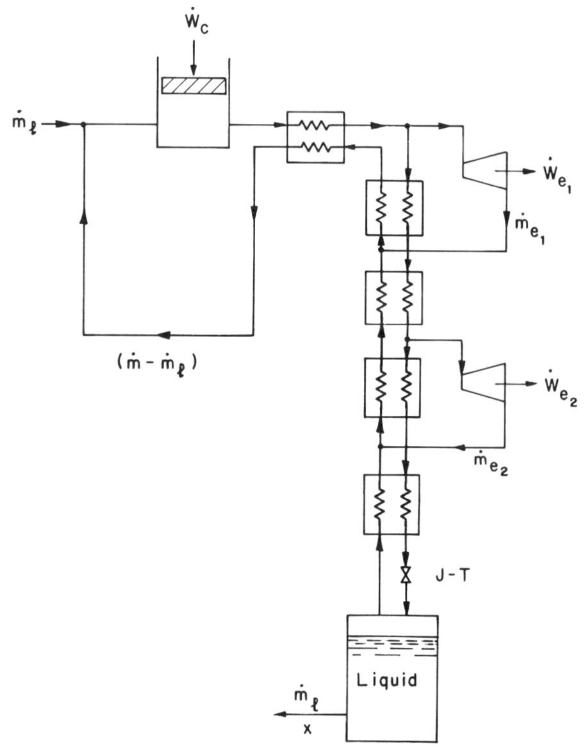

Fig. 8.14. Schematic diagram of Collins liquefaction system.

temperature at the inlet to each expansion engine must be determined to compute the yield. Once these quantities are known it is a straightforward computation to determine the yield for the Collins system,

$$y=\frac{h_1-h_2}{h_1-h_l}+x_1\frac{\Delta h_{e_1}}{h_1-h_l}+x_2\frac{\Delta h_{e_2}}{h_1-h_l}+\cdots \tag{8.44}$$

where Δh_{e_i} is the enthalpy change through the ith expansion engine and x_i is the mass flow ratio through the ith circuit. Conservation of mass requires that $y+\sum_i x_i \leqslant 1$.

The first Collins liquefiers were developed with two expansion engines. Typical working temperature ranges for the expanders are from 60 to 30 K and 15 to 8 K, respectively. During steady-state operation the approximate values for the expander mass flow fractions are $x_1=0.30$ and $x_2=0.55$ at 15 bars. Insertion of these numerical values into (8.44) leads to a yield of only 3.6%. However, it has been demonstrated that the liquid yield could be improved by a factor of 2–3 by liquid nitrogen precooling.

8.7.3. Stirling Cycle

An alternative concept for the production of liquid helium has evolved out of the application of a heat engine cycle invented by Stirling in 1827.[11–15] Operated in reverse, the Stirling cycle produces cooling. The Phillips Company was the first to produce liquefiers that operated on this principle. The cycle consists of two isothermal and two isochoric (constant volume) processes. Cooling is achieved by adiabatic expansion of the working fluid, but instead of counterflow heat exchangers, the Stirling cycle employs a regenerator. The regenerator is a porous material used for

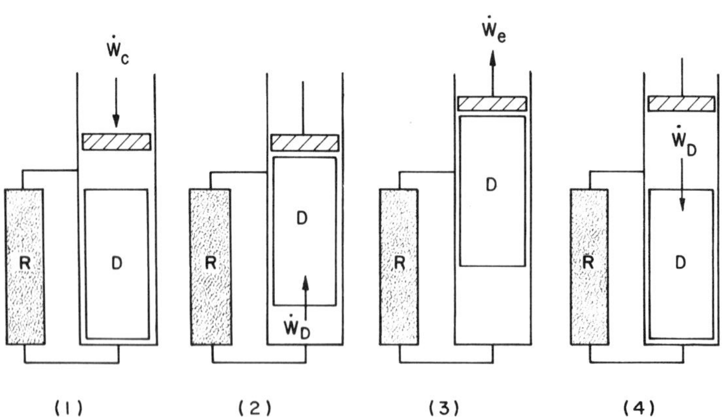

Fig. 8.15. Stirling cycle refrigerator.

storing the thermal inertia of the system. A schematic representation of the Stirling system is shown in Fig. 8.15. There are three main components: a regenerator (R), compressor, and displacer (D). The displacer works in consort with the compressor to provide low-temperature compression and expansion processes.

By separately moving the compressor and displacer, it is possible to achieve compression or expansion of the working fluid in the upper or lower region of the chamber. The four stages displayed in Fig. 8.15 outline the procedure whereby these components produce cooling. Once the process has achieved the steady state it can be described in terms of p-V and T-S diagrams as in Fig. 8.16. At (1) the helium is compressed but resides in the upper compression chamber since the displacer is at its lowest position. From (1) to (2), the displacer is moved to its upper position forcing the fluid through the regenerator into the expansion chamber. This process takes place at constant volume. The pressure decreases because the gas is cooled by the regenerator. Step (2)–(3) is an expansion of the fluid in the lower chamber achieved by moving the displacer upward with the compressor. This process isothermally extracts heat from the regenerator. Between (3) and (4) the displacer is returned to its lowest position, forcing the cold fluid back through the regenerator to the compression chamber. The cold is stored in the regenerator and the working fluid is recompressed to (1) isothermally. Because of the periodic nature of the cycle, cooling is not continuous. However, if the regenerator has sufficient heat capacity, temperature fluctuations at the low end can be minimized. Configured as a liquefier, the regenerator would act as a condenser. The liquid is then siphoned off and only the gaseous phase recycled to the high-pressure point (1). There would be some make-up gas administered at (1) to maintain the flow constant.

The regenerator is a component not present in conventional refrigeration schemes. It performs a similar function to the counterflow

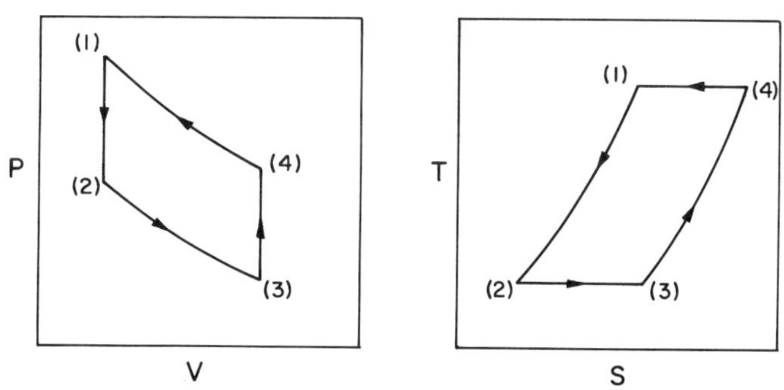

Fig. 8.16. p-V and T-S diagrams for Stirling cycle.

heat exchanger in a continuous refrigerator. A properly designed regenerator should have a number of characteristics:

1. Minimum flow resistance.
2. Minimum longitudinal heat conductance.
3. Maximum heat transfer between the fluid and the packing material.
4. Large specific heat of the packing material.

The desirable characteristics of a regenerator appear similar to those of good heat exchangers, with the exception of the large specific heat. This requirement is particularly difficult in the application of the Stirling cycle to helium liquefaction or refrigeration because most solid materials have rather low specific heats at liquid helium temperatures. Metals become unusuable below about 10 K. Special materials are being developed to allow the Stirling cycle to be used at these lower temperatures. At present, the major interest in this refrigeration technique is to provide closed-cycle cooling at intermediate temperatures, $T \approx 15$ K, using helium gas as a working fluid.

For an ideal Stirling cycle, the total refrigeration resulting from the expansion of a gas from p_1 to p_2 is given by the relationship

$$\dot{Q}_c = nRT_c \nu \ln \frac{p_1}{p_2} \tag{8.45}$$

where T_c is the expansion temperature, n is the total number of moles of the working fluid, and ν is the cycle frequency. Ideally, the cooling rate is linearly proportional to the low temperature and the frequency. Frictional losses and other nonidealities increase with frequency, providing optimum performance parameters for a practical cycle.

8.8. CLOSED-CYCLE REFRIGERATION

Until this point, the discussion has concentrated on systems whose purpose is to liquefy helium. Thus, the principal questions that have been asked are: What is the yield of a particular liquefaction cycle and what is its thermodynamic efficiency? In either case these questions relate only to the production of helium. However, these considerations are not always the most important when designing a helium refrigeration system which is to provide cooling to a particular device. In practice, many large systems have, as part of their standard operation, a closed-cycle refrigerator for continuous cooling. This approach is desirable because it achieves a higher thermodynamic efficiency and dedicated operation, the closed-cycle nature reduces the probability of helium vapor losses. The advantages of a closed-cycle refrigerator in terms of thermodynamic efficiency can be seen most clearly by a single but instructive example.

There are basically two methods of cooling a system to liquid helium temperatures. One method is shown schematically in Fig. 8.17a. Here the cooling of the device requires two stages. First, helium is liquefied and stored in a container. Subsequently, or possibly simultaneously, helium is transferred from the storage container to the system. Heat generated in the system Q_c and absorbed at low temperatures must be removed by the latent heat of the liquid helium. The enthalpy of the evaporated helium is lost because the heat is being generated at low temperatures. Certainly, if the heat Q_c were the result of conduction from the external environment, the situation would be considerably different. Here the vapor can cool the structure continuously, thereby reducing the total heat absorbed by the liquid helium. Liquid helium has a relatively small latent heat λ, being 21 J/g at 4.2 K. This value is equal to 0.73 W · h/L at saturated vapor pressure densities. A useful quantity to keep in mind is that 0.75 W dissipated in liquid helium for 1 h will boil away 1 L. The work required to produce a unit mass of liquid in an ideal Carnot cycle has been computed above to be 6865 J/g. Thus, an efficiency η can be defined in terms of the number of watts which must be dissipated at 300 K to produce 1 W of cooling at low temperatures. For the above open-cycle system, $\eta = \lambda/\dot{W}_i = 1/326$.

Now compare the above computation with a closed-cycle refrigerator, such as shown schematically in Fig. 8.17b. Closed cycle means that the helium gas is recovered cold through the refrigeration system. This approach allows the enthalpy of the return gas to be used to precool the incoming high-pressure stream by means of a heat exchanger. The efficiency of such a device is inherently much higher. For an ideal Carnot

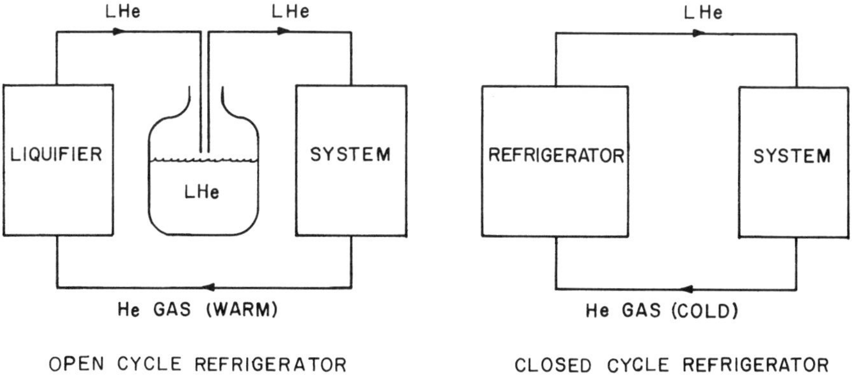

Fig. 8.17. Two liquid helium refrigeration systems.

refrigerator, the efficiency is given in terms of the absolute temperatures T_C and T_H, that is, $\eta = T_C/(T_H - T_C) = 1/70.4$, about a factor of 4.6 higher than the open-cycle system. Real systems which perform at less than Carnot efficiency have a similar ratio of efficiency for closed- versus open-cycle operation.

Listed in Table 8.3 are values for power requirements to open- and closed-cycle refrigeration assuming ideal thermodynamic processes. Comparison of the second and third columns indicates that helium accrues the greatest thermodynamic advantage by operating in the closed cycle. This is because the latent heat of liquid helium is small compared to the enthalpy of the gas between 300 K and T_{bp}. The ratio of columns 2 and 3 decreases as the boiling point of the liquid increases. For the high-temperature cryogenic fluids such as O_2 and N_2, the latent heat is larger than the enthalpy of the gas between T_b and 300 K. The last column in Table 8.3 indicates the minimum power requirements to liquefy a liter assuming ideal processes. This quantity is computed by using (8.34) and dividing by the density of the liquid at saturated vapor pressure.

As is the case for liquefaction systems, real low-temperature refrigerators operate at some fraction of ideal thermodynamic efficiency. That fraction is defined as the figure of merit (FOM). Thus, real refrigerators have a similar makeup to real liquefiers. However, there exists a clear distinction between refrigerators and liquefiers in that the refrigerator is to provide cooling while the liquefier is to condense fluid. For liquid–gas refrigeration systems there are primarily two methods of performing the process which extracts the low-temperature heat. The first is *isothermal refrigeration* where the heat input produces saturated vapor and the bath temperature is fixed by the two-phase coexistence. Alternatively, it may be of interest to perform *isobaric refrigeration* where normally liquid is not present and the gaseous enthalpy is used to provide the cooling. The latter method is not a constant temperature process.

Table 8.3. Power Requirements for Various Ideal Refrigeration Modes[a]

Fluid	T (K)	Closed-cycle refrigeration ($W_{300\,K}/W_{4.2\,K}$)	Open-cycle refrigeration ($W_{300\,K}/W_{4.2\,K}$)	Ideal liquefaction (W · h/L)
He	4.2	70.4	326.0	236
H_2	20.4	13.7	31.7	278
Ne	27.1	10.1	15.5	447
N_2	77.4	2.88	3.87	173
Ar	87.3	2.44	2.95	185
O_2	90.2	2.33	2.89	195

[a] See Ref. 16.

8.8.1. Isothermal Refrigeration

There is not a large difference between the system configuration for an isothermal refrigerator and that of a liquefier. In fact, the only distinction is that heat $\dot{Q}_c$ is applied to an evaporator bath rather than liquid being extracted. Consequently, the mass flow in the isothermal refrigerator is the same throughout the cycle. An example of an isothermal refrigerator based on the Claude cycle is displayed in Fig. 8.18.

Applying the first law of thermodynamics to an isothermal Claude cycle, the rate of heat extraction at low temperatures may be written

$$\frac{\dot{Q}_c}{\dot{m}} = (h_1 - h_2) + x(h_3 - h_e) \tag{8.46}$$

which is related to the yield equation for the liquefaction cycle. In fact, it is generally true for isothermal refrigeration that

$$y = \frac{\dot{Q}_c}{\dot{m}(h_1 - h_l)} \tag{8.47}$$

a quantity which is analogous to the ratio of actual refrigeration to ideal refrigeration.

Apart from this one fact, isothermal refrigeration is related directly to the liquefaction process. Thus, the optimization procedure follows a set of criteria similar to those already outlined. There are a few differences. Since

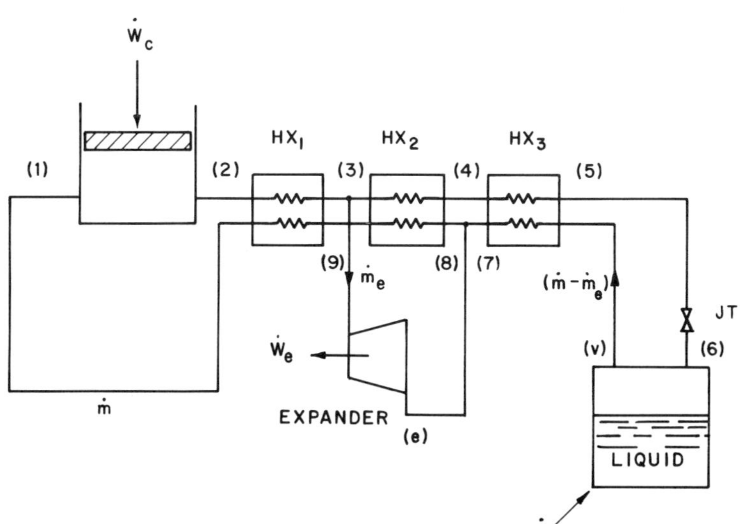

Fig. 8.18. Schematic of isothermal refrigerator based on Claude cycle.

the mass flow is now conserved, the condition applied to the return side heat exchangers are somewhat different. This affects the low-temperature end of the refrigerator and allows more latitude in the selection of mass flow through the various parallel circuits.

8.8.2. Isobaric Refrigeration

Unlike isothermal refrigeration, absorption of heat at constant pressure without liquid present results in an increase of the fluid temperature. Thus, the heat is absorbed in the vapor enthalpy at low temperatures. There are several reasons why this mode of operation is useful as a type of refrigeration. For one, it permits the operating temperature of the low end to be a free parameter adjustable within design constraints and not restricted to the boiling point of the cryogenic fluid. For helium, isobaric refrigerators are particularly useful for systems operating above 4.2 K. Avoiding the two-phase coexistence region permits the refrigerator to be constructed without a Joule–Thomson expansion stage. This approach permits potentially higher thermodynamic efficiency. Numerous applications employ cold gas refrigeration. Among these are superconducting systems using high T_c materials and certain small-scale systems including cryopumping devices.

An example of one cold gas refrigeration system is shown in Fig. 8.19 by a schematic flow chart. The system uses the expansion engine to provide cooling to a low-temperature heat sink. The amount of heat absorbed at

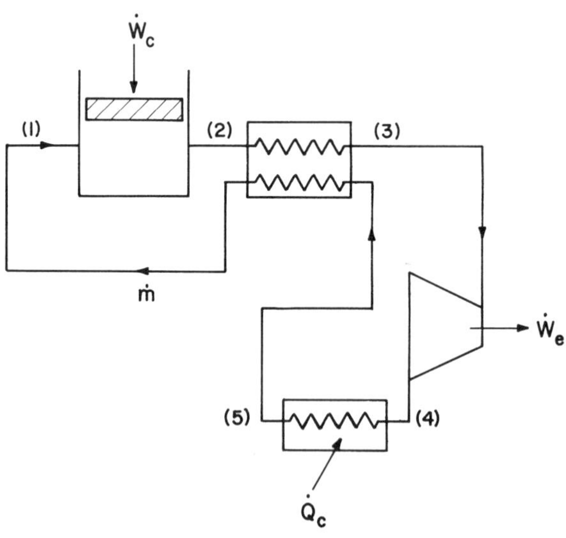

Fig. 8.19. Schematic of isobaric refrigerator.

low temperature is equal to the enthalpy change of the gas as it passes through the low-temperature heat sink. This quantity may be written

$$\frac{\dot{Q}_c}{\dot{m}} = h_5 - h_4 \qquad (8.48)$$

A schematic *T-S* diagram of this cold gas refrigeration cycle is shown in Fig. 8.20. The various locations on the flow chart are also indicated on the *T-S* diagram. Note that (5) is at a temperature which is comparable to (3), the inlet to the expansion engine.

As has been the case for the liquefier design, the isobaric refrigerator is not determined fully by thermodynamic considerations. Applying the first law of steady flows to the entire cycle gives an alternate expression for the refrigeration,

$$\frac{\dot{Q}_c}{\dot{m}} = (h_1 - h_2) + (h_3 - h_4) \qquad (8.49)$$

where the first term is the isenthalpic contribution which is only positive provided $T < T_{\text{inv}}$. The second term in (8.49) is the cooling produced by

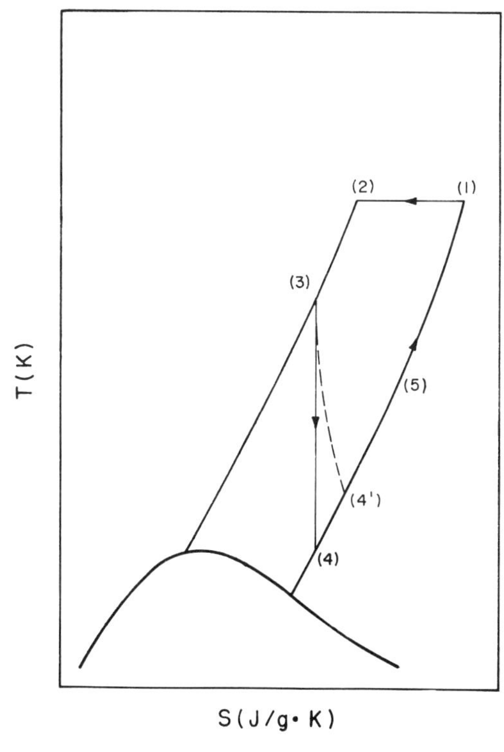

Fig. 8.20. *T-S* diagram of isobaric refrigerator.

the isentropic expansion engine and is always positive. For the presently considered cycle, (8.48) and (8.49) do not determine fully the refrigeration power. However, if one additional constraint is established such as the exit temperature to the heat exchanger, there is enough information to determine the performance.

As an example, assume the above described refrigerator were to be used closed cycle to cool a system, which should have a maximum temperature not to exceed 100 K. Therefore, in addition to (8.48) and (8.49), it is known that $T_5 = 100$ K. Furthermore, let us establish the usual set of starting conditions: at low pressure (1),

$$\begin{aligned} p_1 &= 1 \text{ bar} \\ T_1 &= 300 \text{ K} \\ s_1 &= 31.4 \text{ J/g} \cdot \text{K} \\ h_1 &= 1573 \text{ J/g} \end{aligned}$$

at high pressure (2),

$$\begin{aligned} p_2 &= 20 \text{ bars} \\ T_2 &= 300 \text{ K} \\ s_2 &= 25.2 \text{ J/g} \cdot \text{K} \\ h_2 &= 1579 \text{ J/g} \end{aligned}$$

and exiting the low-temperature heat exchanger (5),

$$\begin{aligned} p_5 &= 1 \text{ bar} \\ T_5 &= 100 \text{ K} \\ s_5 &= 25.7 \text{ J/g} \cdot \text{K} \\ h_5 &= 534 \text{ J/g} \end{aligned}$$

Combining (8.48) and (8.49), an expression for conservation of enthalpy through the first heat exchanger is obtained:

$$h_3 = h_2 - h_1 + h_5 = 540 \text{ J/g}$$

It then follows that state (3) is established:

$$\begin{aligned} p_3 &= 20 \text{ bars} \\ T_3 &= 100 \text{ K} \\ s_3 &= 19.5 \text{ J/g} \cdot \text{K} \\ h_3 &= 540 \text{ J/g} \end{aligned}$$

Finally, h_4 is fixed by isentropic expansion:

$$p_4 = 1 \text{ bar}$$
$$T_4 = 30 \text{ K}$$
$$s_4 = 19.5 \text{ J/g} \cdot \text{K}$$
$$h_4 = 170 \text{ J/g}$$

Now returning to either (8.48) or (8.49) we obtain the cooling power of the refrigerator:

$$\frac{\dot{Q}_C}{\dot{m}} = 364 \text{ J/g}$$

It is most useful to compare this value with the total work done. This exercise establishes the coefficient of performance. The net work is computed by assuming the expansion engine work can be utilized:

$$\frac{\dot{W}}{\dot{m}} = T_2(s_1 - s_2) - (h_1 - h_2) - (h_3 - h_4) = 1496 \text{ J/g}$$

Since this value only represents an example, it is more useful to establish a coefficient of performance that can then be compared to Carnot efficiencies. For the above example,

$$\text{COP} = \frac{\dot{Q}_C}{\dot{W}} = 0.243$$

An isothermal Carnot refrigerator operating between 100 and 300 K would have a coefficient of performance

$$\text{COP} = \frac{T_C}{T_H - T_C} = 0.50$$

That the above designed refrigerator operates at a figure of merit less than 50% is due largely to the isobaric operation. Because the system must cool the gaseous helium to 30 K, its efficiency is lower when compared to an isothermal system operating continuously at 100 K.

8.9. NONIDEAL REFRIGERATION COMPONENTS

Practical refrigeration components are not capable of operating reversibly. Heat exchangers always transfer heat across a finite temperature difference and isotropic expanders always allow some heat transfer between the working fluid and the device. The net effect of irreversible operation is to increase the work required per unit of refrigeration produced. The optimum operating parameters (e.g., pressure ratio, inlet temperatures, and mass flow fractions for expanders) for a particular cycle are critically dependent on the degree of reversibility in each component in the system.

The degree of reversibility of a component is usually expressed as a ratio of actual performance to the ideal performance obtained by a reversible process. The performance parameter of interest for an isentropic compressor or expander is the change in enthalpy experienced by the fluid. The efficiency is defined as the actual change in enthalpy divided by the maximum enthalpy change for a reversible process:

$$\eta = \frac{\Delta h_{\text{actual}}}{\Delta h_{\text{reversible}}} \tag{8.50}$$

The efficiency of the expander in the cycle depicted in Fig. 8.20 can be expressed as

$$n_e = \frac{h_3 - h_{4'}}{h_3 - h_4} \tag{8.51}$$

where $h_3 - h_4$ is the enthalpy change produced by an isentropic expansion. Typical expansion efficiencies are in the range of 70% for state-of-the-art machinery.

The thermal effectiveness is a similar measure of performance for heat exchangers. According to the first law of thermodynamics for a flowing system, enthalpy must be conserved in a heat exchanger. The second law requires that the outlet temperature cannot be lower than the inlet temperature at the cold end of a counterflow heat exchanger. At the hot end, the outlet temperature cannot be higher than the inlet temperature. Using the points shown in Figs. 8.19 and 8.20, these three constraints can be expressed as

$$\dot{m}_i(h_2 - h_3) = \dot{m}_o(h_1 - h_5) \tag{8.52a}$$

$$T_5 \leqslant T_3 \tag{8.52b}$$

and

$$T_1 \leqslant T_2 \tag{8.52c}$$

where $\dot{m}_i$ and $\dot{m}_o$ are the inlet and outlet mass flow rates. The maximum possible enthalpy transferable is always the lesser of the enthalpy difference on the two sides of the heat exchanger. The largest possible enthalpy difference occurs when the fluid is at the maximum temperature occuring within the heat exchanger at one end and at the minimum temperature at the other end. Note that a difference in mass flow rates between the two heat exchanger legs can produce a large difference in the maximum enthalpy change possible in each of the two legs. The thermal effectiveness of the heat exchanger is defined as the ratio of the actual enthalpy change in a leg of the heat exchanger divided by the maximum possible enthalpy change,

$$\frac{\Delta h_{actual}}{\Delta h_{max}} \tag{8.53}$$

As an example, consider the heat exchanger in the cycle depicted in Fig. 8.20. For the high-pressure leg of the heat exchanger, the maximum possible enthalpy drop would occur if the fluid entered at temperature T_2 and left at temperature T_5. Therefore, the maximum possible enthalpy differences in the two legs of the heat exchanger are

$$\Delta h_{max\ 2\text{-}3} = h(p_2, T_2) - h(p_3, T_5) \tag{8.54a}$$

and

$$\Delta h_{max\ 5\text{-}6} = h(p_1, T_2) - h(p_5, T_5) \tag{8.54b}$$

The maximum possible enthalpy change for either fluid stream is the lesser of these two quantities. The thermal effectiveness can be written in terms of the actual enthalpy change by inserting (8.54) into (8.53). The thermal effectiveness for heat exchangers used in refrigeration systems typically ranges around 90%. Heat exchangers with larger thermal effectiveness are large and expensive and must be made with care to minimize the pressure drop.

8.10. He II REFRIGERATORS

Refrigeration using the low-temperature phase of liquid helium, He II, is an area of current development. Unlike normal helium refrigerators, there are very few commercial systems available and most are one of a kind.[10,17] However, He II refrigeration is a growing area and there soon should be a number of large systems in operation.

It is important to realize that most He II refrigeration systems exist as an addendum to a normal helium closed-cycle refrigeration.[18] Thus, the starting point for the low-temperature ($T \approx 1.8$ K) refrigeration circuit is usually an isothermal bath of liquid helium near 4.2 K. This approach is not essential because it is equally possible to expand directly to low pressures ($p \approx 50$ mbar). However, the existence of a normal helium bath allows its use in other parts of the system or to intercept heat leaks at higher temperature with improved thermodynamic efficiency.

A schematic of a He II refrigeration system using this concept is shown in Fig. 8.21. A second J-T valve operates between the normal helium bath and the low-temperature heat exchanger, $p \approx 20$ mbar. The return flow of helium vapor is then available to precool the incoming fluid. The heat exchanger is unconventional technology because the low-density vapor has considerably different heat transfer and fluid flow characteristics. Above the heat exchanger there are basically two approaches. The simplest and most common is to recompress the helium vapor at ambient temperature with high-capacity vacuum pumps. The disadvantage here is that the vapor density is too low to allow good heat exchange with incoming fluids, which works against thermodynamic efficiency. Alternatively, it is possible to insert a cold compressor with a pressure ratio around 10 and operating near helium temperature. This approach is apparently contradictory to thermodynamic arguments, which indicate that it is best to compress at ambient temperature. However, it allows recovery of a larger fraction of the return helium vapor enthalpy, thereby improving the overall efficiency.

As an example of this technology, consider an isenthalpic expansion of helium from 1 bar, 4.2 K to 16 mbar corresponding to 1.8 K. A low-temperature helium *T-S* diagram is shown in Fig. 8.22. The expansion follows along path (1)–(2). Determination of the yield for the process is analogous

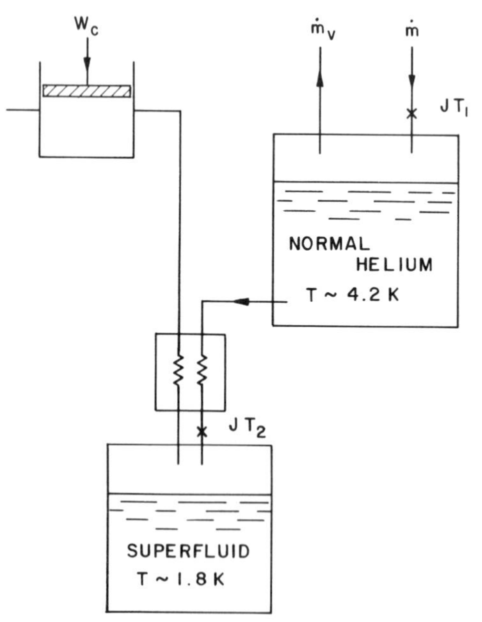

Fig. 8.21. Schematic of He II refrigeration system.

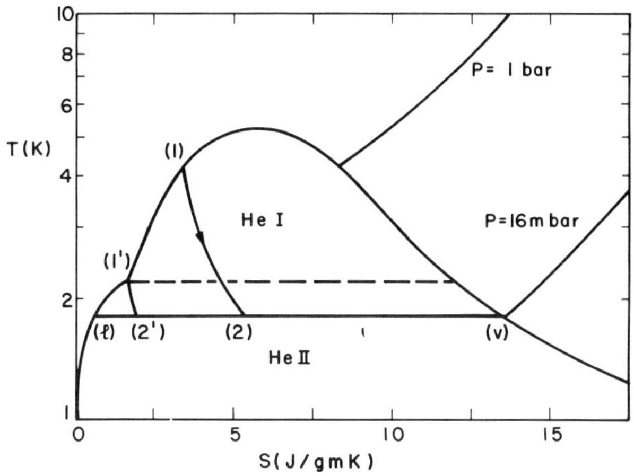

Fig. 8.22. Low-temperature *T-S* diagram for helium showing isenthalpic processes.

to the methods applied above in conventional helium liquefaction. By equating the enthalpies before and after expansion, we obtain an expression for the yield,

$$y = \frac{h_v - h_1}{h_v - h_l} \tag{8.55}$$

where the subscripts apply to the various locations on the *T-S* diagram. For the process under consideration, the yield $y = 62\%$. This yield is fairly high although actual systems are not perfectly isenthalpic. The yield in a real system can be improved considerably by precooling the incoming fluid with the return vapor. Perhaps the best that can be obtained is to have the inlet of the J-T valve at 2.2 K just above the λ-temperature. With this initial condition, the isenthalpic expansion will follow path $(1')$–$(2')$ with a corresponding yield $y = 89\%$.

8.11. REFRIGERATION TECHNOLOGY

When designing and constructing a helium cryogenic system for a particular application, it is necessary to keep in mind a number of factors which are influenced by economics and operational methods. Included among these are:

1. Duty cycle: Is the system to be operated on a short-term basis or is it an on-line component?

2. Owning and operating cost: A system may be costly to operate because of large running costs including power, labor, or materials.
3. Overall size of the system.

Within the context of the above factors there are several choices that can be exercised when providing the helium coolant to a system. One of four methods are appropriate for cooling the system to liquid helium temperatures.

1. Purchase liquid helium and vent the helium vapor that evolves.
2. Purchase helium gas, liquefy it on a batch basis, and vent the vapor.
3. Purchase helium gas, liquefy it on a batch basis, and recover the evolved vapor which is then reliquefied.
4. Purchase helium gas and a closed-cycle refrigerator. Run the system on a continuous basis, recovering the gas cold.

Of the above four methods, the first is suitable only for small systems with low duty cycles. More typically, a combination of the second and third methods is used in small research laboratories, where the percentage of helium recovery depends on the care with which the system is operated. The last method, although most efficient, is actually only suitable for systems that have the cryogenics as an integral part. Clearly, economics is the major factor that determines the cooling method best suited for a particular application.

Actual refrigeration systems operate far from the ideal efficiency. This fact is brought about by the cycles deviating from Carnot and the system components operating at less than ideal efficiency. Some of the nonidealities of components have been described above; however, those components that have the greatest tendency for nonideal performance are the moving parts, including compressors and expansion engines. A usual way of displaying actual refrigerator efficiencies is in terms of percent of Carnot efficiency, $T_C/(T_H - T_C)$. Plotted in Fig. 8.23 are actual efficiencies of various cryogenic systems operating at all temperature ranges from 1.8 to 90 K.[16] It is significant to note that the most important factor in the efficiency appears to be the capacity and not the operating temperature range. The reason for the increase in efficiency with size is twofold. For larger refrigerators, there is more to be gained by making the system more efficient. The total power saved can be a significant cost item. Also, larger systems can justify more stages of refrigeration and expansion, thus allowing the cycle to be tuned closer to ideal conditions. An actual refrigeration system is made up of a number of individual components. A schematic representation[19] of the way these components fit together in the system is shown in Fig. 8.24. Besides the helium refrigerator or liquefier,

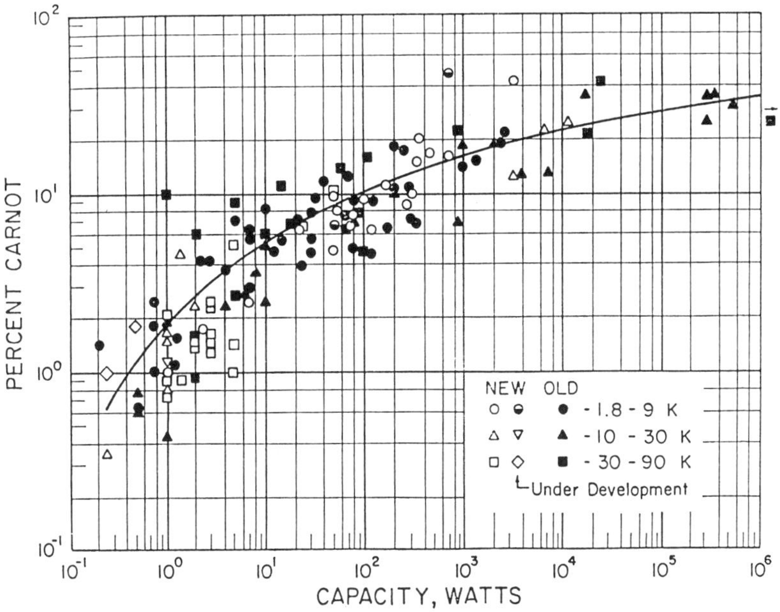

Fig. 8.23. Efficiency of cryogenic refrigeration systems (from Strobridge[16]).

there must be storage of gaseous and liquid helium and nitrogen. Furthermore, in the helium system, there must be some form of purifier to recycle the gas. Finally, there must be a system of vacuum insulated transfer lines to take the fluids from storage or liquefier to the system.

The cost of refrigeration components has been surveyed by Trepp,[20] Strobridge,[16] and Robinson.[19] Plotted in Fig. 8.25 are costs as summarized by Strobridge of only the liquefier or refrigerator. The room temperature power P is a more important parameter in determining the cost than is the refrigeration temperature. The graph suggests a relationship for the cost of refrigeration of the form

$$C = \sigma P^n \qquad (8.56)$$

where σ is an empirical coefficient and n is an exponent less than unity. For C to be the cost in dollars and P the input power in kilowatts at room temperature, in 1974 Strobridge found $\sigma = 6000$ and $n = 0.7$ to be the best-fit parameters. This relationship appears to fit all the data to within a factor of 2. More recent surveys give similar values for the exponent but have revised the coefficient σ upward. Furthermore, as standard units become available, the empirical relationship suggested by Fig. 8.25 has an exponent closer to unity. This fact is the result of certain fixed costs such as

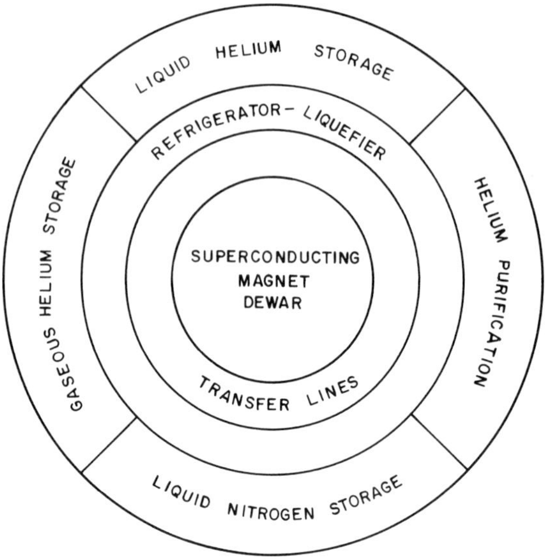

Fig. 8.24. Components making up a cryogenic system (from Robinson[19]).

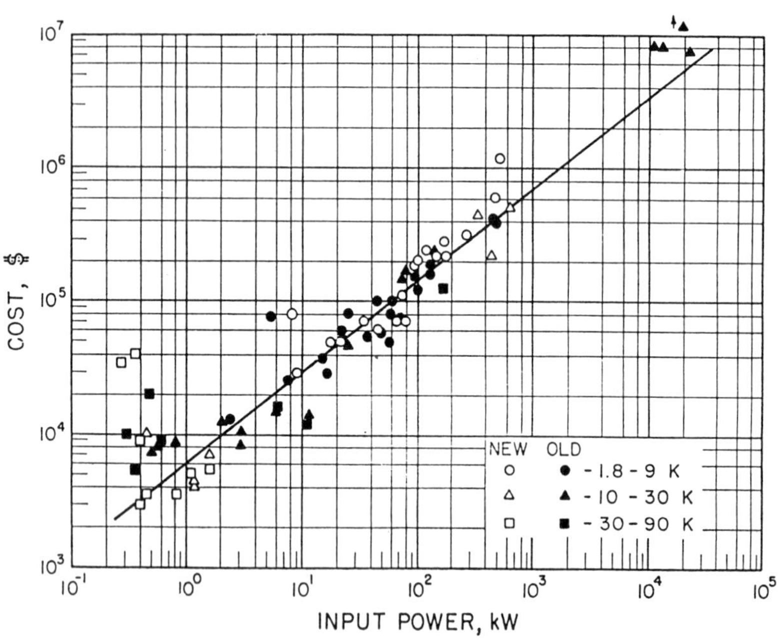

Fig. 8.25. Cost of low-temperature refrigerators and liquefiers as a function of installed power (from Strobridge[16]).

engineering and design which enter in the development of a new system. To arrive at a reasonable estimate for the cost of an actual system, the values from (8.56) should be multiplied by about a factor of 2 to account for peripheral equipment and installation. The nonlinearity of capital cost coupled with the costs of operation and maintenance are factors which must be considered when determining which cooling scheme is appropriate for a particular system.

The foregoing chapter presents a survey of the principles of refrigeration and liquefaction applied to helium. Greater detail can be considered particularly with respect to real systems and components. However, as the goal of this book is the emphasis of basic principles rather than hardware, no further discussion of this subject is given here. For the reader who is particularly interested in this subject, several of the references[7,10–15] listed at the end this chapter will provide greater detail.

PROBLEMS

8.1. An alternate equation of state for real gases is described by the relationship

$$pv = \frac{RT}{(1 - b/4v)^4} - \frac{a}{v}$$

a. Using the universal behavior of gases near the critical point, determine the coefficients a and b as functions of critical parameters. Evaluate a and b for helium.
b. Derive a relationship for the inversion temperature T_{inv} and calculate its reduced maximum value (T^m_{inv}/T_c).
c. Derive an expression for the Joule–Thomson coefficient μ_j and determine its value for helium at the critical point.

8.2. a. Calculate the ideal liquefaction work for helium which is initially at liquid nitrogen temperature, 77 K.
b. Determine the minimum quantity of liquid nitrogen required to precool a unit mass of helium gas from 300 to 77 K. ($\lambda_{LN_2} = 198$ J/g.)
c. Compare the total work required for the liquid nitrogen precooling method to that of an ideal liquefier.

8.3. Calculate the temperature corresponding to the maximum inversion pressure for a van der Waals gas. Determine the corresponding temperature and pressure for helium gas.

8.4. Derive the following relationship for the entropy change for a van der Waals gas:

$$s_2 - s_1 = C_p \ln\left(\frac{T_2}{T_1}\right) + R \ln\left(\frac{v_2 - b}{v_1 - b}\right)$$

Begin from a total differential of the entropy. Find the relationship between temperatures and specific volumes for an isentropic change in state.

8.5. A particular piston compressor can achieve a maximum pressure ratio of $p_2/p_1 = 100$ by either isentropic or isothermal compression of a gas. The compressor is to be used in one of two processes within a refrigeration cycle: (1) isentropic compression followed by isochoric cooling of the gas in the cylinder, or (2) isothermal compression. Calculate the entropy change, heat removed from the gas, and the maximum gas temperature for each process. Assume an ideal gas which begins at $p = 1$ bar, $T = 300$ K and ends up at $T = 300$ K.

Given that the pressure ratio and that the maximum operating temperature will always be limited for real compressors, which process should be used in the refrigeration cycle?

8.6. The Collins helium liquefaction system uses five heat exchangers and two expanders, one cooling heat exchanger #2 and one cooling heat exchanger #4. Draw the refrigeration cycle for this system showing the locations of all components. Label the various stages of cooling on a helium *T-S* diagram. Assume the following:

$$p_1 = 1 \text{ bar} \qquad x_1 = 0.25 \qquad T_{e_1} = 150 \text{ K}$$
$$p_2 = 20 \text{ bars} \qquad x_2 = 0.50 \qquad T_{e_2} = 10 \text{ K}$$
$$T_H = 300 \text{ K}$$

Calculate the yield, compressor work, and figure of merit for an ideal liquefier.

8.7. Consider the ideal isothermal helium refrigerator based on the Claude cycle with the first heat exchanger fixed by liquid nitrogen precooling at 77 K.

a. Calculate the cooling per unit mass flow. Determine an optimum based on varying m_e.
b. Estimate the liquid nitrogen consumption.
c. Determine the effective FOM for the entire system. Assume that $T_1 = T_2 = 300$ K and $p_1 = 1$ bar, $p_2 = 20$ bars.

8.8. For the simple He II refrigerator shown schematically in Fig. 8.21, assume the inlet temperature to JT_2 is at 2.2 K, 1 bar and the system is in steady-state operation.

a. Calculate the refrigeration capacity per unit mass flow.
b. Estimate the temperature of the vapor which is exiting the heat exchanger on the return side.

8.9. A Stirling refrigerator is used to cool an infrared sensor. The working fluid is helium gas which may be treated as ideal. If the pressure and temperature at point 1 (see Fig. 8.16) are 5 bars and 100 K, the volume ratio (V_3/V_2) is 10 and p_2 is 1 bar, Find:

a. The pressure, temperature, and volume at points 1 through 4.
b. The temperature of the regenerator after steps 1 and 2 if it consists of 100 g of copper and has an initial temperature of 100 K.
c. Repeat part b if the regenerator consists of 100 g of lead.
d. On the basis of thermodynamics alone, which of the above materials is preferable for use in the regenerator?
e. Find the coefficient of performance (COP).

8.10. An isobaric refrigerator is to be designed with an inlet temperature $T_1 = T_2 = 300$ K and an inlet pressure $p_1 = 1$ bar to provide cooling at temperatures at or below 18.5 K. Assume the refrigerator is too small to use the work produced by the expander.

a. For a heat exchanger effectiveness and an expander efficiency of unity, calculate the compressor work per unit of cooling produced $(\dot{W}/\dot{Q}_c)$ for isothermal compressor outlet pressures of 5, 10, 20, 40, and 100 bars. Plot $\dot{W}/\dot{Q}_c$ as a function of the compression ratio p_2/p_1 and fit the curve with an equation.
b. Redo the calculation in part a assuming a heat exchanger effectiveness of 0.95 and expander efficiency of 0.7.

REFERENCES

1. M. W. Zemansky, *Heat and Thermodynamics*, 5th ed., McGraw-Hill, New York, 1968.
2. N. Kurti, Thermodynamics and Statistical Mechanics and Their Relevance to Cryogenics, in *Advanced Cryogenics*, C. A. Bailey (Ed.), Chap. 1, Plenum Press, London, 1971.
3. E. R. Lady, Liquefaction of Gases, in *Cryogenic Engineering*, E. R. Lady (Ed.), Chap. 3, Engineering Summer Conference, University of Michigan, 1965.
4. R. D. McCarty, *Thermophysical Properties of Helium-4 from 2 to 1500 K with Pressures to 1000 Atmospheres*, NBS Technical Note 631, U.S. Government Printing Office, Washington, DC, 1972.
5. R. M. Gibbons, Thermophysical Data for Cryogenic Materials, in *Cryogenic Fundamentals*, G. G. Haselden (Ed.), Academic Press, London, 1971.
6. D. B. Mann, Properties of Helium, in *Technology of Liquid Helium*, R. H. Kropschot, B. W. Birmingham, and D. B. Mann (Eds.), Chap. 2, National Bureau of Standards Monograph 111, U.S. Government Printing Office, Washington, DC, Oct. 1968.
7. R. F. Barron, *Cryogenic Systems*, 2nd Ed., Clarendon Press, Oxford, 1985.
8. E. A. Guggenheim, *Thermodynamics*, North-Holland Publishing, Amsterdam, 1977.
9. F. E. Hoare, L. C. Jackson, and N. Kurti, *Experimental Cryophysics*, Butterworth, London, 1961.
10. R. W. Johnson, Liquefaction, in *Liquid Cryogens*, Vol. I, *Theory and Equipment*, K. D. Williamson, Jr. and F. J. Edeskuty (Eds.) Chap. 2, CRC Press, Boca Raton, FL, 1983.
11. S. C. Collins, Liquefaction Techniques, in *Technology of Liquid Helium*, R. H. Kropschot, B. W. Birmingham, and D. B. Mann (Eds.), Chap. 3, National Bureau of Standards Monograph 111, U.S. Government Printing Office, Washington, DC, 1968.
12. R. W. Vance, Small Refrigerator Systems (Space Applications), in *Liquid Cryogenics*, Vol. II, K. D. Williamson, Jr. and F. J. Edeskuty (Eds.), Chap. 5, CRC Press, Boca Raton, FL, 1983.
13. G. G. Haselden, Refrigeration and Liquefaction Cycles, in *Cryogenic Fundamentals*, G. G. Haselden (Ed.), Chap. 2, Academic Press, New York, 1971.
14. A. J. Croft, Liquefiers and Refrigerators, in *Advanced Cryogenics*, C. A. Bailey (Ed.), Chap. 7, Plenum Press, New York, 1971.
15. G. Walker, *Cryocoolers, Part 1: Fundamentals*, Plenum Press, New York, 1983.
16. T. R. Strobridge, *Cryogenic Refrigerators—An Updated Survey*, NBS Technical Note 655, U.S. Government Printing Office, Washington, DC, 1974.
17. W. Daus and R. Ewald, A Refrigeration Plant with 300 W Capacity at 1.8 K, *Cryogenics* **15**, 591 (1975).
18. G. Claudet, A. Lacaze, P. Roubeau, and J. Verdier, The Design and Operation of a Refrigerator System Using Superfluid Helium, in *Proceedings of the 5th International Cryogenics Engineering Conference*, pp. 265–267, Butterworth, Surrey, U.K., 1974.
19. G. Y. Robinson, Economics of Cryogenic Systems for Superconducting Magnets, *Adv. Cryog. Eng.* **25**, 342 (1981).
20. C. Trepp, Problems of Large Scale Refrigeration, in *Low Temperatures and Electric Power*, pp. 31–41, Pergamon Press, New York, 1970.

Chapter 9

Special Topics in Cryogenics

In the preceding discussion, the emphasis has been on bulk helium properties and systems operating in the temperature range above 1 K. However, there are numerous other topics in cyrogenics which do not fall in this general context but still have considerable relevance to the general subject of helium cryogenics. In the present chapter, four topics outside the main focus of this book are reviewed. These topics include (1) thermal insulation systems, (2) the behavior of liquid ^3He at very low temperatures, (3) helium adsorption, and (4) the practical aspects of magnetic refrigeration. The review of these subjects is not all-inclusive but rather represents a few areas of potential interest to the general subject of low-temperature phenomena important to the useful application of helium cryogenics.

9.1. THERMAL INSULATION

In the design of any cryogenic system, thermal isolation of the low-temperature environment must be achieved effectively. This is particularly true for liquid helium owing to the small value of latent heat and high cost per watt of refrigeration. Various factors contribute to degradation of thermal isolation. These are mainly in the form of heat leaks to the system which can be controlled and reduced by proper design. Heat leaks can be divided into three classes:

1. Heat conduction through support structure and other penetrations to the system.
2. Heat transport through residual gas in the insulating vacuum space.
3. Radiant heat transfer between walls of the insulating container.

9.1.1. Solid Conduction

The first of these modes can be minimized by use of materials with low thermal conductivity and high strength. The relative importance of this mechanism depends on design. If the system demands a great deal of load transfer to ambient temperature, then the conduction heat leak will probably dominate. This quantity can be calculated by application of Fourier's law of heat conduction,

$$q = -k \nabla T \tag{9.1}$$

where k represents the material thermal conductivity. This property has been discussed to a considerable extent in Chapter 2.

Many practical problems invole one-dimensional conduction over a finite temperature difference. A commonly useful quantity is the integrated or mean thermal conductivity $\bar{k}$, which is used for determining the total heat conduction between two fixed temperature reservoirs. The integrated thermal conductivity is determined by the expression

$$\bar{k} = \frac{1}{(T_2 - T_1)} \int_{T_1}^{T_2} k \, dT \tag{9.2}$$

where T_2 and T_1 are the temperature of the reservoirs. In general, k is a complex function of temperature making evaluation of (9.2) difficult. For most materials at low temperatures, k can be approximated by an expression of the form

$$k = AT^n \tag{9.3}$$

where for metals n approaches 1 and for insulators $n \approx 3$. It follows that the integrated thermal conductivity may be written

$$\bar{k} = \frac{A}{(n+1)} \left(\frac{T_2^{n+1} - T_1^{n+1}}{T_2 - T_1} \right) \tag{9.4}$$

As long as the thermal conductivity is a monotonic increasing function of temperature, $\bar{k}$ is dominated by the high-temperature end. There are numerous lists of integrated thermal conductivity available in the literature.[1-6] Table 9.1 gives some typical values for materials useful in low-temperature applications. These are given for commonly encountered reservoir temperatures. To determine the conduction heat leak between reservoir temperatures other than those listed in Table 9.1, it is necessary to evaluate (9.2).

9.1.2. Gas Conduction

The second mode of heat leak is conduction through the residual gas in the vacuum space surrounding the low-temperature environment. In

Table 9.1. Integrated Thermal Conductivity in W/m · K of Several Materials Useful for Low-Temperature Applications[a]

Material	$\bar{k}\|_1^4$	$\bar{k}\|_4^{77}$	$\bar{k}\|_{77}^{300}$
Copper	200	980	410
Brass	1.7	26	81
Stainless steel	0.2	4.5	12.3
Pyrex	0.06	0.25	0.82
Nylon	0.006	0.17	0.31

[a] From White.[1]

principle, this mode can be made arbitrarily small by further reduction of the residual pressure; however, this is not always possible. To estimate the magnitude of the gas conduction heat leak consider an ideal system consisting of two surfaces at different temperatures with a residual gas between the surfaces at pressure p. There are primarily two regimes of heat transport in the residual gas. The first represents the case where the mean free path l is short compared to the spacing between the two surfaces. Under this circumstance, the heat leak is nearly independent of pressure except at low temperatures and high pressures (see Section 3.4). Of course, if the pressure is high, convection heat transfer can enhance the effective conductivity of the gas. However, the short mean free path case generally occurs at high pressures, $p \gtrsim 1$ Pa ($\sim 10^{-2}$ torr). For most applications pertaining to cryogenics, the residual pressure is considerably lower than this value. For lower pressures the heat transfer process is represented by a second regime, where the mean free path is greater than the distance between the two surfaces. The mean free path of a gas can be estimated from the expression

$$l = 8.6 \frac{\eta}{p} \left(\frac{T}{M} \right)^{1/2} \tag{9.5}$$

where η is the viscosity in poise, p is the pressure in torr, and M is the molecular weight in g/mol. The mean free path computed from (9.5) is in centimeters. For a residual pressure of 10^{-3} torr, the mean free path for helium at 4.2 K is about 1 mm.

The regime where the mean free path is greater than the distance between the two surfaces can be evaluated in terms of molecular kinetic theory. The physical picture is that of molecules making several collision-free trips between the two surfaces before interacting with other molecules in the volume. Thus, the rate of heat transfer is more determined by the molecule–wall interaction than intermolecular behavior. In this regime, the heat transfer is approximately proportional to the absolute pressure. The molecule–wall interaction can be considered in terms of an accommodation coefficient α given by[7]

$$\alpha = \frac{T_i - T_e}{T_i - T_w} \tag{9.6}$$

where T_i is the effective temperature of the incident molecule, T_e that of the emitted or reflected molecule, and T_w the wall temperature. The maximum value for α is unity, associated with the molecule coming into complete thermal equilibrium with the wall.

For parallel surfaces at temperatures T_1 and T_2, the heat flux by gas conduction at low pressures is given by the relationship

$$q = \frac{\alpha_0}{4} \frac{\gamma + 1}{\gamma - 1} \left(\frac{2R}{\pi M T} \right)^{1/2} p(T_1 - T_2) \text{ mW/m}^2 \tag{9.7}$$

Here γ is the ratio of the specific heats, and R is the universal gas constant in J/mol · K. The quantity α_0 is an averaged accommodation coefficient dependent on the individual accommodation coefficient and surface areas. For unequal values of α and A, the quantity α_0 must be averaged over the two surfaces, that is,

$$\alpha_0 = \frac{\alpha_1 \alpha_2}{\alpha_2 + (A_2/A_1)(1 - \alpha_2)\,\alpha_1} \tag{9.8}$$

The value of α is known only approximately. For very clean metallic surfaces near room temperature, the accommodation coefficient for helium gas can be quite small, taking on values as low as 0.025.[1] At low temperatures, α increases considerably, being reported[8] to equal 0.6 at 20 K. For many low-temperature applications, helium gas is the major contributor to gaseous conduction heat leak because most other gases condense or adsorb on the cold surfaces. For helium gas, a reasonably good value to assume for average accomodation coefficients is around 0.5.

9.1.3. Radiation Heat Transfer

Thermal radiation represents the mode of heat transfer which is often of primary concern in cryogenic insulation systems. This mechanism occurs independent of residual gas pressure or structural supports. The usual method for reducing radiation heat leaks is to use one of various forms of multilayer shielding. The present section considers both the basic aspects of radiant heat transfer and some practical methods for reducing this mode of heat leak to a system.

The usual starting point for the description of radiant heat transfer is to consider the spectral energy density associated with a black body,

$$e_b(T, \lambda) = \frac{8\pi hc}{\lambda^5}\left(\frac{1}{e^{hc/\lambda k_B T} - 1}\right) \tag{9.9}$$

where λ is the wavelength of the thermal radiation. This relationship has a well-known form shown in Fig. 9.1. At very high temperatures, the peak occurs in the range of visible wavelengths, while at low temperatures, the peak is shifted to longer wavelengths and has smaller amplitude. Wien's displacement law is an exact description of the temperature dependence of the peak wavelength:

$$\lambda_m T = 2898\ \mu\text{m} \cdot \text{K} \tag{9.10}$$

Note that for surfaces at 4.2 K, the corresponding peak wavelength is 0.69 mm, which is in the far infrared. Such long wavelengths can have potentially interesting diffractive effects when the physical spacing between radiant surfaces approaches that of the peak wavelength.[9]

For a black body the total radiant energy flux can be obtained by integration of (9.9) over wavelength. The integration results in the Stefan–Boltzmann law,

$$\begin{aligned} E_b &= \int_0^\infty e_b(T, \lambda)\, d\lambda \\ &= \sigma T^4 \end{aligned} \tag{9.11}$$

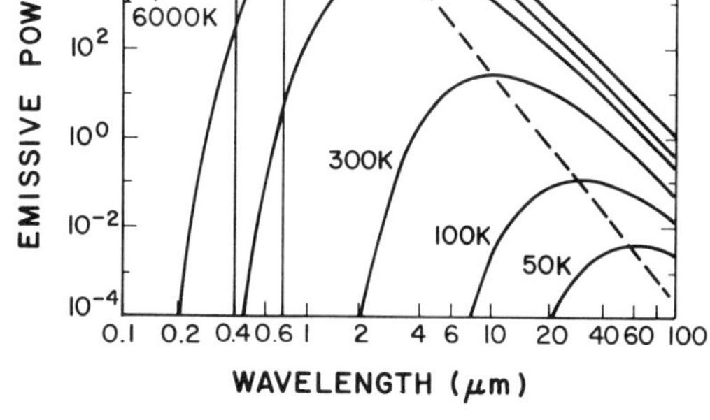

Fig. 9.1. Spectral energy density of black body radiation.

where σ is the Stefan–Boltzmann constant taking the value $2\pi^5 k_B^4/15c^2h^3 = 5.67 \times 10^{-8}$ W/m$^2 \cdot$ K^4. Most surfaces encountered in cryogenics do not approximate black bodies. In fact, the effort in cryogenics is usually a minimization of the thermal radiation heat leak. If a surface is not black, its spectral energy density is smaller in proportion to its emissivity ε_λ. The emissivity is actually dependent on wavelength and is defined by the relationship

$$e_r(\lambda, T) = \varepsilon_\lambda e_b(\lambda, T) \tag{9.12}$$

where ε_λ is less than unity for nonblack surfaces.

It is often possible to make an approximation for radiant heat transfer which allows nonblack surfaces to be handled in a fairly straightforward manner. This assumption is to characterize surfaces as having constant emissivity less than unity, which is independent of wavelength. The *gray body* assumption allows the writing of the total radiant energy flux as

$$E_r = \varepsilon\sigma T^4 \tag{9.13}$$

where the emissivity is determined for the appropriate set of conditions for the system of interest.

There are a number of implicit assumptions that enter into the gray body approximation. The first, already mentioned above, is that the emissivity must be wavelength independent over the range of wavelengths appropriate to the problem. Second, it is necessary to be able to equate the emissivity with the absorptivity a which is a measure of the rate of energy absorption by the gray surface. The total absorptivity is defined by the relationship

$$E_a = aE_i \tag{9.14}$$

where E_i is the incident energy flux usually defined according to (9.13). Finally, for the gray body assumption to be reasonable, all surfaces must be diffuse scatterers of radiation. This is certainly far from the case for highly polished shields that occur with multilayer insulation systems; however, it simplifies calculations considerably.

There have been numerous attempts to measure and tabulate the emissivities of different materials at low temperatures.[10,11] Generally, these values are determined for radiant energy flux between ambient (273 or 300 K), liquid nitrogen (77 K), and liquid helium temperatures (4.2 K). A compiled graphical representation of emissivity measurements[11] is shown in Fig. 9.2. To achieve low values of emissivity, it is necessary to have highly polished, high-conductivity surfaces made from gold, silver, copper, or aluminum. It has also been indicated[1] that since the emissivity is related to

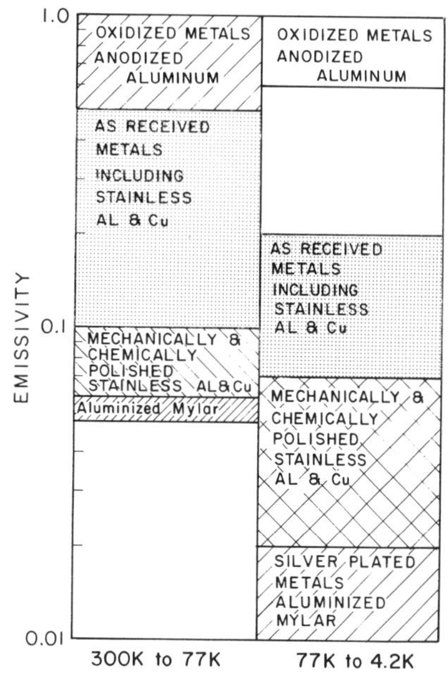

Fig. 9.2. Emissivity of different metallic surfaces (from Obert et al.[11]).

surface conductivity, it is best to polish the surfaces with a strain-free technique such as chemical etches. Finally, to estimate the radiant energy flux between surfaces of unknown emissivity, it is necessary to choose an approximate value for ε. If the surface in question is dielectric, it is reasonable to assume black body heat transfer. If, on the other hand, the surface is a metal and polished, a conservative choice is $\varepsilon \approx 0.1$. However, as can be seen in Fig. 9.2, to achieve emissivities much less than 0.1, special care must be taken with the surface preparation.

The radiant heat transfer between parallel surfaces, shown schematically in Fig. 9.3, is a function of the properties of both surfaces. The rate at which energy is radiated from surface 1 is proportional to the emissivity ε_1 and the fourth power of the temperature T_1. A similar situation occurs for surface 2. In the simple case that the surfaces can be approximated as black bodies, that is, $\varepsilon_1 = \varepsilon_2 = 1$, the net radiant heat flux is the difference between these two values:

$$q_r = \sigma(T_1^4 - T_2^4) \tag{9.15}$$

In the general case where the emissivity is considerably less than unity, the situation is more complicated. This is because the incident radiation from one surface can either be absorbed by the other or reflected back to the original surface. This problem can be solved by summing the infinite series

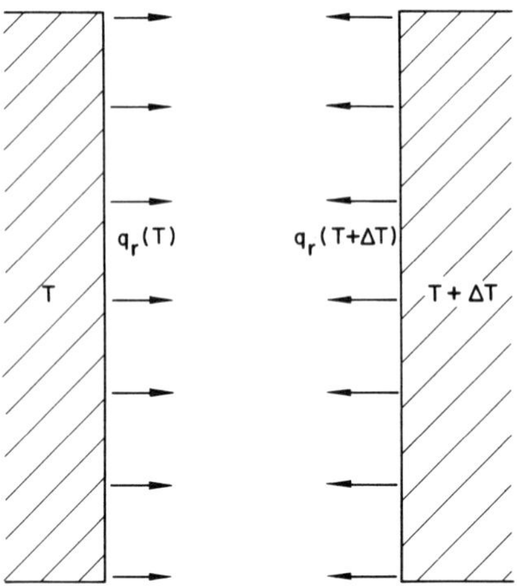

Fig. 9.3. Schematic of radiant heat transfer between two surfaces.

of contributions to the radiant energy flux. The result for the net radiant heat flux for gray surfaces is

$$q_r = \left(\frac{\varepsilon_1 \varepsilon_2}{\varepsilon_1 + \varepsilon_2 - \varepsilon_1 \varepsilon_2} \right) \sigma (T_1^4 - T_2^4) \tag{9.16}$$

Note that in the special case often encountered in cryogenics where $\varepsilon_1 \simeq \varepsilon_2 \simeq \varepsilon \ll 1$, the quantity in brackets reduces to simply $\varepsilon/2$.

As an example, consider the heat leak to a vacuum insulated liquid helium container both with and without liquid nitrogen shielding. Assume that the residual vacuum is sufficiently low to be able to neglect gas conduction heat leak. Taking a fairly optimistic value for $\varepsilon = 0.05$ for both surfaces, (9.16) yields a radiant heat leak between 77 and 4.2 K of $q_r(77\text{ K}) \approx 50\text{ mW/m}^2$, which for most systems is acceptable because it represents about 60 mL/h or liquid helium consumption per square meter of container surface area. If, on the other hand, the exterior surface is maintained at ambient temperature, 300 K, the heat leak is increased by approximately a factor of 300 which represents almost 3 $\text{L/h} \cdot \text{m}^2$ of helium boil-off. Obviously, there is considerable benefit in liquid nitrogen shielding of liquid helium containers.

The above situation can be generalized to include the possibility of multilayers of insulation,[12] which represents a fairly good approximation for aluminized mylar superinsulation at low packing density. For the case of n shields between two parallel surfaces, the radiant heat flux becomes

$$q_r = \left(\frac{\varepsilon_i \varepsilon_k}{(n-1)\,\varepsilon_i + 2\varepsilon_k} \right) \sigma (T_1^4 - T_2^4) \tag{9.17a}$$

where

$$\varepsilon_i \equiv \frac{\varepsilon_0 \varepsilon_s}{\varepsilon_s + \varepsilon_0 - \varepsilon_0 \varepsilon_s} \tag{9.17b}$$

and

$$\varepsilon_k = \frac{\varepsilon_s}{2 - \varepsilon_s} \tag{9.17c}$$

for which ε_0 is the emissivity of the two parallel surfaces and ε_s is that of the shields. For the special case where $\varepsilon_0 = \varepsilon_s = \varepsilon$, (9.17) can be simplified considerably:

$$q_r = \left(\frac{\varepsilon}{(n+1)(2-\varepsilon)} \right) \sigma (T_1^4 - T_2^4) \tag{9.18}$$

Note that for $\varepsilon \ll 1$, (9.18) predicts a heat leak reduced by the factor $(n+1)^{-1}$ from that without the multilayer shields.

9.1.4. Insulating Materials

Superinsulation, which is aluminized mylar with low-density fibrous insulating spacers between layers, represents a special case of a multilayer insulation system. Since there is interlayer material present, it is no longer possible to assume that each of the n shields are isolated from the others except for radiant heat transfer. Thus, in superinsulation there are two contributing heat transfer heat transfer mechanisms which are both functions of density and total number of shields. There is radiant heat transfer that decreases with increasing number of radiation shields, that is, layers of superinsulation. However, as the packing density increases, the heat transferred by conduction though the fibrous insulating spacers begins to make a larger contribution to the total heat leak. Thus, there occurs an optimum layer density for practical multilayer insulations. Note that the solid-state conduction heat leak can be reduced by increasing the spacing between walls for the same packing density while the radiation contribution is only a function of number of layers. Therefore, the optimum number of layers should also be a function of the total insulation thickness.

The existence of a minimum in the layer density dependence of the heat transfer through superinsulation has been demonstrated experimentally.[13,14] Plotted in Fig. 9.4 is one set of results for the heat flux through superinsulation between 4.2 and 77 K as a function of the numbers of layers.[14] The existence of a broad minimum near 0.5 layers/mm indicates the density where the solid-state conduction begins to play a substantial role. Note that the exact position of the minimum is not critical since a fac-

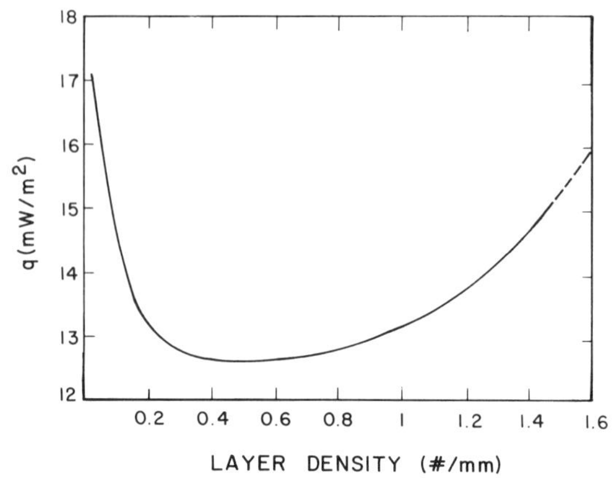

Fig. 9.4. Radiation heat flux for varying layer densities of superinsulation in a 50.8 mm gap between 77 and 4.2 K (from Leung et al.[14]).

tor of 2 change in packing density increases the heat flux by less than 10%. Obviously, these results are not universal because the conduction and radiation contributions scale differently. However, superinsulation at low densities, less than 0.5 layers/mm, can be modeled fairly accurately by pure radiant heat transfer. For these results, the best choice for the emissivity of aluminized mylar is $\varepsilon = 0.011$ at 4.2 K and 0.03 at 77 K.[14]

Glass powder insulation or perlite materials also are used to fill spaces between vacuum walls in insulating cryogenic vessels. These materials generally are comprised of glass microspheres with diameters in the range 100–1000 μm. They have advantages over superinsulation in that the material is easier to install, cheaper, and the residual vacuum requirements are not as high. However, the disadvantages of powder insulation technology are that it requires long and careful pump-down procedures and that the residual effective thermal conductivity is considerably higher than can be achieved with properly installed superinsulation. It is because of these latter two factors that powder insulations are only seriously considered for containers operating in the range of liquid nitrogen temperatures and above.

The apparent thermal conductivity of perlite materials is a reasonably well-known quantity. Measurements have been carried out on the variation of this quantity with residual gas pressure and diameter of glass spheres. Most results are reported in the range of 77–300 K, where it is of interest in liquid nitrogen and oxygen storage container design. Plotted in Fig. 9.5 are typical values for the apparent thermal conductivity of perlite versus residual gas pressure.[15] Note that minimum conductivity is achieved for pressures less than 10^{-2} torr. Also plotted are similar data for super-

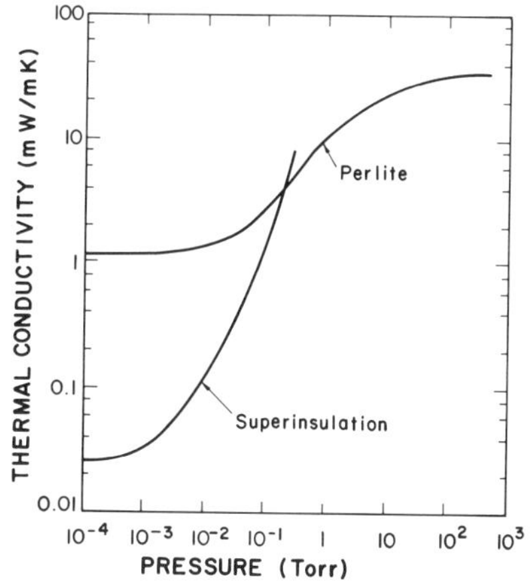

Fig. 9.5. Apparent mean thermal conductivity of glass powder and multilayer insulation versus residual gas pressure (as compiled by Lady[16]).

insulation. In this case, a much lower apparent conductivity is achievable at low pressures ($p \approx 10^{-4}$ torr); however, at intermediate pressures, the benefits of superinsulation are minimal. At high pressures, $p \gtrsim 10^{-1}$ torr, superinsulation actually has higher thermal conductivity than the powders.

As a final comment, an additionally interesting problem concerns the refrigeration of radiation shields. In some large systems where radiation can contribute substantially to the heat leak, it is possible to optimize the system by cooling the shields with closed-cycle refrigerators operating at intermediate temperatures. The advantage here is that improved thermodynamics can be achieved since radiation heat leaks from ambient temperature can be absorbed at higher temperatures, thus taking advantage of higher refrigeration efficiency. The thermodynamic principles needed to carry out this analysis are discussed in Chapter 8. For further details, the reader should consult references in the literature.[17,18]

9.2. THE HELIUM-3 ISOTOPE

By far, the most abundant naturally occurring isotope of helium is ^{4}He. However, there exists one other stable isotope, ^{3}He, which is only about 0.1 ppm of natural helium but can be obtained readily as a by-product of nuclear reactions. The physical behavior of low-temperature ^{3}He is fundamentally quite different from that of the common isotope. ^{3}He has application in a number of areas in cryogenics. Most notable of these is the role it plays in the achievement of temperatures below 1 K. This temperature range is primarily of interest in the studies of basic physical phenomena.

Interest in performing experiments with ^{3}He was not raised significantly until the mid 1940s, when it was determined that ^3He was a by-product of the Li^6 neutron capture reaction,

$$^6\text{Li} + n \rightarrow {}^3\text{H} + \alpha$$
$$\hookrightarrow {}^3\text{He} + \beta$$

$$\tau = 12.5 \text{ yr}$$

and therefore could be isolated in quantities sufficient for low-temperature experimentation. ^6Li is relatively abundant, occurring as 7.5% of natural lithium, and the β-decay reaction has a half-life of 12.5 yr. Consequently, the wide availability of neutrons in experimental fission reactors has made ^3He available to numerous researchers in low-temperature physics. Although the gas is a relatively expensive commodity, costing in the neighborhood of $50 per STP liter, it is still orders of magnitude less costly than it would be to separate ^3He from natural helium.

Some properties of liquid ^3He are listed in Table 9.2. Note that the critical point is below the normal boiling point of ^4He. In practice, ^3He is condensed by heat exchange with a boiling pot of superfluid ^4He ($T \approx 1$ K). The relatively low binding energy of ^3He causes the vapor pressure to be significantly higher than that of the common isotope. This fact is important for distillation of ^3He–^4He mixtures as well as reducing the pumping speed required for very-low-temperature ($T < 1$ K) refrigerators. One other

Table 9.2. Some Properties of Liquid ^3He[a]

Normal boiling point (K)	3.191
Critical constants	
T_c (K)	3.324
p_c (bar)	1.15
ρ_c (kg/m^3)	41.3
Density at 0 K (kg/m^3)	82.3
Compressibility at 0 K (mm^3/J)	361
Heat of vaporization at 0 K (J/mol)	20.56
Surface tension at 0 K (mJ/m^2)	0.16
Velocity of sound at 0 K (m/s)	183
Thermal conductivity at 3.2 K (mW/m · K)	20
Viscosity at 3.2 K (mg · s/m)[b]	1.9
Magnetic moment (nuclear magnetons)	−2.127
Vapor pressure at 1.7 K (kPa)[c]	10.9
Specific heat at 1.0 K(J/mol · K)	4.222
Superfluid transition (mK)	2.6

[a] Compiled by Lounasmaa.[3]
[b] 1 mg · s/m = 10 μP.
[c] 1 kPa = 10^{-2} bar = 7.5 torr.

important feature of ^3He is that it possesses a nonzero magnetic moment due to the unpaired nuclear spin. The magnetic moment is responsible for a large spin entropy which contributes a major fraction of the fluid entropy over much of its temperature range. For example, the ratio of specific heats of ^3He and ^4He at 1.0 K is about 40, a ratio that increases with decreasing temperature.

^3He is observed to experience a superfluid transition at 2.6 mK, about three orders of magnitude in temperature below the λ-point of ^4He. This ratio should not be surprising owing to the wholly different statistical character of the two fluids. Listed in Table 9.3 is a statistical comparison between ^3He and ^4He. ^4He has integral nuclear spin and obeys Bose–Einstein statistics. ^3He, on the other hand, has half-integer spin and a nuclear magnetic moment, thus obeying Fermi–Dirac statistics. These two statistical models predict very different physical behavior. Consider, for example, the condensation mechanisms associated with superfluid transitions in ^3He and ^4He. The superfluid transition from ^4He is described approximately by Bose–Einstein (BE) condensation. In the case of ^3He, the condensation is more analogous to that accurring in a superconductor which requires pairing of the Fermi particles.

9.2.1. ^3He–^4He Mixtures

Before continuing the discussion of pure ^3He, it is important to consider the behavior of mixtures of ^3He and ^4He. Mixtures have technically important characteristics which allow for improved methods of cooling at very low temperatures. The equilibrium phase diagram of the ^3He–^4He system is shown in Fig. 9.6. The λ-point, which determines the phase transition between He I and He II, is depressed by the addition of small quantities of ^3He. Much more significant for refrigeration is the existence of a phase-separation region below 1 K. ^3He–^4He mixtures are the only systems that experience a separation of phase due to isotopic mass differences only! The point where the λ-line intersects the phase separation is called the tricritical point (TCP). In the temperature–concentration plane, the TCP occurs at $T = 0.872$ K and $x = 0.669$. The TCP is physically quite

Table 9.3. Statistical Comparison

	^4He	^3He
Nuclear spin	Integral	Half integral
Statistics	Bose–Einstein	Fermi–Dirac
Magnetic moment	None	$-2.127\ \mu_B$
Condensation	BE type	BCS type

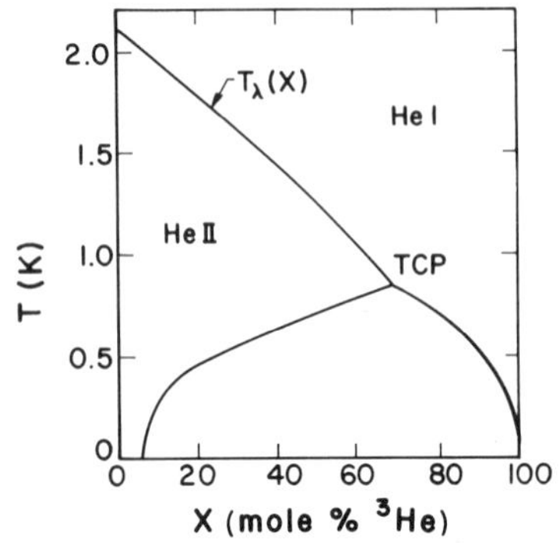

Fig. 9.6. Phase diagram of ^{3}He–^{4}He mixtures.

interesting because it represents the intersection of three phase boundaries, one being a second-order phase transition. Another feature apparent in Fig. 9.6 is the finite solubility of ^{3}He in the ^{4}He-rich phase. At absolute zero the phase boundary intersects the concentration axis at $x = 0.064$. This effect is contrary to classical physical arguments which require a two-phase region in equilibrium at absolute zero to be completely separated. The occurrence of finite solubility in ^{3}He–^{4}He mixtures is the result of the quantum statistical character of the system.

The existence of a finite solubility allows heat to be removed from the system by transferring ^{3}He from the pure component to the dilute component and thereby extracting the heat of mixing which is finite even at absolute zero. In a two-phase system, the entropy of mixing can be written

$$\Delta s = \frac{R}{M}\left[x \ln x + (1-x)\ln(1-x)\right] \tag{9.19}$$

which for the example at hand at $T \simeq 0$ K, $x = 0.064$, $\Delta s = 0.66$ J/g · K. At any finite temperature, a very sizable amount of heat can be removed by this mechanism. The process forms the basis for a well-established, very-low-temperature refrigeration method known as dilution refrigeration. The principles of dilution refrigeration are described here only briefly. For the reader who is interested in further detail, there is a very good review of the subject in Lounasmaa.[3]

The principal components of a continuously operating dilution refrigerator are shown in Fig. 9.7. The system consists of a closed-loop circulation system in which ^{3}He is usually the circulating component. Refrigeration is obtained by locating the phase boundary between the two components in the mixing chamber at the lower end of the system. Then by

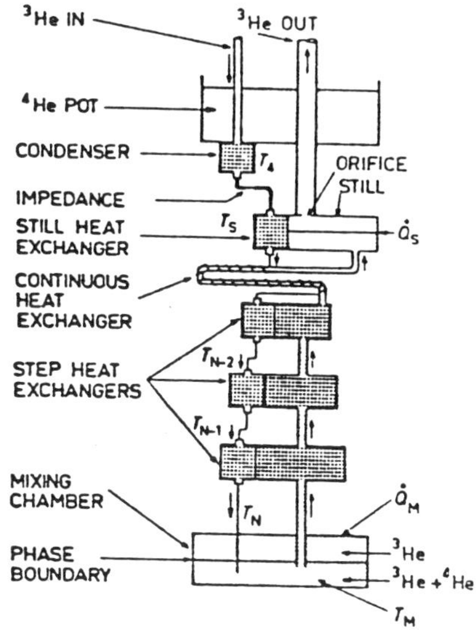

Fig. 9.7. Principal parts of a continuously operating dilution refrigerator.

circulating the ^{3}He, heat of mixing is removed from this region. The rate of cooling, independent of nonideal loss mechanisms, is

$$Q = \frac{\dot{m}}{M} T \Delta S$$

$$= \frac{\dot{m}}{M} RT[x \ln x + (1-x)\ln(1-x)] \tag{9.20}$$

where M, the molecular weight of ^{3}He, is equal to 3 g/mol. Based on (9.20) a mass flow rate of 0.3 mg/s is required to obtain a cooling power of 10 μW at 50 mK, reasonable values for a moderate-size dilution refrigeration system. The state-of-the-art in this technology is continuous cooling in the range of a few millikelvin.

In addition to the mixing chamber, there are several other important components to the dilution refrigerator. The ^{4}He pot, which operates around 1.2 K, condenses the circulating ^{3}He and provides a low-temperature heat sink for heat leaks into the system. Either concentric counterflow or discrete heat exchangers are used in dilution refrigerators. The latter type are more effective at obtaining the lowest temperatures. The still is the point in the cooling circuit where the ^{3}He is evaporated to produce the circulation. Typically, the still operates around 0.5 K where the ^{3}He vapor pressure is high enough to obtain sizable mass flow easily. Often a special ^{4}He film flow inhibitor is installed in the still so that the superfluid film will not creep up the pumping line and cause significant ^{4}He evaporation.

Circulation of ^{4}He can result in degradation of the performance of a ^{3}He–^{4}He dilution refrigerator.

Dilution refrigerators are the best method of obtaining continuous cooling a temperatures below 0.4 K. Other options that exist include pure ^{3}He evaporation which can produce more cooling power than dilution refrigerators above 0.4 K but becomes exponentially less effective with decreasing temperature. Also, adiabatic demagnetization can be employed to reach the millikelvin regime; however, it is generally not possible to obtain continuous cooling with this method.

9.2.2. Properties of Pure ^{3}He

By analogy to the treatement of pure ^{4}He in Chapter 4, it is appropriate to introduce the behavior of ^{3}He in terms of deviations from the ideal quantum gas model. In this case, it is necessary to consider quantum statistical mechanics applied to particles obeying Fermi-Dirac statistics. This problem is basically the same as the free-electron model in metals although the relative temperature range is quite different.

As is the case for the ideal Bose gas model, the properties of an ideal Fermi gas are established in terms of a summation over discrete energy levels. The properties of an ideal quantum gas can be obtained by taking these discrete levels in the density of states to be infinitesimal and converting the summation to an integral. The resulting expression for the number density can be written

$$\frac{N}{V}=\frac{2\pi}{h^3}(2m)^{3/2}\int_0^\infty \frac{\varepsilon^{1/2}\,d\varepsilon}{z^{-1}e^{\beta\varepsilon}+1} \tag{9.21}$$

where, as in the case of the Bose gas, $z=e^{\beta\mu}$ is the fugacity which is a function of temperature. Equation (9.21) is an application of (4.9) for the special case of a system of particles obeying Fermi–Dirac statistics. Note that an important corollary to Fermi statistics is that the system of particles must obey the Pauli exclusion and therefore can have energy level occupation numbers of 0 or 1. This fact is important because there is no need to be concerned with substantial occupation numbers in the ground state as was the case in Bose statistics. By further analogy with the discussion in Chapter 4, the energy density of a set of Fermi particles is

$$\frac{E}{V}=\frac{2\pi}{h^3}(2m)^{3/2}\int_0^\infty \frac{\varepsilon^{3/2}\,d\varepsilon}{z^{-1}e^{\beta\varepsilon}+1} \tag{9.22}$$

which can be used in consort with (9.21) to derive the properties of the gas. Note that for a set of Fermi particles having spin s, the energy levels have a

$(2s+1)$ degeneracy, which is only removed by splitting the levels in a magnetic field. At $T=0$, all levels are occupied up to the Fermi energy which is defined by the expression

$$\varepsilon_F = \frac{\hbar^2}{2m}\left(\frac{6\pi^2}{(2s+1)}\frac{N}{V}\right)^{2/3} = k_B T_F \tag{9.23}$$

Introducing numerical values appropriate to ^{3}He with spin $\frac{1}{2}$, the Fermi temperature T_F is computed to be 4.9 K. Therefore, considerable effects of degeneracy in ^{3}He are expected at low temperatures. For a finite temperature, a fraction of the particles are excited above the Fermi energy, the result of which is a smearing of the distribution function ε_F.

The specific heat of an ideal Fermi gas is obtained from the derivative of the internal energy per particle:

$$\frac{C_v}{Nk_B} = \frac{1}{Nk_B}\left(\frac{\partial E}{\partial T}\right)_v \tag{9.24}$$

The result of this calculation[19] is plotted in Fig. 9.8 for a range of energies including ε_F. Note that at high temperatures, $k_B T > \varepsilon_F$; the specific heat asymptotically approaches the Boltzmann limit of $\frac{3}{2}$. At low temperatures, $k_B T \ll \varepsilon_F$; the behavior of a degenerate Fermi gas is evident with a linear temperature dependence of the specific heat,

$$\frac{C_v}{Nk_B} \simeq \frac{\pi^2}{2}\frac{T}{T_F} \tag{9.25}$$

which is the same result obtained for the free-electron model in Chapter 2.

The magnetic susceptibility of an ideal Fermi gas is derived similarly. At high temperatures, $T \gg T_F$, the system approaches the classical limit obtained by Curie–Langevin,

$$\chi = \frac{\mu^2}{k_B T}\left(\frac{N}{V}\right) \tag{9.26}$$

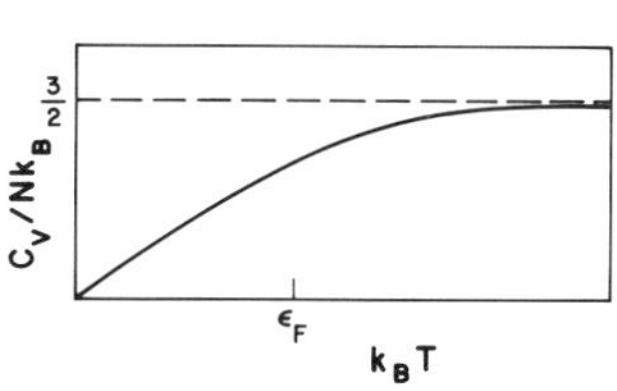

Fig. 9.8. Specific heat of an ideal Fermi gas.

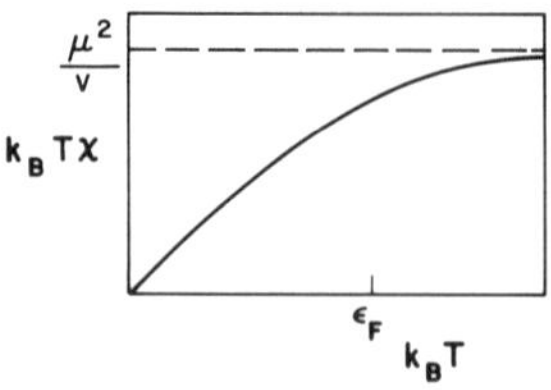

Fig. 9.9. Susceptibility of an ideal Fermi gas.

while at low temperatures the susceptibility approaches a constant modified by a quadratic term,

$$\chi = \frac{3\mu^2}{2k_B T_F}\left(\frac{N}{V}\right)\left[1 - \frac{\pi^2}{12}\left(\frac{T}{T_F}\right)^2 + \cdots\right] \tag{9.27}$$

where in both equations μ is the magnetic moment of the particle. The susceptibility of an ideal Fermi gas over a range of temperatures including T_F is plotted in Fig. 9.9.

The behavior of liquid ^{3}He is considerably different from that of an ideal Fermi gas. This is primarily the result of the sizable interparticle interactions which occur in the fluid. Landau's theory of Fermi liquids is a fairly straightforward approach to this problem.[20,21] This theory, which is primarily a modification to the ideal quantum gas model, calls for introducing an effective mass m^*, which in turn modifies the relationship for the physical properties of the Fermi liquid. The effective mass should then be calculable and compared with experimental results for ^{3}He.

Consider first the behavior of the specific heat at low temperatures. The Landau theory predicts that

$$\frac{C_v}{Nk_B} = \frac{m^*}{m}\left(\frac{C_v}{Nk_B}\right)_{\text{ideal}} \simeq \frac{\pi^2}{2}\left(\frac{T}{T_F^*}\right) \tag{9.28}$$

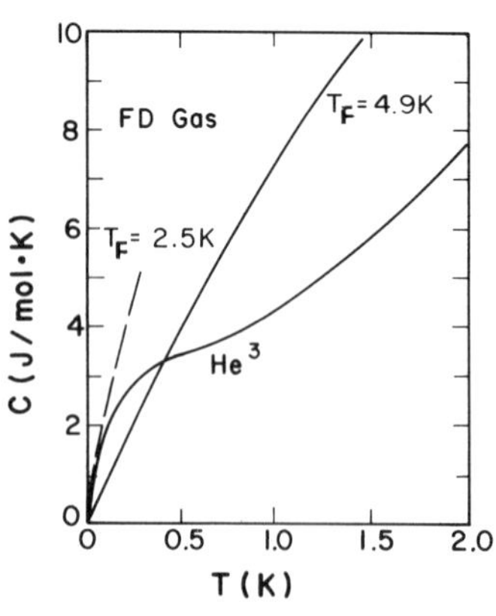

Fig. 9.10. Specific heat of liquid ^{3}He compared to curves for ideal Fermi gas.

where in this case the modified Fermi temperature is given by the expression

$$T_F^* = \left(\frac{m}{m^*}\right) T_F \tag{9.29}$$

By comparison of (9.28) with experiment it is possible to determine the effective mass appropriate to ^{3}He. Plotted in Fig. 9.10 is the specific heat of liquid ^{3}He compared to that of an ideal Fermi gas of liquid density, $T_F = 4.9$ K. Clearly, the ideal gas model is a poor representation of the experimental data. A better fit to the low-temperature specific heat is obtained with a choice of effective mass about twice that of the ^{3}He nucleus, which indicates the Fermi temperature appropriate to the Landau model is closer to 2.5 K.

The other property to compare with theory is that of the magnetic susceptibility at low temperatures. According to Landau's theory the low-temperature susceptibility should take the form

$$\chi = \frac{3\mu^2}{2k_B T_F^{**}}\left(\frac{N}{V}\right)\left[1 - \frac{\pi^2}{12}\left(\frac{T}{T_F^{**}}\right)^2 + \cdots\right] \tag{9.30}$$

where now the modified Fermi temperature T_F^{**} is not necessarily the same as that appropriate to the specific heat. The corresponding experimental results are shown in Fig. 9.11. Note that the general form of the susceptibility is in agreement with theory. Listed in the figure are the best-fit values for T_F^{**} as a function of pressure. Over the entire range of pressure, the effective temperature T_F^{**} is considerably less than T_F^* obtained from specific heat data. Also, the Fermi temperature T_F^{**} decreases with increasing pressure, an effect contrary to the predictions of the ideal Fermi gas model.

Fermi liquid theory can be extended to describe the transport properties of ^{3}He. The theoretical development of this problem stems from a solution to the semiclassical Boltzmann equation. In this model the viscous

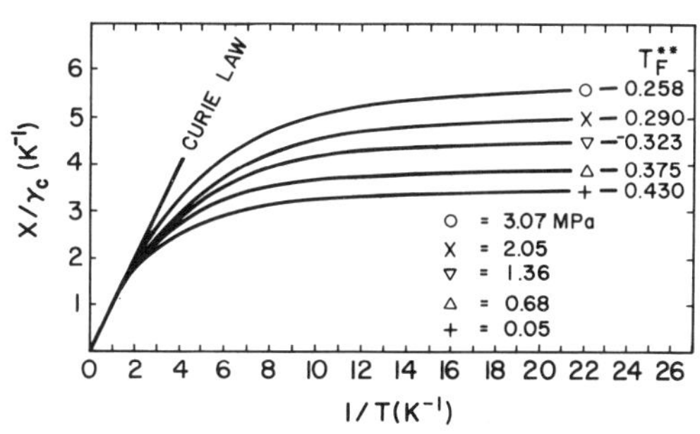

Fig. 9.11. Susceptibility of liquid ^{3}He versus temperature and pressure (as compiled by Keller[19]).

and thermal interactions are associated with the scattering of quasiparticles at the Fermi surface. For an isotropic fluid, the relationship for the viscosity may be written in the general form

$$\eta = \frac{A(P_F, m^*)}{T^2} \tag{9.31}$$

where A is a function of the effective mass and P_F is the momentum at the Fermi level. Similar analysis for the thermal conductivity yield a form for this property,

$$k = \frac{B(P_F, m^*)}{T} \tag{9.32}$$

where B is a similar function of P_F and m^*. Note that these relationships are similar to those derived for helium gas in Chapter 3. At that time it was noted that the ratio of the thermal conductivity to the viscosity took on a value of approximately $\frac{5}{2}C_v$ which is at least approximately correct for Fermi liquids.

At low temperatures, $T \lesssim 0.1$ K, the viscosity of ^{3}He has been shown to be consistent with the inverse square relationship in (9.31). There appears to be some discrepancies between experiments but over the low-temperature range η can be fit to the relationship[22]

$$\eta = (0.38 \pm 0.06)\, T^{-2}\ \mathrm{mg \cdot s/m} \tag{9.33}$$

The thermal conductivity according to theory should be proportional to T^{-1}. The inverse temperature dependence of the thermal conductivity can be seen from (9.32) as well as the fact that a degenerate Fermi liquid has C_v proportional to T which when coupled with the form of the viscosity yields $k \approx T^{-1}$. Experimental results generally support this conclusion although only at very low temperatures, $T < 20$ mK, where C_v is linear with temperature. As a point of reference, the thermal conductivity of liquid ^{3}He at 3 mK is equal approximately to that of pure copper at the same temperature.

9.2.3. Submillikelvin Refrigeration

To study the behavior of ^{3}He or any other material below a few millikelvin, it is necessary to use an additional cooling mechanism to that of dilution refrigeration. There are primarily two methods to achieve lower temperatures, one being adiabatic demagnetization and the other Pomeranchuk cooling.

Adiabatic demagnetization has been used extensively for very-low-temperature refrigeration. It is usually a single-cycle process that does not provide continuous cooling. Some methods, available for continuous magnetic refrigeration at higher temperatures, are discussed in Section 9.4. To achieve cooling by adiabatic demagnetization it is necessary to have a working magnetic material which has weakly interacting spins so that the ordering temperature is below the minimum temperature to be achieved. For cooling of ^3He the most effective material is cerium magnesium nitrate (CMN) because its magnetic ordering temperature is below 1 mK (see Table 9.5).

Adiabatic demagnetization of CMN has been used extensively to study ^3He in the range of a few millikelvin. The method has advantages in that it is indirect, not affecting the thermodynamic state of the ^3He except perhaps by the imposition of a magnetic field. However, because cooling is indirect the thermal relaxation time for the transfer of entropy from the liquid to the CMN can be very long, ultimately affecting the minimum achievable temperature. Even with this limitation, adiabatic demagnetization of CMN is the most widely applied method for achieving ^3He temperatures below 1 mK. If lower temperatures than this are desired, the only presently successful method is demagnetization of nuclear spins in materials such as Cu or $PrNi_5$.[23,24]

An alternative method used to cool ^3He to about 1 mK is adiabatic compression of ^3He along its melting line. The method was first suggested by Pomeranchuk,[25] who predicted the existence of a minimum in the solid–liquid phase boundary for pure ^3He ($T_{min} = 319$ mK, $p_{min} = 2.913$ MPa). The minimum exists because the spin ordering in the liquid occurs at higher temperatures than that in the solid. Therefore, between the ordering temperature of the liquid and that of the solid, the entropy of the solid is greater and cooling can be achieved by adiabatically compressing the liquid through the solid–liquid phase transition.

The thermodynamics of Pomeranchuk cooling are reasonably straightforward given the existence of the minimum in the solid–liquid phase boundary in ^3He. Using the Clausius–Clapeyron equation for the phase transition,

$$\frac{dp}{dT} = \frac{s_l - s_s}{v_l - v_s} \tag{9.34}$$

for temperatures below the minimum the liquid state has a greater specific volume than that of the solid. Below about 40 mK, the denominator in (9.34) takes on approximately a constant value of 1.31 cm^3/mol. The entropy of each phase is dominated by the spin contribution. The liquid has an entropy that is approximately linear with temperature,

$$s_l = \frac{R}{M}\left(\frac{T}{T_F^{**}}\right) \ln 2 \tag{9.35}$$

where T_F^{**} is the pressure-dependent magnetic Fermi temperature shown in Fig. 9.11. The entropy of the solid is nearly constant until its magnetic ordering temperature, $T_N = 1.1$ mK. The relationship for the solid entropy is more complex, involving the nuclear exchange interaction, J, that is,

$$s_s = \left(\frac{R}{M}\right) \ln 2 - \frac{1}{M} \sum_n A_n \left(\frac{J}{k_B T}\right)^{n+1} \tag{9.36}$$

where A_n are weighting factors in the summation. The behavior of the entropy of each phase is shown in Fig. 9.12.

Pomeranchuk cooling is achieved by first bringing the system to thermodynamic equilibrium at a temperature below the minimum in the solid–liquid phase boundary. This is best carried out by using a continuously operating dilution refrigerator. With the fluid compressed to just below the melting line, the Pomeranchuk cell is then isolated from the dilution refrigerator and adiabatically compressed across the phase boundary. Temperature reductions of at least an order of magnitude are possible. For example, starting with the system at 30 mK, point ① in Fig. 9.12, adiabatic compression along line ① to ② achieves a minimum temperature of about 1 mK. Obviously, as with adiabatic demagnetization, this method is not suited to continuous refrigeration at the lower temperature. Furthermore, the end product of the cooling stage is a two-phase mixture of solid and liquid ^3He, which has the added difficulty of separating the physical phenomena occurring in each phase. This situation can lead to confusion when one of the phases is experiencing most unusual behavior as occurs in ^3He below 3 mK. As a result of these complications and the interest in studying the magnetic field dependence of ^3He, the most

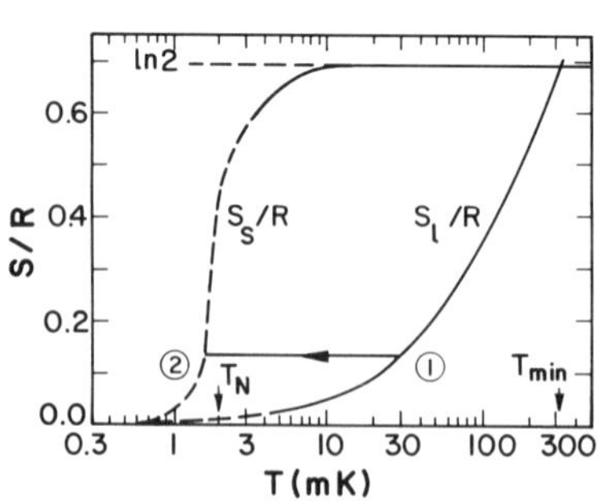

Fig. 9.12. Entropy of solid and liquid ^3He along the melting curve. Temperatures indicated include T_{min}, the minimum in the melting curve and T_N, the spin ordering temperature in the solid.

popular method for cooling ^{3}He to submillikelvin temperatures is adiabatic demagnetization.

9.2.4. Superfluid ^{3}He

Predictions as early as the 1960s indicated that ^{3}He would become superfluid by a mechanism similar to that occurring in superconductors. The uncertainty was based mainly on determining the transition temperature. The physical basis known as BCS-type interaction has a well-established form for the transition temperature,

$$T_c = 2.28 \frac{\hbar^2 P_F^2}{k_B m^*} \exp\left(\frac{\pi}{2}\frac{h^2}{m^*}\frac{P_F}{\langle\phi|\, U\,|P_F\rangle}\right) \tag{9.37}$$

where U is the exchange interaction and ϕ is the particle wave function. It is difficult to predict T_c from first principles because m^*, the effective mass, is primarily an experimental quantity. Since m^* enters in the denominator of the exponential argument, small changes in its value can have profound effect on T_c. However, with liquid helium research perpetually pursuing lower temperatures, it was only a matter of time before superfluid ^{3}He would be discovered.

The first observations attributed to a new phase transition in ^{3}He were made by Osheroff et al.[26] in 1972. Their experiment was an investigation of the pressure variations along the liquid–solid phase boundary using a Pomeranchuk cell. Two phase transitions were observed. The first, referred to as the A-transition, occurred at 2.65 mK and was associated with a slope change in the pressure variation of the cell, suggesting a higher-order phase transition. The lower-temperature phase transition B, occuring at 1.8 mK, was seen as a first-order transition with an associated latent heat. Since that time, an enormous amount of research has been conducted on these new phases of ^{3}He. In this brief review, it is not possible to do justice to the state-of-the-art in superfluid ^{3}He research. Fortunately, a number of excellent reviews are available in the literature.[22,27,28] The present section presents some of the interesting properties of these new phases of ^{3}He.

To begin, consider the phase diagram of ^{3}He in the very-low-temperature regime as it depends on pressure, temperature, and magnetic field; see Fig. 9.13. The dependence on magnetic field is quite unique for pure liquids. The diagram indicates the existence of primarily three phases in addition to the normal Fermi liquid occurring above T_c. The intersection point between the A, B, and normal Fermi liquid phases at zero magnetic field is referred to as the polycritical point (PCP). Note that the A phase takes precedence over the B phase with increasing magnetic field, indicating A is more strongly magnetic than B.

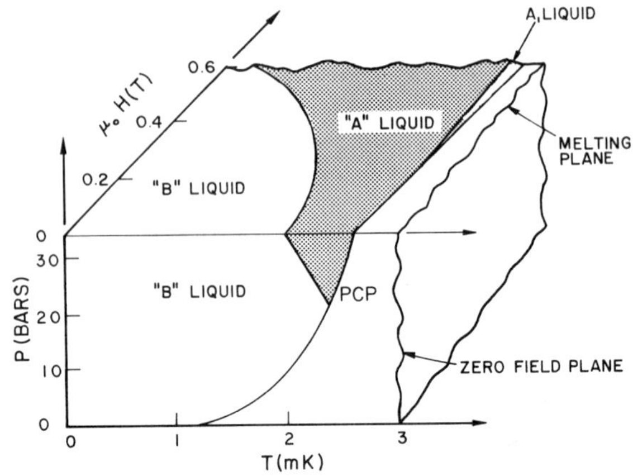

Fig. 9.13. *P-T-H* phase diagram of very-low-temperature ^{3}He.

The theoretical description of superfluid ^{3}He is extremely complex and beyond the scope of the present discussion. Evidence suggests that three phases occur in ^{3}He because it is a magnetic superfluid that pairs in a triplet state. This fact establishes an important difference between the interactions occuring in superfluid ^{3}He and those in superconductors. In the latter case the pairing is in the singlet state and consequently only one phase is present. In ^{3}He three orientations are possible for the nuclear spins, two with parallel and one with antiparallel alignment. Since the A phases are more strongly magnetic, it is expected that they contain mostly parallel spin orientations. The *B* phase being less magnetic must contain a significant fraction of antiparallel aligned spins. This effect has been shown experimentally through susceptibility measurements. The A and A_1 phases have magnetic properties approximately the same as the normal Fermi liquid with a temperature-independent susceptibility. The B phase has a decreasing susceptibility with temperature suggestive of an antiferromagnetic ordering.

Perhaps the most interesting aspects of ^{3}He in this low-temperature regime are those normally associated with superfluidity, that is, anomalous behavior in the energy transport mechanisms. As discussed in Chapter 4, superfluid ^{4}He (He II) has a number of very unique transport properties. Among these are new sound propagation mechanisms (second sound), viscosity that depends on the mechanism used for measurement, the existence of vortex lines in the bulk fluid, and extremely high heat transport capability. The latter feature makes He II a technically useful fluid. Generally, these properties are explained successfully in terms of the reasonably simple two-fluid model. It is therefore of interest to consider similar characteristics of superfluid ^{3}He.

One property that makes a superfluid very different from ordinary fluids is its unique sound propagation mechanisms. In addition to ordinary sound propagation, there is second sound and fourth sound which is propagation through a superleak. This latter sound propagation mechanism is one of several methods used to measure the superfluid density. In fact, it can be shown from the two-fluid model that the superfluid density ρ_s is related directly to the fourth and first sound velocities:

$$\frac{\rho_s}{\rho} = \left(\frac{c_4}{c_1}\right)^2 \tag{9.38}$$

Since ρ_s is an important parameter in the theory of superfluidity, experimental determinations of ρ_s can be compared with calculation. Such an analysis, based on a renormalized theoretical calculation,[29] is well beyond the scope of the present discussion. However, it is important to be aware of the substantial difference between the behavior of ρ_s in ^{3}He and that for the same parameter in ^{4}He. These properties have different temperature dependencies owing to the different condensation mechanisms. In the BCS theory, which is appropriate for superfluid ^{3}He, the superfluid density goes as the order parameter squared which is approximately linear with temperature near T_c:

$$\frac{\rho_s}{\rho} \approx 1 - \frac{T}{T_c} \tag{9.39}$$

This equation is roughly consistent with experiment. By contrast, the temperature dependence of ρ_s for ^{4}He varies as $1 - (T/T_\lambda)^{5.6}$.

Now turn to the problem of viscous flow in superfluid ^{3}He. Above T_c, the viscosity of ^{3}He is that of a normal Fermi liquid, which increases with decreasing temperature as $1/T^2$. Therefore, in the millikelvin range, the viscosity is very large, being of the order of 10^{-2} kg · s/m (10^{-1} P), a value comparable to the viscosity of light machine oil. The viscosity of ^{3}He decreases for temperatures below T_c; it never becomes vanishingly small. In fact, there is a minimum in the viscosity occurring around 1.6 mK for which the ratio to that of the Fermi liquid at the A transition is approximately 0.2. Therefore, although superfluid ^{3}He has a decreasing viscosity, it is still orders of magnitude more viscous than ^{4}He near T_λ.

Heat transport in superfluid ^{3}He can be discussed in terms of two-fluid hydrodynamics. Therefore, the mechanism which is appropriate for describing heat flow is internal convection between the normal and superfluid components. For fine capillary tubes of diameter d, at low heat fluxes and laminar flow conditions, the dominant mechanism is the normal fluid viscous interaction with the tube walls. In this case, an effective thermal conductivity k_e may be defined as

$$k_e = \frac{d^2(\rho s)^2 T}{32\,\eta_n} \tag{9.40}$$

where η_n is the normal fluid viscosity. However, in the case of ^3He the value of η_n is about four orders of magnitude larger than η_n for ^4He around 1.5 K. Therefore, the effective thermal conductivity of superfluid ^3He in small-diameter channels should be much less than occurs in ^4He. To establish some rough numerical values for this property, note that the thermal conductivity of the normal Fermi liquid increases with decreasing temperature until it reaches a value of 0.1 W/m · K just above T_c. Below T_c, the effective thermal conductivity in a channel of diameter 3 mm, is about 10 times higher or 1 W/m · K. Although superfluid ^3He does not possess the phenomenally high heat conductivity capability evident in superfluid ^4He, it exists at much lower temperatures where its heat conductivity is comparable to that of pure metals.

In the above brief description of the properties of ^3He, both normal and superfluid, a number of interesting features have been reviewed. A great deal more is known about ^3He, however, space and theoretical limitations do not permit further analysis of this interesting system. There are a number of other unique properties of superfluid ^3He not described heretofore. These properties include the anisotropic nature of the system. Because of magnetic properties, the fluid can be oriented in a magnetic field. Also, superfluid ^3He is believed to possess some spatial correlations called "textures," referring to correlated behavior of ^3He over dimensions long compared to the cohenrence length. Needless to say, these are very new theoretical and experimental topics, which explains the substantial research effort in this very exciting field.

9.3. HELIUM ADSORPTION

Another interesting and technically significant topic related to the field of helium cryogenics is that of physical adsorption. Physical adsorption is a general term used to describe the process whereby an inert molecule comes in contact with and adheres to an inert surface or substrate. The inert character of each component is important because if substantial chemical reaction occurs, the process is referred to as chemisorption. Since helium is inert and does not bond chemically to any other element, the adsorption process is described by physical mechanisms similar to those encountered in bulk condensation. It is often possible to treat the adsorbed film as an independent system, with the substrate forming the noninteractive boundaries much like walls of a container form the boundaries to bulk fluids.

Studies of adsorbed gases over the past few decades have expanded the understanding of the basic physical processes. Growth in the field can be attributed primarily to two factors. First, recent developments have introduced a number of experimental techniques for surface investigation. Several of these techniques are sensitive to one atomic layer, allowing investigations of monolayer and submonolayer films. Second, scientists have discovered substrates that are regular and homogeneous and permit nearly ideal film behavior. Due to surface morphology, real surfaces tend to have considerable variability in adsorption potential which can result in inhomogeneous films. Although this situation is realistic in practice, it gives little insight into the detailed physical behavior of the film itself.

Physical adsorption has considerable technical importance in the field of cryogenics. One application is in cryopumping, a process whereby a gas is pumped from a system by exposing it to a large surface at low temperatures. This technology is used in numerous cryogenic devices. When properly designed and used, cryopumps provide clean, oil-free vacuum systems with high capacities and essentially no moving parts. Another area of technical application for adsorption is in the separation of rarified gases. In this method, more strongly interacting molecules are adsorbed onto the substrate leaving the vapor phase enriched with the lighter more weakly bound molecules. Subsequently, the substrate can be "regenerated" by raising its temperature to desorb the heavier species. A major sector of the technology of air separation is based on this concept. In addition to its use in cryopumping and separation systems, the existence of physically adsorbed layers has considerable indirect impact on the behavior of cryogenic systems. One example not fully understood is the variation of surface heat transfer rate caused by the existence of adsorbed films. This can affect both solid–liquid helium heat transfer processes such as Kapitza conductance as well as radiative heat transfer, where variation of emissivity with cryodeposits can occur. All these factors make knowledge of the physical adsorption process important for proper understanding of the behavior of low-temperature systems.

9.3.1. Adsorption Thermodynamics

To begin, consider a simple thermodynamic description of the adsorption process. A more extensive treatment of this type of analysis is given in Dash.[30] A closed container, shown in Fig. 9.14a, has N_v molecules in the vapor state at temperature T_v and an equilibrium pressure p. Assume that T_v is high enough that the molecules do not adhere in any significant numbers to the container walls, which are also at T_v. One of the container

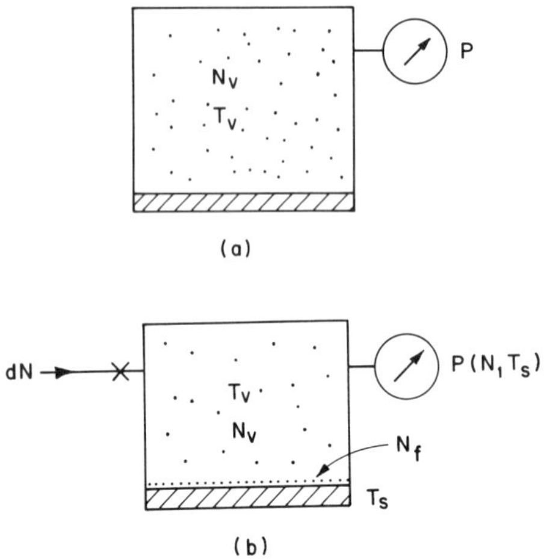

Fig. 9.14. Schematic representation of adsorption. (a) A closed volume with N_v molecules in the vapor phase. The number of molecules adsorbed, N_f, is small compared to N_v because $T_s \gg T_c$. (b) The same volume with $T_s \lesssim T_c$ so appreciable adsorption occurs.

walls forms the adsorption substrate at a temperature T_s that can be independently regulated relative to T_v. Of interest are the physical processes that take place as the substrate temperature is reduced below T_v. Initially, as T_s decreases, the pressure will decrease also in rough proportion because the vapor temperature will be some thermodynamic average of T_s and T_v. However, once T_s reaches a temperature of the order of the critical temperature of the vapor species, 5.2 K in the case of ^{4}He, a substantial quantity will begin to adhere to the substrate. The mechanism whereby this adherence occurs is the van der Waals interaction between the oscillating electric dipoles. A number of approximate potentials exist to model this interaction. One of these is the Lennard–Jones (LJ) 12-6 potential, which was introduced to describe interactions in nonideal gases in Chapter 3. In the case of an ideal two-dimensional planar surface comprised of substrate atoms, it is possible to write a modified form of the LJ potential for a gas atom a distance d above the plane,

$$U(d) = 4\pi\varepsilon n_s \left(\frac{\sigma^{12}}{45d^9} - \frac{\sigma^6}{6d^3} \right) \tag{9.41}$$

where n_s represents the density of substrate atoms and σ and ε are the Lennard–Jones parameters for the gas–substrate atom interaction.

Now consider the impact of introducing a small quantity of gas dN to the system. This situation is illustrated in Fig. 9.14b. If T_s is sufficiently low, some of the gas will be adsorbed, increasing the total particle number on the substrate by dN_f. The remainder will stay in the vapor such that $dN_v = dN - dN_f$. For a closed system in which this extra gas is introduced reversibly, the chemical potentials of each phase are equal, $\mu_f = \mu_v$. This condition establishes a relationship between the entropy of each phase:

$$\left(\frac{\partial S_f}{\partial N_f}\right)_{E,A,V} = \left(\frac{\partial S_v}{\partial N_v}\right)_{E,A,V} \tag{9.42}$$

The behaviors of the two phases are related inherently through thermodynamics. For example, detailed information about the state of the film can be extracted from measurements of pressure of the equilibrium vapor phase.

In bulk matter, the behavior of the system is determined by an equation of state, a relationship between pressure, temperature, and volume. By analogy, in a film system an equation of state exists relating the three surface relevant parameters: spreading pressure ϕ, temperature, and coverage of capacity x. However, under most conditions it is not possible to measure the film spreading pressure directly so that it must be inferred from bulk vapor pressure determinations. The most common situation is to construct an equation of state relating bulk vapor pressure, temperature, and capacity. Such a relationship is shown schematically in Fig. 9.15 which represents a surface equilibrium of an adsorption system. Note that processes occuring at constant capacity are referred to as "isosteric" by analogy with isochoric or constant volume processes in bulk systems.

On of the most important parameters in an adsorption system is the isosteric heat of adsorption, q_{st}. This quantity is defined as the amount of

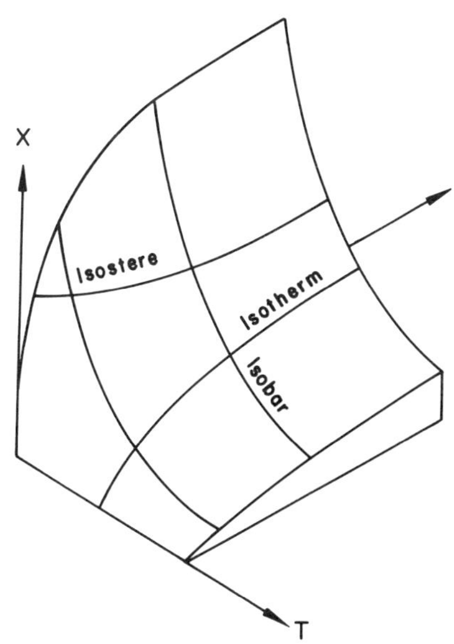

Fig. 9.15. An adsorption equilibrium surface.

energy required to bring a molecule from the film into its equilibrium vapor. In the limit of zero coverage and at absolute zero, the isosteric heat is simply the single-particle binding energy. As the film grows, this value decreases until for very thick films $q_{st} \to \lambda$, the latent heat of vaporization of a bulk condensate comprised of film molecules. These values form the limits for q_{st}. However, the detailed structure of this quantity can give considerable information about the thermodynamic state of the film.

The thermodynamic definition of q_{st} is given in terms of the entropy change at constant coverage:

$$q_{st} = -T\left(\frac{\partial S}{\partial N_f}\right)_{T,p,A} \tag{9.43}$$

By holding T, p, and A constant the variation of q_{st} is essentially along a constant coverage isostere; see Fig. 9.15. Substituting the relationship for equilibrium between the film and vapor, (9.42), we can rewrite (9.43) as a function of individual component entropies,

$$q_{st} = -T\left[\left(\frac{\partial S_f}{\partial N_f}\right)_{T,A} - \left(\frac{\partial S_v}{\partial N_v}\right)_{T,p}\right] \tag{9.44}$$

An alternate definition of the isosteric heat is given in terms of a temperature derivative of the equilibrium vapor pressure. This expression, which can be derived explicitly from (9.44), is a function of easily measurable quantities:

$$q_{st} \equiv k_B T^2 \left(\frac{\partial(\ln p)}{\partial T}\right)_{N_f,A} \tag{9.45}$$

In the simplest case where q_{st} is a constant, integration of (9.45) establishes an experimental form of the equilibrium vapor pressure,

$$\frac{p}{p_0} = e^{-q_{st}/k_B T} \tag{9.46}$$

where p_0 is a constant of integration. This expression identifies the importance that the isoteric heat plays in determining the ultimate pressure of an adsorption system.

The traditional method used for measurement of q_{st} is by vapor pressure isotherms. The behavior of the isotherms depends strongly on the type of substrate material and adsorbed gas. Substrate materials considered range from nearly ideal single-crystal surfaces to more practical inhomogeneous surfaces composed by polycrystalline metals or insulators. In addition, a given surface can be modified by first preplating the substrate with a more strongly adsorbed gas which then forms the substrate for subsequent adsorption. Experimental data exist for numerous systems including almost all commonly available gaseous elements and a number of compounds. The temperature range of investigation in these studies is determined mostly by the strength of the gas–substrate interaction.

Most substrates encountered in practical adsorption problems are inhomogeneous, composed of polycrystalline or amorphous materials. Because of their nonuniformity, only qualitative models are available to describe their behavior. For inhomogeneous substrates the vapor pressure isotherm follows the general form shown in Fig. 9.16. Initially, the vapor pressure is quite low because the first layer is bound strongly. However, as the film builds thickness, the pressure increases rapidly until it asymptotically approaches that of the bulk liquid, p_s. An expression can be derived to compare with the results of Fig. 9.16 by assuming a layer-dependent form of q_{st} in (9.46). Considering only the attractive term in the van der Waals expression (9.41), we obtain the Frenkel–Halsey–Hill equation[30]

$$\frac{p}{p_s} = \exp\left(\frac{-\alpha}{k_B T \bar{d}^3}\right) \tag{9.47}$$

where α represents the strength of the attractive potential on the bare substrate and should be approximately equal to q_{st} at low coverage, and $\bar{d}$ is the number of atomic layers. A comparison between the adsorption of ^{4}He on a mylar surface and that derived from (9.47) is made in Fig. 9.16. Theory and experiment are in qualitative agreement although the theory is not particularly sensitive to the detailed nature of q_{st}.

The limiting value of q_{st} is determined from the slope of low-coverage isotherms. Plotted in Fig. 9.17 is the absolute pressure versus T^{-1} for ^{4}He adsorbed on a copper sponge substrate.[31] The general form of (9.43) is

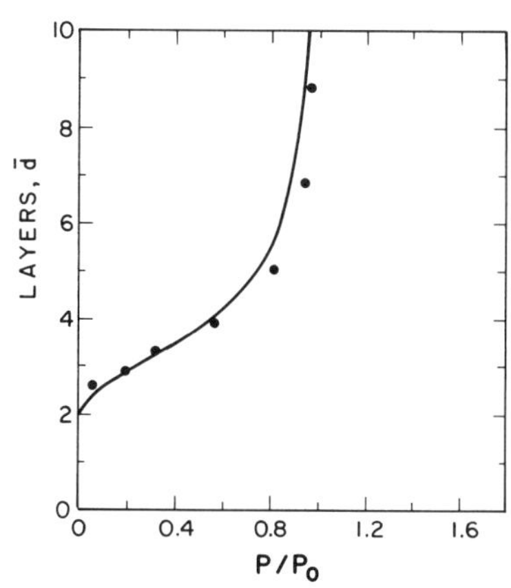

Fig. 9.16. An adsorption isotherm for a mylar substrate representing typical inhomogeneous character. The curve is based on the Frenkel–Halsey–Hill expression (9.44) (reprinted from Dash[30]).

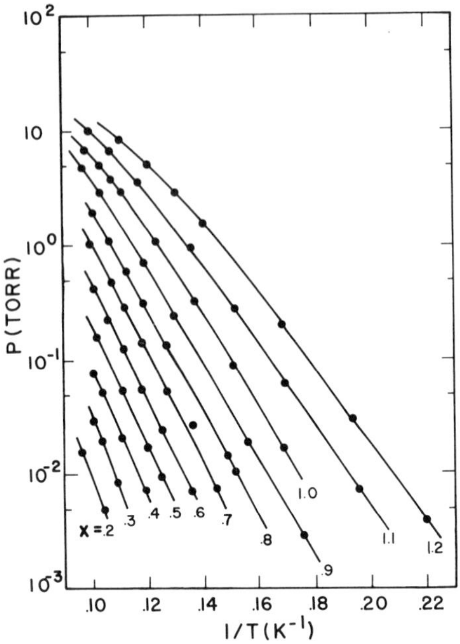

Fig. 9.17. Temperature dependence of the vapor pressure for constant coverage films on copper sponge. x refers to fractional monolayer coverage (reprinted from Princehouse[31]).

obeyed. Furthermore, by extrapolation to zero coverage, it is possible to deduce q_{st} for the substrate. A number of different measurements of q_{st} for ^{4}He at low coverage are listed in Table 9.4. Note that the values of q_{st} for submonolayer ^{4}He are at least an order of magnitude larger than the latent heat of vaporization. As a result, there can be considerable adsoption of ^{4}He even at temperatures in the vicinity of T_c. However, the total quantities adsorbed are strong functions of temperature and are small because the process has an exponential dependence.

As more gas is adsorbed on the surface the vapor pressure increases quite strongly. This effect can be seen most clearly in Fig. 9.16. Initially, the equilibrium vapor pressure is quite low due to the large value of q_{st}. However, as N_f increases, q_{st} decreases until for films having an equivalent

Table 9.4. Isosteric Heat of Adsorption at Low Coverage for ^{4}He on Various Substrates

Substrate	q_{st}/k_B (K)	Reference
Copper	160	31,32
Argon plated TiO_2	230	33
Exfoliated graphite	143–156	34, 35
Zeolite (13X)	220	36

thickness of 4–6 atomic layers the vapor pressure is essentially that of the bulk. The quantity of adsorbed gas necessary to complete one monolayer is an important parameter. Not only is this parameter an effective method of determining the surface area of the substrate but it also can be used as a fiducial point for estimating the density of submonolayer films. A number of methods exist for measuring monolayer completion. The most direct is to measure the vapor pressure isotherm. Above layer completion, the average position of the next absorbed gas molecule must be further from the substrate, and thus the molecule experiences a weaker binding. In an ideal system the value of q_{st} would decrease discontinuously. However, most substrates are inhomogeneous and the layers are compressible so that, more commonly, q_{st} undergoes a change in slope at layer completion. The precise value of layer completion is also quite temperature dependent, owing to thermal excitations in the film, and therefore is defined only approximately for any substrate. Listed in Table 9.5 are measured values for monolayer capacities of different substrates. Note that the capacity per unit area of substrate does not vary greatly from substrate to substrate. This is because the surface area is determined primarily by the hard sphere radius of the adsorbed molecule. On the other hand, there are wide variations of effective surface areas per gram of substrate material. Considerable benefit for cryopumping can be obtained by use of one of the high-surface-area materials. It should be pointed out that high adsorption capacity does not necessarily imply good cryopumping characteristics because the latter is a rate-dependent process. The rate at which adsorption occurs in high-surface-area materials is mostly a function of gas flow hydrodynamics within the material, a characteristic not necessarily desirable for high-speed cryopumps.

Vapor pressure isotherms are most commonly of the type described above, having a smoothly varying relationship between N_f and p. However, for some substrates that are very uniform this dependence is not observed. In these systems a discontinuous slope change in the vapor pressure isotherm occurs at monolayer completion. Further adsorption occurs on

Table 9.5. ^{4}He Capacities of Various Adsorbent Materials

Substrate	Monolayer capacity ($STP \cdot cm^3/m^2$)	Area (m^2/g)	Reference
Sintered copper sponge ($2 \times 2 \times 0.5\ \mu m^3$ particles)	0.24	0.41	31
Zeolite 13X	0.29	527	32
Exfoliated graphite	0.36	25	35, 37
Vycor glass	0.41	130	38

the second layer with a correspondingly lower value of q_{st}. This process continues until layer completion where a second step in the isotherm occurs. In the ideal case this mechanism for layer growth would continue indefinitely until the thermal excitations of the molecule of the order of $k_B T$ blur the distinction between individual layers.

Perhaps the best example of a near-uniform adsorption surface is exfoliated graphite. Graphite possesses a regular crystal structure with only slight spatial variations in adsorption potential. As a result, layer by layer adsorption occurs, which is reflected in the vapor pressure isotherms. Plotted in Fig. 9.18 is an absorption isotherm for krypton on exfoliated graphite. The discrete nature of the layer formation is apparent until well into the fourth layer. Such regularity of adsorption has also been observed for ^{4}He adsorbed on graphite, although in this case only two distinct layers are evident.[35]

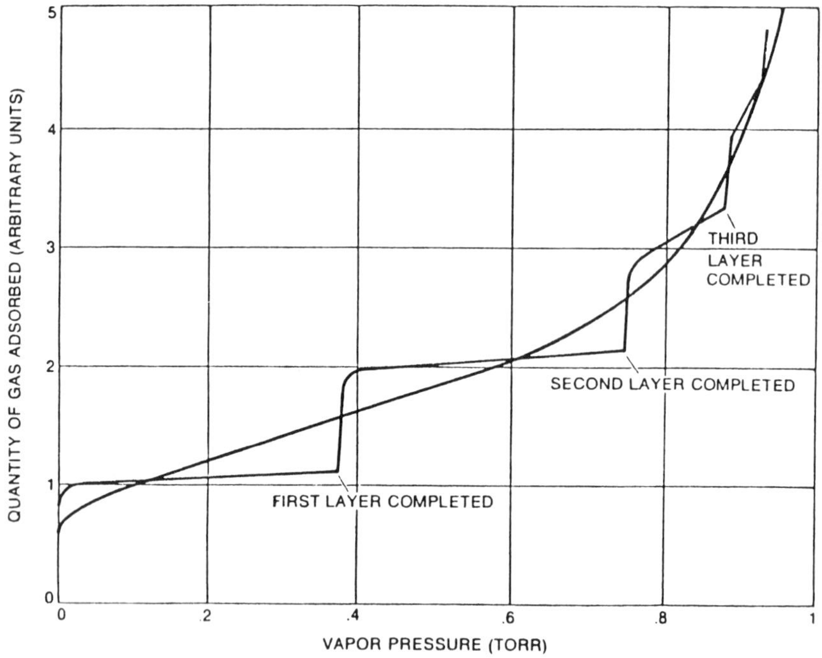

Fig. 9.18. Vapor pressure isotherm for krypton on exfoliated graphite showing individual layer behavior characteristic of uniform substrates. Also shown is a typical isotherm for non-uniform substrate (from Dash[30]).

9.3.2. Physical Properties of Helium Films

Through the introduction of large-surface-area uniform substrates, it has become possible to investigate a number of interesting quasi-two-dimensional phases occurring mostly in submonolayer films. The models that describe the physical processes emphasize the two-dimensional nature of the system. At low densities, much less than one atomic layer, helium adsorbed on graphite behaves much as a two-dimensional quantum gas. At high temperatures, its specific heat asymptotically approaches a constant value of $C \approx Nk_B$, while at lower temperatures, quantum effects brought on by statistical correlations reduce the specific heat in a manner similar to the bulk fluids. At intermediate surface coverages, behavior occurs which is correlated with regularity in the adsorbed gas. This effect is brought on by the adsorbed gas being commensurate with the periodicity of the lattice. At still higher coverages up to layer completion, the adsorbed gas behaves much as a two-dimensional solid with low-temperature specific heat varying as T^2, consistent with a two-dimensional Debye model. For these high coverages, transitions in the specific heat are observed at intermediate to high temperatures above which fluidlike behavior occurs. An interesting artifact of surface phases in adsorption systems is the occurrence of two-dimensional solids even though bulk helium does not solidify unless considerable external pressure is applied. The adsorption interaction compresses the molecules on the substrate to densities similar to those that occur in the bulk solid. These high densities allow the formation of crystal structure.

Although submonolayer helium films exhibit a number of unique phases, there does not appear to be any clear evidence for the existence of superfluidity. This is a point of considerable debate. The simplest theoretical models based on Bose–Einstein condensation mechanisms, introduced for bulk helium in Chapter 4, indicate that the ground state does not have a substantial occupation number at any temperature above absolute zero. For a two-dimensional Bose gas, the total number of particles is given by an integral over the density of states,

$$N = \int_0^\infty n(\varepsilon)\, g(\varepsilon)\, d\varepsilon \tag{9.48}$$

where $n(\varepsilon)$ is the single-particle occupation number given by

$$n(\varepsilon) = \frac{1}{e^{\beta(\varepsilon - \mu)} - 1} \tag{9.49}$$

where $\beta = 1/k_B T$. In two dimensions, the density of states is on a per unit area basis such that

$$g(\varepsilon) = \frac{2\pi m A}{h^2} \tag{9.50}$$

where A is the total area by analogy to the total volume for the relationship in Chapter 4. Substitution into the equation for the total particle number produces a closed-form solution,

$$N = \frac{2\pi m A}{h^2} \int_0^{\infty} \frac{d\varepsilon}{e^{\beta(\varepsilon-\mu)} - 1}$$

$$= -\frac{A}{\lambda_T^2} \ln(1 - e^{\beta\mu}) \tag{9.51}$$

where $\lambda_T = (h^2/2\pi m k_B T)^{1/2}$ is the thermal de Broglie wavelength. Bose–Einstein condensation implies that there is a finite occupation number in the ground state. In other words, the ground-state occupation number must be some fraction of the total density N, which can only occur when the chemical potential vanishes. Solving for μ in (9.51). we obtain

$$\mu = k_B T \ln(1 - e^{-n\lambda_T^2}) \tag{9.52}$$

where $n = N/A$. Therefore, to have significant condensation, it is necessary to have either $T = 0$ or $n \to \infty$.

The above analysis has been used extensively to argue against the possibility of any ordering occurring in helium films. Part of the avoidance of this dilemma can be achieved by noting that the divergence in (9.52) is logarithmic, therefore allowing for order over finite dimensions. In addition, more recent theoretical evidence[39] has redefined the ordering process in terms of a "topological order" no longer requiring that (9.52) be obeyed.

For the case of helium films of more than a few atomic layers, different behavior occurs. Perhaps the most important phenomenon is the onset of superfluidity. This event has tremendous physical and practical significance. In a He II refrigeration system it can produce numerous unwanted effects, including excessive mass and heat transfer. The first work published on the variation of superfluid character with film thickness was by Frederikse,[40] who studied the dependence of the specific heat on layer thickness. These results are reprinted in Fig. 9.19. The substrate consisted of a fine powder of jewelers rouge (Fe_2O_3). Even for coverages up to 12 layers the λ anomaly in the specific heat deviated considerably from that of the bulk liquid.

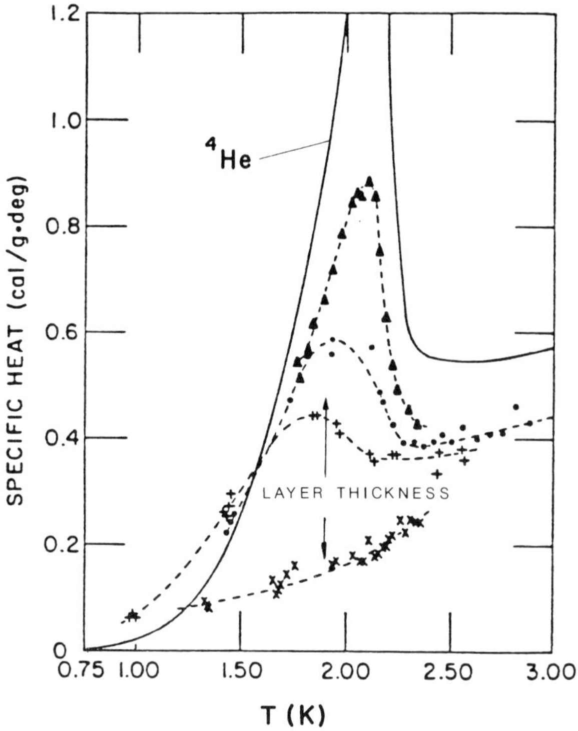

Fig. 9.19. Specific heat of ^{4}He films adsorbed on jeweler's rouge substrate as measured by Frederikse.[40] Data are for increasing coverage: (×) 3–4 layers, (+) 5–6 layers, (●) 7–9 layers, and (△) 9–12 layers. Comparison is shown with the bulk ^{4}He λ-anomaly.

Perhaps more interesting from the viewpoint of applications is the temperature dependence of the onset of superfluidity in thin films. The appearance of superfluidity can be measured by a number of techniques including mass transport through superleaks, heat transport, flow of persistent currents, and third sound. The latter phenomenon refers to the transport of second sound in thin films where the normal fluid is not free to move due to viscous interactions with the substrate. The onset of superfluidity as measured by these techniques is shown in Fig. 9.20 versus the fraction of saturated vapor pressure, p/p_0.[41] There is a fairly universal dependence on this curve with the exception of the results on the graphite substrate. There is still some question as to the origin of the higher onset temperature for the graphite substrate.[42]

As was stated early on in this section, gas separation and cryopumping are the main applications for physical adsorption in cryogenic system. In the first case the desired characteristics of the adsorbent are light weight,

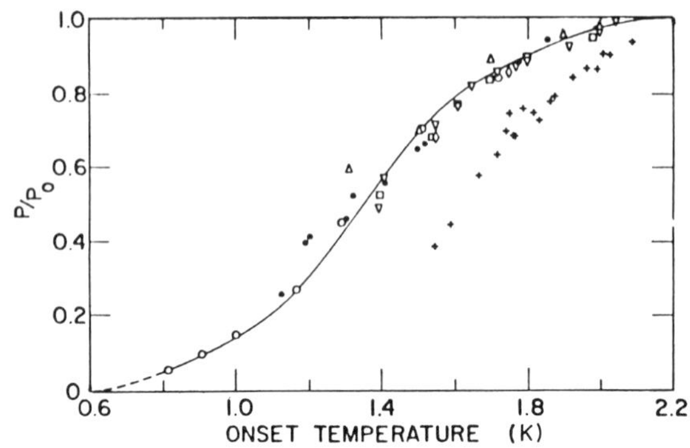

Fig. 9.20. Superfluid onset temperatures of ^{4}He films on different substrates and measured by different techniques (from Herb and Dash[41]).

small heat capacity, and large surface area. Typical materials used are activated charcoal, silica gel, activated alumina, or molecular sieves. The specific area of these materials are in the range 300–1200 m^2/g and bulk densities less than 1 g/cm^3. No particular concern is given to surface uniformity. Applications generally call for designing an adsorption "bed" which provides good adsorption characteristics yet still allows for bulk flow. This application depends strongly on details of design. The reader who is interested in a review of the technology of adsorption bed design should consult one of several reviews in the literature.[43,44]

In terms of cryopumping applications of adsorption, the choice of substrate material and configuration is more dependent on specific application. If high-speed pumping is required, the surface must not restrict the flow of gas. Therefore, the most common design is to use a planar surface. To achieve substantial total pumping capacity before regeneration, this type of surface is necessarily quite large, a fact that leads to problems with reducing the radiation heat leak to warm surrounding surfaces. A number of standard techniques are available for providing radiation shields that allow high flow rates of gas. If high pumping speed is not necessary, use can be made of numerous high-surface-area materials, called "getters," thermally well anchored to a low-temperature heat sink. Getters of this type often are sealed into closed helium dewars to maintain good vacuum quality over an extended period of time.

9.4. MAGNETIC REFRIGERATION

Throughout this book emphasis has been placed on the physical description and engineering applications of low-temperature helium. It has been an implicit assumption that these low temperatures are achieved by the methods outlined in Chapter 8. In most cases this assumption is correct in that the vast majority of helium refrigeration systems work on cycles using helium as the working media. However, there are a limited and growing number of special refrigeration systems that do no use helium to achieve cooling. One example, which is gaining considerable interest because of its improved efficiency over conventional helium cylcles, is that of magnetic refrigeration. In this system, refrigeration is achieved by adiabatic demagnetization of paramagnetic materials. There are basically two distinct types of magnetic system which can be used in adiabatic demagnetization. The strongest of the two, associated with uncompensated spins in the orbital electrons, can be used for cooling to below 1 K. To achieve much lower temperatures it is necessary to take advantage of the nuclear magnetic moment which is weaker but aligns at much lower temperatures.

It was first suggested in 1925 by Debye[45] and separately by Giauque[46] that cooling could be achieved by adiabatic demagnetization of a paramagnetic salt. Shortly thereafter, in 1933, Giauque and MacDougall[47] demonstrated this principle by successfully cooling a sample of paramagnetic material from 3.5 to 0.5 K. Until the separation of ^{3}He and development of dilution refrigeration, adiabatic demagnetization was used almost exclusively as the only effective means to cool below 1.0 K. During this period, most systems operated in the open cycle, providing a simple cooling step with relatively low cooling powers. In 1954 Herr and coworkers[48] were the first to operate a closed-cycle magnetic refrigerator. The system, which used iron ammonium alum $Fe_2(SO_4)_3 \cdot (NH_4)_2SO_4 \cdot 24\,H_2O$ as a working material, achieved a cooling capacity of 40 μW at 0.5 K. With the advent of continuously operating dilution refrigerators, the millikelvin regime became accessed more easily, with adiabatic demagnetization delegated to achieving submillikelvin temperatures. At the same time, and particularly in the last decade, there has been substantial effort directed toward development of closed-cycle magnetic refrigeration with large cooling powers. The goal of this work is to develop an alternate method of achieving temperatures in the range of the ^{4}He boiling point without resorting to helium gas cycles, with their inherent inefficiency.

The following discussion surveys the thermodynamic aspects of magnetic refrigeration. This includes a description of adiabatic demagnetization or open-cycle cooling as well as closed-cycle systems. The emphasis of the discussion is placed on the use of paramagnetic salt demagnetization systems, since their range of operation is of more practical significance. The performance of state-of-the-art closed-cycle magnetic refrigerators also is reviewed. Throughout this discussion the assumption is made that the reader is familiar with the fundamental physics of magnetic ion subsystems, a subject surveyed in Section 2.5.

9.4.1. Principles of Magnetic Cooling

The principles by which magnetic cooling can be achieved are seen easily by analogy with a fluid system. For the sake of simplicity, assume that the magnetic system is composed of a solid material so that pressure–volume work is negligible. In this case, the combination of the first and second laws of thermodynamics can be written

$$T\,dS = dE - \mu_0 H\,dM \tag{9.53}$$

where M is the magnetization of the material. Thus, the second term on the right-hand side is magnetic work done on the system by direct analogy to $p\,dV$, the work done in a fluid system.

Adiabatic demagnetization is analogous to isentropic expansion in a fluid system. The important parameter that controls this process is the isentropic coefficient, which for a magnetic system is the derivative of the temperature with respect to applied field at constant entropy:

$$\mu_m = \frac{1}{\mu_0}\left(\frac{\partial T}{\partial H}\right)_s \tag{9.54}$$

Note the similarity between (9.54) and the isentropic expansion coefficient defined in Chapter 8. Normally, μ_m is referred to as the magnetocaloric coefficient. Table 9.6 represents a comparison between parameters and coefficients relevant to magnetic and fluid refrigeration systems.

To derive an expression for μ_m analogous to μ_s, it is necessary to invoke the chain rule:

$$\frac{1}{\mu_0}\left(\frac{\partial T}{\partial H}\right)_s = -\frac{1}{\mu_0}\left(\frac{\partial T}{\partial S}\right)_H\left(\frac{\partial S}{\partial H}\right)_T \tag{9.55}$$

Then applying the Maxwell relation for a magnetic system,

$$\frac{1}{\mu_0}\left(\frac{\partial S}{\partial H}\right)_T = \left(\frac{\partial M}{\partial T}\right)_H \tag{9.56}$$

Table 9.6. Comparison of Parameters and Coefficients in Magnetic and Fluid Refrigeration Systems

	Fluid	Magnetic
Extensive variable	V	M
Intensive variable	p	$\mu_0 H$
Work	$p\,dV$	$-\mu_0 H\,dM$
Isentropic coefficient	$\mu_s = \left(\frac{\partial T}{\partial p}\right)_s$	$\mu_m = \frac{1}{\mu_0}\left(\frac{\partial T}{\partial H}\right)_s$

where the constant field heat capacity is

$$C_H = T\left(\frac{\partial S}{\partial T}\right)_H \tag{9.57}$$

The combination of (9.56) and (9.57) in the expanded form for μ_m results in a fully general relationship

$$\mu_m = -\frac{T}{C_H}\left(\frac{\partial M}{\partial T}\right)_H \tag{9.58}$$

The application of (9.58) for a particular magnetic refrigeration system requires knowledge of two quantities, C_H and the temperature dependence of M.

In Chapter 2 a relationship for the magnetization of a magnetic ion due to the Zeeman effect was derived. The equation for this effect was shown to take the form

$$M = Ng\mu_B J B_J(a) \tag{9.59}$$

where B_J is the Brillouin function for a magnetic ion of spin J and $a = g\mu_B B/k_B T$ is a measure of the ratio of the magnetic energy to thermal energy. The Brillouin function is shown graphically in Fig. 2.6. In the limit of large a, there is a high degree of magnetic ordering and the magnetization approaches a constant value,

$$M = Ng\mu_B J \quad \text{for } g\mu_B B \gg k_B T \tag{9.60}$$

while at small values of a, the magnetic system is weakly ordered and the Brillouin function is linear with a. In this regime it is shown easily that the magnetization can be expressed as

$$M = \frac{\mu_0 \gamma_C H}{T} \quad \text{for } g\mu_B B \ll k_B T \tag{9.61}$$

where γ_C is the Curie constant defined by expansion of the Brillouin function,

$$\gamma_C = \frac{Ng^2\mu_B^2 J(J+1)}{3k_B} \tag{9.62}$$

Returning to the relationship for the magnetocaloric coefficient and considering a weakly interacting magnetic system obeying the Curie law, we find that μ_m takes on a simplified form,

$$\mu_m = \frac{M}{C_H} \tag{9.63}$$

which is analogous to the isentropic expansion coefficient for an ideal gas.

The other term needed to establish the behavior of a magnetic refrigeration system is C_H, the specific heat at constant magnetic field. A relationship for this quantity can be derived through application of the first and second laws of thermodynamics for a magnetic system (9.53). Defining the constant magnetization specific heat as

$$C_M = T\left(\frac{\partial S}{\partial T}\right)_M = \left(\frac{\partial E}{\partial T}\right)_M \tag{9.64}$$

then (9.53) can be recast into the form

$$T\,dS = C_M\,dT - \mu_0 H\,dM \tag{9.65}$$

Further restricting (9.65) to a constant magnetic field process, the expansion of the temperature differential defines the specific heat at constant field:

$$C_H = T\left(\frac{\partial S}{\partial T}\right)_H \tag{9.66}$$

It follows that C_H may be written

$$C_H = C_M - \mu_0 H\left(\frac{\partial M}{\partial T}\right)_H \tag{9.67}$$

The constant magnetization specific heat C_M is made up to several independent contributions. The relative importance of each contribution depends on the type of magnetic material involved and the temperature range over which cooling is to be achieved. In paramagnetic ion systems operating at low temperatures, there are primarily two contributors. One is the lattice specific heat due to phonon excitations which at low temperatures may be approximated by the Debye model,

$$C_{\text{ph}} = 234\ Nk_B\left(\frac{T}{\Theta_D}\right)^3 \quad \text{for } T \ll \Theta_D \tag{9.68}$$

This contribution is particularly important at high temperatures, $T \approx 10$ K.

Because this term varies as T^3, it has little consequence for very-low-temperature refrigeration.

The other major contribution to the specific heat of a magnetic ion system is due to the ordering of the spins which are nondegenerate because of crystal field splitting. In Chapter 2 the Schottky heat capacity due to this effect was shown to take the form

$$C_M = \frac{N_m \delta^2}{k_B T^2} \frac{(g_0/g_1)\, e^{\delta/k_B T}}{[1 + (g_0/g_1)\, e^{\delta/k_B T}]^2} \tag{9.69}$$

where δ is the separation energy between the two levels and g_0 and g_1 are degeneracies of each state. Note that N_m is the number of magnetic ions, which can be much less than N, the total particle number. Typically, $\delta/k_B \approx 100$ mK so that this term dominates the zero field heat capacity at lower temperatures, $T \lesssim 1$ K. For many applications, refrigeration occurs substantially above the Schottky anomaly temperature δ/k_B. In this regime it is possible to take a high-temperature limiting form of (9.69) which yields the equation

$$C_M = \frac{A}{T^2} \tag{9.70}$$

where A, the specific heat coefficient, has a value

$$A = N_m k_B \frac{g_0/g_1}{(1 + g_0/g_1)^2} \left(\frac{\delta}{k_B}\right)^2 \tag{9.71}$$

Listed in Table 9.7 are specific heat coefficients for several paramagnetic materials along with other relevant properties. For most magnetic ion systems the specific heat at constant magnetization is represented accurately by a combination of (9.68) and (9.70). To determine C_H it is

Table 9.7. Selected Properties of Magnetic Refrigerant Materials[a]

Material	Specific volume (cm^3/g · ion)	Curie constant γ_C (J · K/g · ion · T^2)	Debye temperature Θ_D (K)	A (J · K/g · ion)
$2Ce(NO_3)_3 \cdot 3Mg(NO_3)_2 \cdot 24H_2O$	366	0–3.17[b]	60	5.1×10^{-5}
$Cr_2(SO_4)_3 \cdot K_2SO_4 \cdot 24H_2O$	273	18.4	—	0.150
$Fe_2(SO_4)_3(NH_4)_2SO_4 \cdot 24H_2O$	282	43.8	250	0.108
$Gd_2(SO_4)_3 \cdot 8H_2O$	124	78	105	2.91
$Gd_3Ga_5O_{12}$	48	78	203	0.130

[a] See Refs. 49–51.
[b] Anisotropic material.

necessary to evaluate the second term in (9.67). For the simplest case of a material obeying the Curie equation, the constant field specific heat at $T \ll \Theta_D$ can be approximated by

$$C_H = 234\ NK_B \left(\frac{T}{\Theta_D}\right)^3 + \frac{A + \gamma_C \mu_0^2 H^2}{T^2} \tag{9.72}$$

As an example, consider the adiabatic demagnetization of iron ammonium alum (see Table 9.7) initially at 1 K and in a background field of 0.1 T. First, it is important to determine whether the material under this set of conditions can be represented by the Curie law. For this material $g = 2$ so the ratio

$$\frac{g\mu_B B}{k_B} = \frac{2(0.927 \times 10^{-23}\ \mathrm{J/T})(0.1\ \mathrm{T})}{1.38 \times 10^{-23}\ \mathrm{J/K}}$$
$$= 0.134\ \mathrm{K}$$

which is significantly less than the ambient temperature so that the error introduced by using the Curie law is small. Next, compare the relative importance of the phonon and Schottky heat capacities. Assuming that $g_0/g_1 = 0.5$ and taking the ratio of (9.68) to (9.70), we obtain

$$\frac{C_{\mathrm{ph}}}{C_M} = 936 \left(\frac{N}{N_m}\right)\left(\frac{T}{\Theta_D}\right)^3 \left(\frac{k_B T}{\delta}\right)^2 \tag{9.73}$$

For a magnetic material like iron ammonium alum with $\delta/k_B = 242$ mK, $\Theta_D = 250$ K, and $N/N_m \simeq 50$, the above ratio takes on the numerical value $C_{\mathrm{ph}}/C_M = 0.052$. Therefore, the phonon contribution to the specific heat at 1 K is only about 5% and becomes smaller as the temperature decreases, since the ratio goes as T^5. As a consequence, it is possible to neglect the phonon term here. For many systems it must be included, particularly at high temperatures, $T \gtrsim 10$ K.

Neglecting the phonon term in (9.72), we obtain a simplified expression for the magnetocaloric coefficient:

$$\mu_m = \frac{\mu_0 \gamma_C H T}{A + \gamma_C \mu_0^2 H^2} \tag{9.74}$$

This expression can be employed to determine the final temperature achieved by adiabatic demagnetization of a magnetic material. To approach this problem, integrate (9.74) over a finite change in field and temperature,

$$\int_{T_i}^{T_f} \frac{dT}{T} = \int_{\mu_0 H_i}^{\mu_0 H_f} \frac{\mu_0 \gamma_C H T}{A + \gamma_C \mu_0^2 H^2}\, d(\mu_0 H) \tag{9.75}$$

which by demagnetization to zero applied field, $H_f = 0$, results in the expression

$$\frac{T_i}{T_f} = \left(1 + \frac{\gamma_C \mu_0^2 H_i^2}{A}\right)^{1/2} \tag{9.76}$$

Returning to the example, by inserting the appropriate values $A = 0.108\ \mathrm{J \cdot K/mol}$ and $\gamma_C = 43.8\ \mathrm{J \cdot K/mol}$ into (9.76), the ratio of temperatures for $\mu_0 H_i = 0.1$ T becomes $T_i/T_f = 2.25$. Therefore, with the material initially at 1 K the final temperature as a result of adiabatic demagnetization would be $T_f = 0.45$ K. In practice, demagnetizing fields much larger that 0.1 T allow lower final temperatures, but the simplified expression derived above (9.76) does not apply in that case.

Now turn to the issue of entropy in magnetic ion systems. This quantity is related to the magnetic Gibbs function,

$$G^* = E - TS - \mu_0 HM \tag{9.77}$$

which is also defined according to the partition function

$$G^* = -nRT \ln Z \tag{9.78}$$

The entropy is given in terms of a derivative of the magnetic Gibbs function,

$$S = RT\left(\frac{\partial \ln Z}{\partial T}\right)_H + R \ln Z \tag{9.79}$$

Inserting the expression for the partition function in a magnetic ion system derived in Chapter 2, we can determine the entropy by a straightforward calculation. The resulting expression is difficult to assess, but by taking limits at high and low fields we can simplify the form considerably.

In the limit of small magnetic field, $g\mu_B B \ll k_B T$ and for $T \gg \delta/k_B$ the spin entropy can be shown to obey the relationship

$$S = R\left(\ln(2J+1) - \frac{2(A + \mu_0 \gamma_C H^2)}{RT^2}\right) \tag{9.80}$$

where J is the total spin quantum number. Generally, except at very low temperatures, the first term in (9.80) is the dominant contribution to the spin entropy. This situation is desirable for magnetic cooling because the entropy change between the magnetized and demagnetized materials must be large.

In the high magnetic field limit, $g\mu_B B \gg k_B T$, a considerably different situation occurs. Here the magnetic moments of the dipoles are approaching complete alignment with the magnetic field. The magnetic field is imposing a greater degree of order on the system, thus lowering the entropy. The spin entropy approaches zero, leaving the only remaining term associated with the Stark effect and the lattice (see Chapter 2), the latter of which can be neglected at low temperatures. In this limit, the principal behavior of the entropy is obtained by integration of the heat capacity (9.69). This analysis leads to an exponentially decaying entropy for the limit where $\delta/k_B \gg T$,

$$S = N_m k_B \left(1 + \frac{\delta}{k_B T}\right) \frac{g_0}{g_1} e^{-\delta/k_B T} \tag{9.81}$$

where S is the total entropy associated with N_m magnetic ions.

As the exact calculation of the entropy in a magnetic ion subsystem is dependent on choice of materials and operating range, it will not be carried out here. However, the principles have been established so that an estimate of the useful range of a magnetic refrigerator should be possible. A schematic representation of the entropy of a magnetic ion subsystem is shown in Fig. 9.21. At high temperatures, $T \gg \delta/k_B$ and $T \gg g\mu_B \mu_0 H/k_B$, the entropy difference between the magnetized and unmagnetized state decreases because the magnetic field is not very effective at magnetizing the sample. On the other hand, at very low temperatures, $T \ll \delta/k_B$ and $T \gg g\mu_B \mu_0 H/k_B$, the entropy difference decreases because the crystal field orders the system independent of the applied magnetic field. There is a range, indicated by the dashed lines in Fig. 9.21, over which the entropy difference approaches $Nk_B \ln(2J+1)$, thereby allowing for efficient magnetic cooling. Obviously, the exact temperature range over which this region occurs depends on the particular paramagnetic material in question.

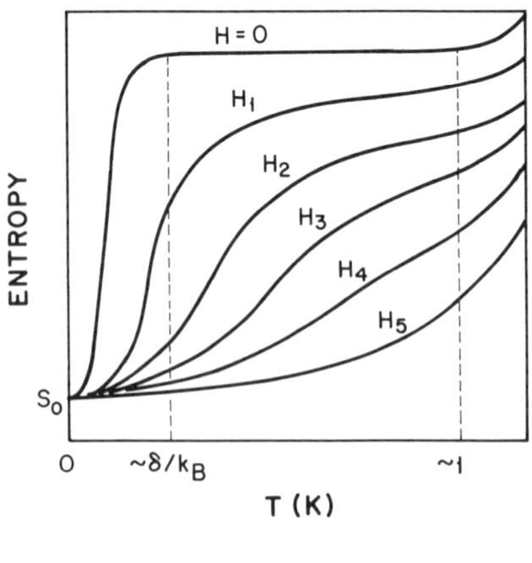

Fig. 9.21. Entropy of a magnetic-ion subsystem as a function of temperature for various values of the magnetic field.

9.4.2. Continuous Magnetic Refrigerators

With the above survey of the thermodynamics of magnetic cooling, it is instructive to turn to the practical prolems associated with achieving continuous refrigeration with magnetic materials. In this discussion emphasis is placed on continuous refrigeration rather than single-cycle adiabatic demagnetization which was outlined above. The material that goes into a magnetic refrigerator must have a number of unique characteristics which set it apart from materials for other applications. Some of these characteristics have been introduced above, but a complete summary includes the following:

1. Magnetic refrigeration materials should have a small electronic and lattice specific heat. The energy used to cool the electrons and lattice is wasted and only reduces the efficiency of the cycle.
2. The magnetic level splitting due to the crystal field Stark effect, δ/k_B, must be below the range of operating temperatures. Otherwise, the entropy change with magnetization is reduced.
3. The magnetic contribution to the entropy and specific heat should be large to allow more thermal energy to be cycled through the demagnetization process.
4. To achieve good thermal exchange with the systems to be refrigerated, it is desirable that the magnetic material have good heat transfer characteristics. Whenever possible, magnetic refrigeration materials should have a high thermal diffusivity.
5. Any material to be used in a device must be able to be fabricated into the configurations necessary for effective operation.

The principles of idealized closed-cycle magnetic refrigeration are similar to those for systems using fluids as working media. The most thermodynamically efficient cycle is the Carnot cycle which is a combination of isothermal and isentropic processes. This cycle can be achieved in a magnetic ion subsystem by a method shown schematically in Fig. 9.22, which is an actual *T-S* diagram for gadolinium sulfate. In this example, the Carnot cycle is shown as operating between two isothermal reservoirs a 15 and 1.7 K. This particular temperature range is attractive for magnetic refrigeration because helium gas–liquid cycles are limited to rather low thermodynamic efficiencies compared to the Carnot cycle. Also the maximum magnetization field of 10 T is a practical limitation because it allows for the use of state-of-the-art superconducting magnet technology.

Consider the methods whereby Carnot refrigeration in a magnetic system can be achieved. A schematic diagram of the system is shown in Fig. 9.23. Apart from the reservoirs, the refrigerator consists of three principal components: the working material made of a magnetic salt, a magnet

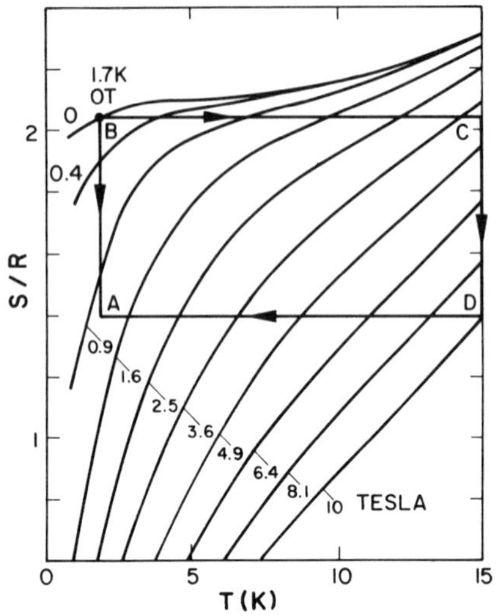

Fig. 9.22. Thermodynamic cycle executed by $Gd_2(SO_4)_3 \cdot 8H_2O$ in a magnetic Carnot cycle (from Steyert[52]).

for aligning the spins of this material, and two thermal switches, one to either reservoir for exchange between the working material and isothermal baths. The Carnot cycle is a four-step process of magnetization, demagnetization, and heat exchange to the isothermal reservoirs.

The methods by which the Carnot cycle can be achieved are seen best by referring to the cycle ABCD in Fig. 9.22. With the thermal switch 1 (TS 1) closed and the working material in good contact with the heat reservoir (HR), the magnetic field is applied up to a maximum of 10 T represented by point D. Here the spins have the maximum alignment at this tem-

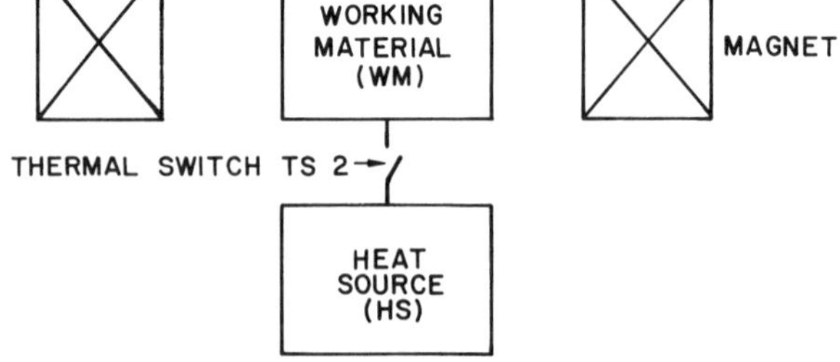

Fig. 9.23. Schematic diagram of magnetic Carnot refrigerator (from Steyert[52]).

perature. Next, TS 1 is opened, isolating the magnetic material, and the field is decreased slowly, cooling the working material to point A at 1.7 K. Note that this point is not at zero field. Thermal switch 2 (TS 2) is then closed and the working material comes into the thermal equilibrium with the heat source. This step, which occurs isothermally, must be accompanied by a further slight demagnetization of the material to point B. At this point all the heat has been transferred. Thermal switch 2 is then opened and the working material is magnetized slowly back to the high temperature represented by point C. Once the working material reaches the heat reservoir temperature, TS 1 is closed and further magnetization occurs isothermally to point D, completing the cycle. This process, as represented in Fig. 9.22, is able to pump 27 J per liter of material per cycle, which if achievable at modest frequencies around 1 Hz could provide a rather high cooling rate in a very small volume. Of course, the difficulty in achieving this kind of cooling power lies in the practical aspects of creating a magnetic refrigerator.

In recent years there have been several devices built that are intended to bring the above concepts to realization on a practical scale.[53–56] Since this field is developing quite rapidly, there will undoubtedly be several more devices in the near future. In most cases the devices are being developed to operate in the low-temperature region, $T < 4.2$ K, and to provide cooling power at the low end of the order of 1 W. Designs intend to scalable to larger systems. Materials employed are either gadolinium sulfate or gadolinium gallium garnet (GGG), $Gd_3Ga_5O_{12}$. The latter material has shown superior thermal properties. In one particular case, a prototype refrigerator has been operated between 1.8 and 4.2 K with a cooling power of 1.2 W and an achieved figure of merit in the range of 45%.[56]

To produce much lower temperatures with magnetic refrigeration it soon becomes inefficient to use paramagnetic salts as the working media because their spins will already be aligned. At these temperatures, which can span into the submillikelvin regime, it is possible to use the unpaired magnetic moment associated with the nuclei. The nuclear magnetic moment μ_N is smaller than the Bohr magneton by the ratio of nucleon to electron mass ($m_n/m_e = 1840$). Thus, the nuclear spins can be aligned only by very high magnetic fields at low temperatures. This method was first applied in 1956 by Simon and Kurti at Oxford where spin temperatures in the neighborhood of 10 μK were achieved through adiabatic demagnetization of the nuclei of a paramagnetic salt. However, the spin temperatures are not the same as the lattice temperatures and entropy must be transferred from one system to the other. The time constant that controls this process, called the spin–lattice relaxation time, can be large, providing a severe hindrance to the ultimately achievable temperature. In recent times, it has become more advantageous to achieve nuclear demagnetization using nuclei of metallic elements. In those systems the density of magnetic moments is much greater and the spin–lattice relaxation time shorter, allowing for minimum actual bulk temperatures. Record minimum bulk temperatures of $\sim 100\,\mu$K have been achieved by multistage devices for wich the last stage is nuclear demagnetization.[24]

REFERENCES

1. G. K. White, *Experimental Techniques in Low Temperature Physics*, 3rd ed., Clarendon Press, Oxford, 1979.
2. F. E. Hoare, L. C. Jackson, and N. Kurti, *Experimental Cryophysics*, Butterworths, London, 1961.
3. O. V. Lounasmaa, *Experimental Principles and Methods Below 1 K*, Academic Press, London, 1974.
4. R. B. Scott, *Cryogenic Engineering*, Princeton Press, Princeton, NJ, 1959.
5. R. F. Barron, *Cryogenic Systems*, 2nd Ed., Clarendon Press, Oxford, 1985.
6. A. C. Rose–Innes, *Low Temperature Techniques*, English Universities Press, London, 1973.
7. E. H. Kennard, *Kinetic Theory of Gases*, McGraw–Hill, New York, 1938.
8. R. J. Corrucini, Gaseous Heat Conduction at Low Pressures and Temperatures, *Vacuum* **7–8**, 19 (1959).
9. E. G. Cravalho, C. L. Tien, and R. P. Caren, Effect of Small Spacings on Radiative Transfer Between Two Dielectrics, *Transactions of ASME*, paper 67-HT-21 (1967).
10. P. F. Dickson and M. C. Jones, Infrared Spectral Reflectances of Metals at Low Temperatures, *Cryogenics* **8**, 24 (1968).
11. W. Obert, J. R. Coupland, D. P. Hammond, T. Cook, and K. Harwood, Emissivity Measurements of Metallic Surfaces Used in Cryogenics Applications, *Adv. Cryog. Eng.* **27**, 293 (1982).
12. H. Brechna, *Superconducting Magnet Systems*, Chap. 5, Springer–Verlag, Berlin, 1973.
13. K. D. Timmerhaus, Conductive Heat Transfer, in *Heat Transfer at Low Temperatures*, W. Frost (Ed.), Chap. 2, Plenum Press, New York, 1975.
14. E. M. Leung, R. W. Fast, H. L. Hart, and J. R. Heim, Techniques for Reducing Radiation Heat Transfer Between 77 K and 4.2 K, *Adv. Cryog. Eng.* **25**, 489 (1980).
15. R. H. Kropschot and R. W. Burgess, Perlite for Cryogenic Insulation, *Adv. Cryog. Eng.* **8**, 425 (1963).
16. E. R. Lady, Cryogenic Insulation, in *Cryogenic Engineering*, E. R. Lady (Ed.), Chap. 7, Engineering Summer Conference, University of Michigan, 1965.
17. Y. M. Eyssa and O. Okasha, Thermodynamic Systemization of Thermal Radiation Shields for a Cryogenic Apparatus, *Cryogenics* **18**, 305 (1978).
18. M. A. Hilal and G. E. McIntosh, Cryogenic Design for Large Superconductive Energy Storage Magnets, *Adv. Cryog. Eng.* **21**, 69 (1976).
19. W. F. Keller, *Helium* 3 *and Helium* 4, Plenum Press, London, 1969.
20. L. D. Landau, The Theory of a Fermi Liquid, *Sov. Phys. JETP* **3**, 920 (1957); **5**, 101 (1957); **8**, 70 (1959).
21. J. Wilks, *The Properties of Liquid and Solid Helium*, Clarendon Press, Oxford, 1967.
22. D. M. Lee and R. C. Richardson, Superfluid ^{3}He, in *The Physics of Liquid and Solid Helium, Part II*, K. H. Bennemann and J. B. Ketterson (Eds.), Chap. 4, Wiley, New York, 1978.
23. K. Andres and O. V. Lounasmaa, Recent Progress in Nuclear Cooling, *Prog. Low Temp. Phys.* **8**, 221 (1982).

24. O. V. Lounasmaa, The Pursuit of Absolute Zero, *Adv. Cryog. Eng.* **29**, 551 (1984).
25. I. Pomeranchuk, On the Theory of Liquid ^{3}He, *Zh. Eksp. Teor. Fiz.* **20**, 919 (1950).
26. D. D. Osheroff, R. C. Richardson, and D. M. Lee, Evidence for a New Phase of Solid ^{3}He, *Phys. Rev. Lett.* **28**, 885 (1972).
27. J. C. Wheatly, Three Lectures on the Experimental Properties of Liquid ^{3}He, in *The Helium Liquids*, J. G. M. Armitage and I. E. Farquar (Eds.), Academic Press, London, 1975.
28. J. C. Wheatly, Superfluid Phases of Helium Three, *Phys. Today* **29**, 32 (1976).
29. L. Corruccini and D. D. Osheroff, Unusual Magnetic Relaxation in Superfluid ^{3}He, *Phys. Rev. Lett.* **34**, 564 (1975).
30. J. G. Dash, *Films on Solid Surfaces*, Academic Press, New York, 1975.
31. D. W. Princehouse, High-Resolution Heat Capacity Study of ^{4}He Adsorbed on Bare Copper, *J. Low Temp. Phys.* **8**, 287 (1972).
32. J. G. Daunt and E. Lerner, Adsorption of ^{3}He and ^{4}He on Various Substrates Below 30 K, *J. Low Temp. Phys.* **8**, 79 (1972).
33. W. A. Steele and J. G. Aston, Heats of Adsorption of Hydrogen and Helium on Prepared Surfaces, *J. Am. Chem. Soc.* **79**, 2393 (1957).
34. A. A. Antoniou, Adsorption of ^{4}He on Graphitized Carbon in the Submonolayer Region Between 2 and 15 K, *J. Chem. Phys.* **62**, 779 (1975).
35. R. L. Elgin and D. L. Goodstein, Thermodynamic Study of the ^{4}He Monolayer Adsorbed on Grafoil, *Phys. Rev. A* **9**, 2657 (1974).
36. J. G. Daunt and C. Z. Rosen, Multilayer Adsorption of ^{3}He and ^{4}He on Zeolite from 4 to 80 K, *J. Low Temp. Phys.* **3**, 89 (1970).
37. M. Bretz, J. G. Dash, D. C. Hickernell, E. O. McLean, and O. E. Vilches, Phases of ^{3}He and ^{4}He Monolayer Films Adsorbed on Basel-Plane Oriented Graphite, *Phys. Rev. A* **8**, 1589 (1973).
38. D. F. Brewer, Some Thermal, Magnetic and Flow Properties of Adsorbed He and ^{3}He–^{4}He Mixtures, *J. Low Temp. Phys.* **3**, 205 (1970).
39. J. M. Kosterlitz and D. J. Thouless, Ordering, Metastability and Phase Transitions in Two-Dimensional Systems, *J. Phys. C* **6**, 1181 (1973).
40. H. P. R. Frederikse, On the Specific Heat of Adsorbed Helium, *Physica* (*Utrecht*) **15**, 860 (1949).
41. J. A. Herb and J. G. Dash, Mass Transport of ^{4}He Films Adsorbed on Graphite, *Phys. Rev. Lett.* **29**, 846 (1972).
42. A. D. Migone, J. Krim, J. G. Dash, and J. Suzanne, Incomplete Wetting of ^{4}He Films on Ag and Au (111) Surfaces, *Phys. Rev. B* **31**, 7643 (1985).
43. K. Wilson, Adsorption, in *Cryogenic Fundamentals*, G. G. Haselden (Ed.), Academic Press, London, 1971.
44. G. Davey, Cryosorption Pumping, in *Advanced Cryogenics*, C. A. Bailey (Ed.), Plenum Press, New York, 1971.
45. P. Debye, Einige Bemerkungen zur Magnetisieurung bei Taefer Temperature, *Ann. Phys.* **81**, 1154 (1926).
46. W. F. Giauque, Paramagnetism and the Third Law of Thermodynamics. Interpretation of the Low Temperature Specific Heat of Gadolinium Sulfate, *J. Am. Chem. Soc.* **49**, 1870 (1927).
47. W. F. Giauque and D. P. MacDogall, Attainment of Temperatures Below 1° Absolute by Demagnetization of $Gd_2(SO_4)_3 \cdot 8H_2O$, *Phys. Rev.* **43**, 768 (1933).
48. C. V. Herr, C. B. Barnes, and J. G. Daunt, The Design and Operation of a Magnetic Refrigeration Maintaining Temperatures Below 1 K, *Rev. Sci. Instrum.* **25**, 1088 (1954).
49. M. W. Zemansky, *Heat and Thermodynamics*, 5th ed., Chap. 14, McGraw–Hill, New York, 1968.

50. R. P. Hudson, *Principles and Application of Magnetic Cooling*, Chap. V, North Holland Publishing, Amsterdam, 1972.
51. J. A. Barclay and W. A. Steyert, Materials for Magnetic Refrigeration, Los Alamos Scientific Laboratory Report LA-8746-MS (May 1981).
52. W. A. Steyert, Magnetic Refrigeration, in *Liquid Cryogens*, Vol. II, K. D. Williamson, Jr. and F. J. Edeskietz (Eds.), Chap. 6, CRC Press, Boca Raton, FL, 1983.
53. W. P. Pratt, S. S. Rosenblum, W. A Steyert, and J. A. Barclay, A Continuous Demagnetization Refrigerator Operating Near 2 K and a Study of Magnetic Refrigerants, *Cryogenics* **17**, 689 (1977).
54. R. Beranger, G. Bon Mardion, G. Claudet, C. Delpuech, A. F. Lacaze, and A. A. Lacaze, A Gadolinium Gallium Garnet Double Acting Reciprocating Magnetic Refrigerator, *Adv. Cryog. Eng.* **27**, 703 (1982).
55. A. F. Lacaze, A. A. Lacaze, R. Beranger, and G. Bon Mardion, Thermodynamic Aspects of a Double Acting Reciprocating Magnetic Refrigerator, *Proceedings of the 9th International Cryogenics Engineering Conference*, pp. 14–17, Butterworths, London, 1982.
56. A. F. Lacaze, R. Beranger, G. Bon Mardian, G. Claudet, and A. A. Lacaze, Double Acting Reciprocating Magnetic Refrigerator: Recent Improvements, *Adv. Cryog. Eng.* **29**, 573 (1984).

Appendixes

Appendix 1

Thermodynamic Properties of Liquid Helium at Saturated Vapor Pressure

	Temperature (K)	Pressure (atm)	Density (g/cm^3)	Entropy ($J/g \cdot K$)	C_{sat} ($J/g \cdot K$)
	1.4	0.002835	0.1450	0.132	0.780
	1.6	0.007487	0.1451	0.284	1.572
	1.8	0.01641	0.1453	0.535	2.810
	2.0	0.03128	0.1456	0.940	5.18
$T_\lambda =$	2.172	0.04969	0.1462	—	—
	2.2	0.05256	0.1461	1.671	3.16
	2.4	0.08228	0.1453	1.898	2.25
	2.6	0.1219	0.1442	2.068	2.10
	2.8	0.1730	0.1428	2.225	2.22
	3.0	0.2371	0.1411	2.382	2.42
	3.2	0.3156	0.1393	2.542	2.67
	3.4	0.4100	0.1372	2.706	2.95
	3.6	0.5220	0.1348	2.875	3.28
	3.8	0.6528	0.1321	3.050	3.68
	4.0	0.8040	0.1290	3.234	4.19
	4.2	0.9772	0.1254	3.429	4.88
$T_{Nbp} =$	4.224	1.000	0.1250	3.454	4.98
	4.4	1.174	0.1213	3.640	5.86
	4.6	1.397	0.1163	3.873	7.44
	4.8	1.648	0.1101	4.144	10.5
	5.0	1.929	0.1011	4.495	19.1
$T_c =$	5.201	2.245	0.0696	5.589	—

Note: 1 atm = 760 torr = 1.013 bar = 1.013×10^5 Pa.

Appendix 2

Properties of Helium at Constant Pressure

Helium—Pressure is 1 atm ≐ 1.013×10^5 Pa 101 325 Pa ≐ 1 bar

Temperature (K)	Density (kg/m^3)	Enthalpy (J/kg)	Entropy (J/kg·K)
1.0000+000	1.4685+002	7.0642+002	1.4753+001
1.2000+000	1.4674+002	7.4787+002	5.1475+001
1.4000+000	1.4677+002	8.5601+002	1.3386+002
1.6000+000	1.4686+002	1.0870+003	2.8745+002
1.8000+000	1.4704+002	1.5271+003	5.4498+002
2.0000+000	1.4734+002	2.3175+003	9.5975+002
2.1631 T-Lambda	1.4811+002	3.4875+003	1.5177+003
2.2000+000	1.4818+002	3.8677+003	1.6488+003
2.4000+000	1.4739+002	4.3578+003	1.8630+003
2.6000+000	1.4628+002	4.7693+003	2.0277+003
2.8000+000	1.4492+002	5.1856+003	2.1818+003
3.0000+000	1.4330+002	5.6361+003	2.3370+003
3.2000+000	1.4142+002	6.1288+003	2.4961+003
3.4000+000	1.3923+002	6.6716+003	2.6606+003
3.6000+000	1.3668+002	7.2720+003	2.8321+003
3.8000+000	1.3366+002	7.9433+003	3.0135+003
4.0000+000	1.3006+002	8.7043+003	3.2087+003
4.2000+000	1.2560+002	9.5912+003	3.4253+003
4.4000+000	1.5240+001	3.1716+004	8.6552+003
4.6000+000	1.3868+001	3.3320+004	9.0118+003
4.8000+000	1.2807+001	3.4796+004	9.3259+003
5.0000+000	1.1944+001	3.6188+004	9.6102+003
5.2000+000	1.1220+001	3.7522+004	9.8719+003
6.0000+000	9.1492+000	4.2512+004	1.0765+004
7.0000+000	7.5295+000	4.8349+004	1.1666+004
8.0000+000	6.4355+000	5.3976+004	1.2417+004
9.0000+000	5.6358+000	5.9488+004	1.3067+004
1.0000+001	5.0215+000	6.4929+004	1.3640+004
1.2000+001	4.1338+000	7.5673+004	1.4620+004
1.4000+001	3.5190+000	8.6315+004	1.5440+004
1.6000+001	3.0663+000	9.6899+004	1.6147+004
1.8000+001	2.7184+000	1.0744+005	1.6768+004
2.0000+001	2.4423+000	1.1795+005	1.7321+004
2.2000+001	2.2177+000	1.2844+005	1.7821+004
2.4000+001	2.0312+000	1.3891+005	1.8277+004
2.6000+001	1.8739+000	1.4936+005	1.8695+004
2.8000+001	1.7394+000	1.5981+005	1.9082+004
3.0000+001	1.6230+000	1.7025+005	1.9442+004
4.0000+001	1.2167+000	2.2236+005	2.0942+004
5.0000+001	9.7337-001	2.7440+005	2.2103+004
6.0000+001	8.1127-001	3.2639+005	2.3051+004
7.0000+001	6.9551-001	3.7837+005	2.3852+004
8.0000+001	6.0867-001	4.3033+005	2.4546+004
9.0000+001	5.4112-001	4.8229+005	2.5158+004
1.0000+002	4.8708-001	5.3424+005	2.5705+004
1.2500+002	3.8976-001	6.6409+005	2.6864+004
1.5000+002	3.2487-001	7.9394+005	2.7811+004
1.7500+002	2.7850-001	9.2377+005	2.8612+004
2.0000+002	2.4371-001	1.0536+006	2.9305+004
2.2500+002	2.1665-001	1.1834+006	2.9917+004
2.5000+002	1.9500-001	1.3133+006	3.0464+004
2.7500+002	1.7728-001	1.4431+006	3.0959+004
3.0000+002	1.6252-001	1.5729+006	3.1411+004

C_p (J/kg · K)	C_v (J/kg · K)	Viscosity (kg/m · S)	k (W/m · K)
1.0470+002	1.0470+002		
3.3939+002	3.3937+002		
8.1729+002	8.1704+002		
1.5643+003	1.5632+003		
2.9516+003	2.9482+003		
5.2444+003	5.2344+003		
1.2391+005	8.0078+004		
2.9404+003	2.9112+003	3.0696-006	1.3865-002
2.1446+003	2.0221+003	3.3022-006	1.4742-002
2.0215+003	1.8172+003	3.4507-006	1.5519-002
2.1255+003	1.8343+003	3.5421-006	1.6217-002
2.2769+003	1.9097+003	3.5906-006	1.6841-002
2.5659+003	1.9905+003	3.6056-006	1.7389-002
2.7799+003	2.0647+003	3.5920-006	1.7854-002
3.1307+003	2.1336+003	3.5534-006	1.8230-002
3.5365+003	2.2019+003	3.4896-006	1.8500-002
4.0809+003	2.2746+003	3.4009-006	1.8654-002
4.7468+003	2.3568+003	3.2824-006	1.8672-002
8.4646+003	3.2882+003	1.2745-006	9.2992-003
7.6489+003	3.2410+003	1.3127-006	9.6032-003
7.1449+003	3.2077+003	1.3522-006	9.9151-003
6.8014+003	3.1834+003	1.3920-006	1.0228-002
6.5525+003	3.1654+003	1.4320-006	1.0539-002
6.0037+003	3.1280+003	1.5897-006	1.1744-002
5.7085+003	3.1139+003	1.7790-006	1.3168-002
5.5597+003	3.1110+003	1.9589-006	1.4498-002
5.4718+003	3.1114+003	2.1299-006	1.5735-002
5.4145+003	3.1126+003	2.2929-006	1.6894-002
5.3462+003	3.1150+003	2.5981-006	1.9030-002
5.3068+003	3.1167+003	2.8801-006	2.0986-002
5.2803+003	3.1177+003	3.1430-006	2.2815-002
5.2619+003	3.1182+003	3.3901-006	2.4546-002
5.2488+003	3.1184+003	3.6239-006	2.6201-002
5.2390+003	3.1185+003	3.8463-006	2.7794-002
5.2316+003	3.1185+003	4.0588-006	2.9335-002
5.2258+003	3.1185+003	4.2628-006	3.0833-002
5.2212+003	3.1184+003	4.4593-006	3.2292-002
5.2174+003	3.1183+003	4.6491-006	3.3718-002
5.2064+003	3.1178+003	5.5201-006	4.0446-002
5.2013+003	3.1174+003	6.2985-006	4.6679-002
5.1985+003	3.1171+003	7.0153-006	5.2554-002
5.1969+003	3.1169+003	7.6886-006	5.8150-002
5.1959+003	3.1168+003	8.3296-006	6.3519-002
5.1952+003	3.1167+003	8.9458-006	6.8698-002
5.1947+003	3.1166+003	9.5426-006	7.3714-002
5.1940+003	3.1164+003	1.0973-005	8.5664-002
5.1936+003	3.1163+003	1.2344-005	9.6938-002
5.1934+003	3.1162+003	1.3677-005	1.0768-001
5.1933+003	3.1162+003	1.5000-005	1.1798-001
5.1932+003	3.1161+003	1.6287-005	1.2792-001
5.1931+003	3.1161+003	1.7532-005	1.3754-001
5.1931+003	3.1161+003	1.8740-005	1.4688-001
5.1931+003	3.1161+003	1.9916-005	1.5597-001

Helium—Pressure is 2 atm = 2.026×10^5 Pa

Temperature (K)	Density (kg/m³)	Enthalpy (J/kg)	Entropy (J/kg·K)
1.0000+000	1.4852+002	1.3926+003	1.4886+001
1.2000+000	1.4840+002	1.4354+003	5.2252+001
1.4000+000	1.4845+002	1.5448+003	1.3548+002
1.6000+000	1.4856+002	1.7772+003	2.8990+002
1.8000+000	1.4876+002	2.2248+003	5.5196+002
2.0000+000	1.4912+002	3.0169+003	9.6808+002
2.1531 T-Lambda	1.5002+002	4.1267+003	1.4991+003
2.2000+000	1.5011+002	4.4912+003	1.6283+003
2.4000+000	1.4941+002	4.9536+003	1.8305+003
2.6000+000	1.4842+002	5.3464+003	1.9880+003
2.8000+000	1.4722+002	5.7469+003	2.1364+003
3.0000+000	1.4579+002	6.1802+003	2.2859+003
3.2000+000	1.4413+002	6.6526+003	2.4386+003
3.4000+000	1.4222+002	7.1697+003	2.5953+003
3.6000+000	1.4003+002	7.7361+003	2.7569+003
3.8000+000	1.3749+002	8.3594+003	2.9254+003
4.0000+000	1.3456+002	9.0516+003	3.1027+003
4.2000+000	1.3112+002	9.8293+003	3.2922+003
4.4000+000	1.2700+002	1.0721+004	3.4993+003
4.6000+000	1.2187+002	1.1776+004	3.7334+003
4.8000+000	1.1493+002	1.3111+004	4.0174+003
5.0000+000	1.0321+002	1.5166+004	4.4363+003
5.2000+000	3.3059+001	3.0258+004	7.4170+003
6.0000+000	2.1611+001	3.8299+004	8.8664+003
7.0000+000	1.6518+001	4.5338+004	9.9542+003
8.0000+000	1.3664+001	5.1586+004	1.0789+004
9.0000+000	1.1750+001	5.7497+004	1.1486+004
1.0000+001	1.0350+001	6.3221+004	1.2089+004
1.2000+001	8.4109+000	7.4343+004	1.3104+004
1.4000+001	7.1095+000	8.5241+004	1.3944+004
1.6000+001	6.1682+000	9.6029+004	1.4665+004
1.8000+001	5.4537+000	1.0673+005	1.5295+004
2.0000+001	4.8913+000	1.1737+005	1.5855+004
2.2000+001	4.4362+000	1.2795+005	1.6360+004
2.4000+001	4.0600+000	1.3851+005	1.6819+004
2.6000+001	3.7435+000	1.4904+005	1.7240+004
2.8000+001	3.4734+000	1.5955+005	1.7630+004
3.0000+001	3.2401+000	1.7004+005	1.7992+004
4.0000+001	2.4277+000	2.2233+005	1.9496+004
5.0000+001	1.9423+000	2.7447+005	2.0660+004
6.0000+001	1.6191+000	3.2653+005	2.1609+004
7.0000+001	1.3883+000	3.7855+005	2.2411+004
8.0000+001	1.2152+000	4.3055+005	2.3105+004
9.0000+001	1.0805+000	4.8253+005	2.3717+004
1.0000+002	9.7273-001	5.3450+005	2.4265+004
1.2500+002	7.7860-001	6.6438+005	2.5424+004
1.5000+002	6.4908-001	7.9424+005	2.6371+004
1.7500+002	5.5651-001	9.2409+005	2.7172+004
2.0000+002	4.8706-001	1.0539+006	2.7865+004
2.2500+002	4.3302-001	1.1838+006	2.8477+004
2.5000+002	3.8977-001	1.3136+006	2.9024+004
2.7500+002	3.5438-001	1.4434+006	2.9519+004
3.0000+002	3.2488-001	1.5733+006	2.9971+004

C_p (J/kg·K)	C_v (J/kg·K)	Viscosity (kg/m·S)	k (W/m·K)
1.0741+002	1.0741+002		
3.4501+002	3.4496+002		
8.2792+002	8.2754+002		
1.5810+003	1.5795+003		
2.9778+003	2.9731+003		
5.2984+003	5.2836+003		
1.2327+005	7.3778+004		
2.7636+003	2.7367+003	3.3023-006	1.4092-002
2.0385+003	1.9443+003	3.5264-006	1.4998-002
1.9520+003	1.7806+003	3.6686-006	1.5809-002
2.0664+003	1.8144+003	3.7560-006	1.6545-002
2.2178+003	1.8951+003	3.8026-006	1.7212-002
2.4779+003	1.9763+003	3.8179-006	1.7809-002
2.6748+003	2.0483+003	3.8071-006	1.8331-002
2.9621+003	2.1135+003	3.7741-006	1.8772-002
3.2990+003	2.1771+003	3.7200-006	1.9123-002
3.6835+003	2.2431+003	3.6468-006	1.9377-002
4.1057+003	2.3150+003	3.5533-006	1.9526-002
4.8671+003	2.3962+003	3.4374-006	1.9566-002
6.1317+003	2.4905+003	3.2928-006	1.9508-002
8.5062+003	2.6076+003	3.1019-006	1.9397-002
1.6957+004	2.7831+003	2.8016-006	1.9414-002
1.5762+004	3.3172+003	1.6651-006	1.2983-002
7.8427+003	3.1731+003	1.7262-006	1.2802-002
6.5148+003	3.1235+003	1.8801-006	1.3855-002
6.0417+003	3.1108+003	2.0415-006	1.5058-002
5.8034+003	3.1088+003	2.2003-006	1.6220-002
5.6625+003	3.1099+003	2.3546-006	1.7323-002
5.5091+003	3.1138+003	2.6479-006	1.9381-002
5.4263+003	3.1173+003	2.9221-006	2.1285-002
5.3705+003	3.1194+003	3.1795-006	2.3075-002
5.3322+003	3.1204+003	3.4227-006	2.4777-002
5.3052+003	3.1210+003	3.6534-006	2.6410-002
5.2853+003	3.1211+003	3.8734-006	2.7985-002
5.2703+003	3.1212+003	4.0841-006	2.9513-002
5.2586+003	3.1211+003	4.2866-006	3.0998-002
5.2493+003	3.1209+003	4.4819-006	3.2448-002
5.2418+003	3.1207+003	4.6707-006	3.3865-002
5.2196+003	3.1197+003	5.5384-006	4.0567-002
5.2094+003	3.1190+003	6.3150-006	4.6786-002
5.2040+003	3.1184+003	7.0307-006	5.2653-002
5.2007+003	3.1180+003	7.7031-006	5.8245-002
5.1987+003	3.1177+003	8.3434-006	6.3611-002
5.1973+003	3.1175+003	8.9590-006	6.8787-002
5.1963+003	3.1173+003	9.5552-006	7.3801-002
5.1949+003	3.1170+003	1.0984-005	8.5749-002
5.1941+003	3.1168+003	1.2354-005	9.7022-002
5.1937+003	3.1166+003	1.3686-005	1.0776-001
5.1935+003	3.1165+003	1.5007-005	1.1806-001
5.1933+003	3.1165+003	1.6293-005	1.2800-001
5.1932+003	3.1164+003	1.7537-005	1.3762-001
5.1932+003	3.1164+003	1.8744-005	1.4696-001
5.1931+003	3.1163+003	1.9919-005	1.5605-001

Helium—Pressure is 5 atm = 5·065 × 10^5 Pa

Temperature (K)	Density (kg/m^3)	Enthalpy (J/kg)	Entropy (J/kg · K)
1.0000+000	1.5295+002	3.4101+003	1.6377+001
1.2000+000	1.5287+002	3.4559+003	5.4939+001
1.4000+000	1.5296+002	3.5695+003	1.4150+002
1.6000+000	1.5313+002	3.8128+003	3.0306+002
1.8000+000	1.5347+002	4.2779+003	5.7538+002
2.0000+000	1.5411+002	5.1121+003	1.0132+003
2.1205 T-Lambda	1.5503+002	6.0095+003	1.4482+003
2.2000+000	1.5500+002	6.3503+003	1.5800+003
2.4000+000	1.5447+002	6.7471+003	1.7540+003
2.6000+000	1.5374+002	7.0984+003	1.8953+003
2.8000+000	1.5283+002	7.4643+003	2.0313+003
3.0000+000	1.5174+002	7.8621+003	2.1687+003
3.2000+000	1.5049+002	8.2937+003	2.3084+003
3.4000+000	1.4907+002	8.7618+003	2.4503+003
3.6000+000	1.4745+002	9.2675+003	2.5947+003
3.8000+000	1.4563+002	9.8149+003	2.7423+003
4.0000+000	1.4360+002	1.0408+004	2.8939+003
4.2000+000	1.4133+002	1.1053+004	3.0506+003
4.4000+000	1.3879+002	1.1758+004	3.2135+003
4.6000+000	1.3595+002	1.2531+004	3.3843+003
4.8000+000	1.3274+002	1.3387+004	3.5651+003
5.0000+000	1.2906+002	1.4343+004	3.7585+003
5.2000+000	1.2481+002	1.5419+004	3.9680+003
6.0000+000	1.0259+002	2.0986+004	4.9572+003
7.0000+000	6.0768+001	3.3441+004	6.8743+003
8.0000+000	4.2359+001	4.3301+004	8.1957+003
9.0000+000	3.3689+001	5.1054+004	9.1104+003
1.0000+001	2.8444+001	5.7900+004	9.8326+003
1.2000+001	2.2135+001	7.0319+004	1.0969+004
1.4000+001	1.8306+001	8.2010+004	1.1871+004
1.6000+001	1.5678+001	9.3425+004	1.2634+004
1.8000+001	1.3753+001	1.0461+005	1.3293+004
2.0000+001	1.2272+001	1.1563+005	1.3873+004
2.2000+001	1.1093+001	1.2652+005	1.4392+004
2.4000+001	1.0129+001	1.3733+005	1.4863+004
2.6000+001	9.3247+000	1.4807+005	1.5293+004
2.8000+001	8.6423+000	1.5876+005	1.5689+004
3.0000+001	8.0555+000	1.6941+005	1.6056+004
4.0000+001	6.0275+000	2.2223+005	1.7576+004
5.0000+001	4.8232+000	2.7468+005	1.8746+004
6.0000+001	4.0226+000	3.2694+005	1.9699+004
7.0000+001	3.4511+000	3.7910+005	2.0503+004
8.0000+001	3.0223+000	4.3120+005	2.1199+004
9.0000+001	2.6885+000	4.8325+005	2.1812+004
1.0000+002	2.4213+000	5.3527+005	2.2360+004
1.2500+002	1.9395+000	6.6525+005	2.3520+004
1.5000+002	1.6178+000	7.9517+005	2.4468+004
1.7500+002	1.3877+000	9.2505+005	2.5269+004
2.0000+002	1.2149+000	1.0549+006	2.5962+004
2.2500+002	1.0804+000	1.1848+006	2.6574+004
2.5000+002	9.7270-001	1.3146+006	2.7121+004
2.7500+002	8.8453-001	1.4444+006	2.7616+004
3.0000+002	8.1101-001	1.5743+006	2.8068+004

C_p (J/kg · K)	C_v (J/kg · K)	Viscosity (kg/m · S)	k (W/m · K)
1.1517+002	1.1515+002		
3.6171+002	3.6147+002		
8.5969+002	8.5857+002		
1.6449+003	1.6404+003		
3.0940+003	3.0783+003		
5.5592+003	5.5071+003		
1.2124+005	5.8186+004		
2.3412+003	2.3213+003	4.0137-006	1.4674-002
1.8195+003	1.7558+003	4.1954-006	1.5653-002
1.8120+003	1.6862+003	4.3052-006	1.6547-002
1.9273+003	1.7562+003	4.3687-006	1.7373-002
2.0689+003	1.8481+003	4.3982-006	1.8138-002
2.2780+003	1.9295+003	4.4023-006	1.8841-002
2.4229+003	1.9974+003	4.3857-006	1.9479-002
2.6460+003	2.0569+003	4.3521-006	2.0049-002
2.8677+003	2.1136+003	4.3034-006	2.0546-002
3.1274+003	2.1715+003	4.2422-006	2.0968-002
3.3659+003	2.2334+003	4.1693-006	2.1312-002
3.7350+003	2.3005+003	4.0858-006	2.1575-002
4.1469+003	2.3736+003	3.9920-006	2.1758-002
4.6676+003	2.4528+003	3.8869-006	2.1859-002
5.2638+003	2.5372+003	3.7695-006	2.1882-002
5.8719+003	2.6017+003	3.6387-006	2.1840-002
9.3085+003	2.7406+003	3.0901-006	2.1141-002
1.1717+004	3.0878+003	2.4575-006	1.7999-002
8.4881+003	3.1033+003	2.3955-006	1.7770-002
7.1973+003	3.0999+003	2.4670-006	1.8393-002
6.5865+003	3.1008+003	2.5732-006	1.9186-002
6.0435+003	3.1075+003	2.8130-006	2.0874-002
5.8018+003	3.1164+003	3.0569-006	2.2547-002
5.6483+003	3.1228+003	3.2947-006	2.4173-002
5.5456+003	3.1261+003	3.5241-006	2.5753-002
5.4749+003	3.1277+003	3.7448-006	2.7290-002
5.4238+003	3.1284+003	3.9572-006	2.8789-002
5.3855+003	3.1285+003	4.1619-006	3.0253-002
5.3559+003	3.1284+003	4.3596-006	3.1686-002
5.3327+003	3.1281+003	4.5510-006	3.3091-002
5.3139+003	3.1277+003	4.7366-006	3.4470-002
5.2590+003	3.1254+003	5.5940-006	4.1043-002
5.2338+003	3.1236+003	6.3651-006	4.7191-002
5.2202+003	3.1223+003	7.0772-006	5.3014-002
5.2122+003	3.1213+003	7.7469-006	5.8576-002
5.2070+003	3.1205+003	8.3849-006	6.3922-002
5.2036+003	3.1200+003	8.9986-006	6.9084-002
5.2012+003	3.1195+003	9.5930-006	7.4087-002
5.1976+003	3.1187+003	1.1018-005	8.6016-002
5.1958+003	3.1182+003	1.2383-005	9.7278-002
5.1948+003	3.1179+003	1.3712-005	1.0801-001
5.1942+003	3.1176+003	1.5030-005	1.1831-001
5.1938+003	3.1174+003	1.6313-005	1.2824-001
5.1935+003	3.1173+003	1.7553-005	1.3785-001
5.1933+003	3.1171+003	1.8757-005	1.4719-001
5.1932+003	3.1170+003	1.9929-005	1.5628-001

Helium—Pressure is 10 atm 1.013 × 10^6 Pa

Temperature (K)	Density (kg/m³)	Enthalpy (J/kg)	Entropy (J/kg·K)
1.0000+000	1.5907+002	6.6582+003	1.8826+001
1.2000+000	1.5903+002	6.7120+003	6.2784+001
1.4000+000	1.5916+002	6.8365+003	1.5757+002
1.6000+000	1.5944+002	7.1041+003	3.3534+002
1.8000+000	1.5997+002	7.6171+003	6.3584+002
2.0000+000	1.6099+002	8.5634+003	1.1320+003
2.0598 T-Lambda	1.6182+002	9.0348+003	1.3661+003
2.2000+000	1.6140+002	9.3967+003	1.5244+003
2.4000+000	1.6104+002	9.7195+003	1.6663+003
2.6000+000	1.6052+002	1.0024+004	1.7894+003
2.8000+000	1.5987+002	1.0352+004	1.9115+003
3.0000+000	1.5908+002	1.0711+004	2.0359+003
3.2000+000	1.5816+002	1.1100+004	2.1621+003
3.4000+000	1.5711+002	1.1521+004	2.2896+003
3.6000+000	1.5593+002	1.1972+004	2.4183+003
3.8000+000	1.5460+002	1.2456+004	2.5487+003
4.0000+000	1.5315+002	1.2975+004	2.6811+003
4.2000+000	1.5155+002	1.3532+004	2.8161+003
4.4000+000	1.4981+002	1.4132+004	2.9543+003
4.6000+000	1.4792+002	1.4776+004	3.0961+003
4.8000+000	1.4587+002	1.5470+004	3.2420+003
5.0000+000	1.4365+002	1.6216+004	3.3925+003
5.2000+000	1.4124+002	1.7015+004	3.5475+003
6.0000+000	1.3107+002	2.0476+004	4.1652+003
7.0000+000	1.1475+002	2.6029+004	5.0185+003
8.0000+000	9.4470+001	3.3485+004	6.0133+003
9.0000+000	7.5498+001	4.2154+004	7.0391+003
1.0000+001	6.2401+001	5.0239+004	7.8952+003
1.2000+001	4.6887+001	6.4332+004	9.1867+003
1.4000+001	3.7887+001	7.7106+004	1.0174+004
1.6000+001	3.1956+001	8.9416+004	1.0997+004
1.8000+001	2.7755+001	1.0133+005	1.1698+004
2.0000+001	2.4604+001	1.1292+005	1.2309+004
2.2000+001	2.2142+001	1.2429+005	1.2851+004
2.4000+001	2.0155+001	1.3549+005	1.3338+004
2.6000+001	1.8515+001	1.4656+005	1.3781+004
2.8000+001	1.7135+001	1.5753+005	1.4188+004
3.0000+001	1.5955+001	1.6842+005	1.4564+004
4.0000+001	1.1917+001	2.2209+005	1.6108+004
5.0000+001	9.5393+000	2.7504+005	1.7290+004
6.0000+001	7.9628+000	3.2763+005	1.8249+004
7.0000+001	6.8376+000	3.8002+005	1.9057+004
8.0000+001	5.9929+000	4.3227+005	1.9754+004
9.0000+001	5.3350+000	4.8445+005	2.0369+004
1.0000+002	4.8078+000	5.3656+005	2.0918+004
1.2500+002	3.8563+000	6.6670+005	2.2080+004
1.5000+002	3.2196+000	7.9670+005	2.3028+004
1.7500+002	2.7636+000	9.2664+005	2.3829+004
2.0000+002	2.4208+000	1.0565+006	2.4523+004
2.2500+002	2.1537+000	1.1864+006	2.5135+004
2.5000+002	1.9397+000	1.3163+006	2.5682+004
2.7500+002	1.7644+000	1.4461+006	2.6177+004
3.0000+002	1.6181+000	1.5759+006	2.6629+004

C_p (J/kg·K)	C_v (J/kg·K)	Viscosity (kg/m·S)	k (W/m·K)
1.3419+002	1.3411+002		
3.9465+002	3.9398+002		
9.4451+002	9.4142+002		
1.8150+003	1.8023+003		
3.4106+003	3.3674+003		
6.3571+003	6.1892+003		
1.1748+005	4.1141+004		
1.8487+003	1.8390+003	5.3020-006	1.5447-002
1.5657+003	1.5275+003	5.3575-006	1.6524-002
1.6304+003	1.5569+003	5.3748-006	1.7524-002
1.7656+003	1.6621+003	5.3688-006	1.8462-002
1.9118+003	1.7634+003	5.3451-006	1.9344-002
2.0772+003	1.8447+003	5.3079-006	2.0168-002
2.2076+003	1.9097+003	5.2597-006	2.0934-002
2.3883+003	1.9661+003	5.2023-006	2.1637-002
2.5589+003	2.0204+003	5.1365-006	2.2276-002
2.7185+003	2.0762+003	5.0644-006	2.2849-002
2.9074+003	2.1361+003	4.9865-006	2.3356-002
3.1308+003	2.2007+003	4.9035-006	2.3795-002
3.4060+003	2.2702+003	4.8163-006	2.4168-002
3.6939+003	2.3441+003	4.7245-006	2.4472-002
3.9617+003	2.4205+003	4.6284-006	2.4708-002
4.2097+003	2.4729+003	4.5287-006	2.4878-002
4.7633+003	2.5109+003	4.1782-006	2.5250-002
6.3621+003	2.6712+003	3.7368-006	2.4745-002
8.1137+003	2.8577+003	3.3454-006	2.3625-002
8.2892+003	3.0009+003	3.1204-006	2.2968-002
7.6281+003	3.0578+003	3.0633-006	2.3011-002
6.7562+003	3.0857+003	3.1464-006	2.3909-002
6.3277+003	3.1048+003	3.3140-006	2.5126-002
6.0558+003	3.1216+003	3.5067-006	2.6435-002
5.8669+003	3.1309+003	3.7069-006	2.7779-002
5.7343+003	3.1357+003	3.9071-006	2.9131-002
5.6376+003	3.1379+003	4.1045-006	3.0479-002
5.5647+003	3.1389+003	4.2978-006	3.1818-002
5.5084+003	3.1390+003	4.4865-006	3.3144-002
5.4638+003	3.1387+003	4.6706-006	3.4458-002
5.4280+003	3.1381+003	4.8502-006	3.5758-002
5.3222+003	3.1344+003	5.6891-006	4.2050-002
5.2733+003	3.1311+003	6.4503-006	4.8030-002
5.2469+003	3.1285+003	7.1560-006	5.3744-002
5.2311+003	3.1266+003	7.8209-006	5.9231-002
5.2209+003	3.1252+003	8.4549-006	6.4522-002
5.2141+003	3.1241+003	9.0651-006	6.9643-002
5.2093+003	3.1232+003	9.6563-006	7.4615-002
5.2022+003	3.1216+003	1.1074-005	8.6491-002
5.1986+003	3.1206+003	1.2433-005	9.7721-002
5.1965+003	3.1199+003	1.3755-005	1.0843-001
5.1953+003	3.1194+003	1.5068-005	1.1871-001
5.1945+003	3.1190+003	1.6345-005	1.2863-001
5.1939+003	3.1187+003	1.7580-005	1.3824-001
5.1936+003	3.1185+003	1.8779-005	1.4757-001
5.1933+003	3.1183+003	1.9945-005	1.5665-001

Helium—Pressure is 15 atm 1.5195×10^6 Pa

Temperature (K)	Density (kg/m³)	Enthalpy (J/kg)	Entropy (J/kg·K)
1.0000+000	1.6419+002	9.7947+003	2.1694+001
1.2000+000	1.6415+002	9.8573+003	7.1869+001
1.4000+000	1.6432+002	9.9975+003	1.7861+002
1.6000+000	1.6468+002	1.0296+004	3.7719+002
1.8000+000	1.6536+002	1.0872+004	7.1403+002
1.9928 T-Lambda	1.6749+002	1.1952+004	1.2858+003
2.0000+000	1.6733+002	1.1989+004	1.3042+003
2.2000+000	1.6662+002	1.2371+004	1.4837+003
2.4000+000	1.6633+002	1.2644+004	1.6040+003
2.6000+000	1.6592+002	1.2917+004	1.7146+003
2.8000+000	1.6541+002	1.3217+004	1.8268+003
3.0000+000	1.6478+002	1.3549+004	1.9417+003
3.2000+000	1.6404+002	1.3908+004	2.0583+003
3.4000+000	1.6319+002	1.4296+004	2.1759+003
3.6000+000	1.6222+002	1.4711+004	2.2943+003
3.8000+000	1.6114+002	1.5156+004	2.4140+003
4.0000+000	1.5996+002	1.5632+004	2.5352+003
4.2000+000	1.5867+002	1.6141+004	2.6585+003
4.4000+000	1.5727+002	1.6687+004	2.7841+003
4.6000+000	1.5578+002	1.7271+004	2.9123+003
4.8000+000	1.5417+002	1.7895+004	3.0434+003
5.0000+000	1.5246+002	1.8562+004	3.1775+003
5.2000+000	1.5063+002	1.9266+004	3.3140+003
6.0000+000	1.4317+002	2.2247+004	3.8463+003
7.0000+000	1.3219+002	2.6665+004	4.5259+003
8.0000+000	1.1930+002	3.2057+004	5.2452+003
9.0000+000	1.0499+002	3.8532+004	6.0089+003
1.0000+001	9.0917+001	4.5857+004	6.7845+003
1.2000+001	6.9787+001	6.0516+004	8.1393+003
1.4000+001	5.6915+001	7.3544+004	9.1482+003
1.6000+001	4.7981+001	8.6264+004	9.9978+003
1.8000+001	4.1559+001	9.8628+004	1.0726+004
2.0000+001	3.6744+001	1.1064+005	1.1359+004
2.2000+001	3.2995+001	1.2237+005	1.1918+004
2.4000+001	2.9987+001	1.3389+005	1.2419+004
2.6000+001	2.7514+001	1.4524+005	1.2874+004
2.8000+001	2.5440+001	1.5645+005	1.3289+004
3.0000+001	2.3674+001	1.6756+005	1.3672+004
4.0000+001	1.7668+001	2.2201+005	1.5240+004
5.0000+001	1.4151+001	2.7543+005	1.6432+004
6.0000+001	1.1823+001	3.2833+005	1.7397+004
7.0000+001	1.0161+001	3.8093+005	1.8208+004
8.0000+001	8.9134+000	4.3335+005	1.8908+004
9.0000+001	7.9407+000	4.8564+005	1.9524+004
1.0000+002	7.1605+000	5.3785+005	2.0074+004
1.2500+002	5.7506+000	6.6814+005	2.1237+004
1.5000+002	4.8057+000	7.9823+005	2.2185+004
1.7500+002	4.1279+000	9.2822+005	2.2987+004
2.0000+002	3.6178+000	1.0582+006	2.3681+004
2.2500+002	3.2199+000	1.1880+006	2.4293+004
2.5000+002	2.9010+000	1.3179+006	2.4840+004
2.7500+002	2.6396+000	1.4478+006	2.5335+004
3.0000+002	2.4214+000	1.5776+006	2.5787+004

C_p (J/kg · K)	C_v (J/kg · K)	Viscosity (kg/m · S)	k (W/m · K)
1.5633+002	1.5619+002		
4.3847+002	4.3725+002		
1.0548+003	1.0489+003		
2.0250+003	2.0019+003		
3.8345+003	3.7535+003		
1.1337+005	3.0580+004		
2.5000+003	2.4960+003	6.9863-006	1.4820-002
1.5157+003	1.5083+003	6.7862-006	1.6085-002
1.3889+003	1.3624+003	6.6358-006	1.7244-002
1.5004+003	1.4506+003	6.5107-006	1.8330-002
1.6523+003	1.5734+003	6.4008-006	1.9359-002
1.7878+003	1.6779+003	6.2985-006	2.0332-002
1.9396+003	1.7584+003	6.2004-006	2.1250-002
2.0568+003	1.8226+003	6.1040-006	2.2110-002
2.2072+003	1.8792+003	6.0083-006	2.2911-002
2.3475+003	1.9347+003	5.9119-006	2.3649-002
2.5100+003	1.9927+003	5.8152-006	2.4324-002
2.6437+003	2.0551+003	5.7180-006	2.4937-002
2.8538+003	2.1224+003	5.6202-006	2.5486-002
3.0310+003	2.1944+003	5.5219-006	2.5973-002
3.2856+003	2.2704+003	5.4228-006	2.6397-002
3.4619+003	2.3481+003	5.3226-006	2.6759-002
3.6538+003	2.4013+003	5.2219-006	2.7062-002
3.9913+003	2.4366+003	4.8816-006	2.7955-002
4.8377+003	2.5738+003	4.4873-006	2.8227-002
5.8524+003	2.7201+003	4.1331-006	2.7843-002
6.8163+003	2.8532+003	3.8417-006	2.7207-002
7.2983+003	2.9599+003	3.6438-006	2.6728-002
6.8913+003	3.0588+003	3.5286-006	2.6817-002
6.5333+003	3.0848+003	3.6055-006	2.7684-002
6.2872+003	3.1129+003	3.7424-006	2.8735-002
6.0854+003	3.1301+003	3.9064-006	2.9879-002
5.9291+003	3.1394+003	4.0819-006	3.1069-002
5.8082+003	3.1442+003	4.2615-006	3.2282-002
5.7137+003	3.1466+003	4.4414-006	3.3505-002
5.6387+003	3.1475+003	4.6198-006	3.4731-002
5.5784+003	3.1476+003	4.7957-006	3.5957-002
5.5293+003	3.1472+003	4.9686-006	3.7178-002
5.3812+003	3.1426+003	5.7870-006	4.3182-002
5.3111+003	3.1381+003	6.5374-006	4.8978-002
5.2727+003	3.1346+003	7.2362-006	5.4565-002
5.2496+003	3.1319+003	7.8959-006	5.9961-002
5.2346+003	3.1298+003	8.5257-006	6.5184-002
5.2245+003	3.1281+003	9.1322-006	7.0254-002
5.2174+003	3.1268+003	9.7201-006	7.5185-002
5.2068+003	3.1245+003	1.1130-005	8.6991-002
5.2014+003	3.1230+003	1.2483-005	9.8177-002
5.1983+003	3.1219+003	1.3799-005	1.0886-001
5.1964+003	3.1212+003	1.5105-005	1.1912-001
5.1952+003	3.1206+003	1.6377-005	1.2902-001
5.1944+003	3.1202+003	1.7606-005	1.3862-001
5.1938+003	3.1198+003	1.8799-005	1.4794-001
5.1934+003	3.1195+003	1.9960-005	1.5702-001

Helium—Pressure is 20 atm = 2.026 × 10^6 Pa

Temperature (K)	Density (kg/m³)	Enthalpy (J/kg)	Entropy (J/kg · K)
1.0000+000	1.6865+002	1.2842+004	2.5942+001
1.2000+000	1.6860+002	1.2916+004	8.3582+001
1.4000+000	1.6882+002	1.3074+004	2.0371+002
1.6000+000	1.6929+002	1.3411+004	4.2822+002
1.8000+000	1.7025+002	1.4075+004	8.1664+002
1.9196 T-Lambda	1.7242+002	1.4805+004	1.2178+003
2.0000+000	1.7208+002	1.4991+004	1.3133+003
2.2000+000	1.7115+002	1.5275+004	1.4489+003
2.4000+000	1.7088+002	1.5513+004	1.5547+003
2.6000+000	1.7053+002	1.5764+004	1.6563+003
2.8000+000	1.7009+002	1.6043+004	1.7608+003
3.0000+000	1.6956+002	1.6353+004	1.8683+003
3.2000+000	1.6892+002	1.6689+004	1.9772+003
3.4000+000	1.6819+002	1.7051+004	2.0870+003
3.6000+000	1.6736+002	1.7438+004	2.1976+003
3.8000+000	1.6643+002	1.7853+004	2.3093+003
4.0000+000	1.6541+002	1.8298+004	2.4224+003
4.2000+000	1.6430+002	1.8774+004	2.5375+003
4.4000+000	1.6311+002	1.9284+004	2.6547+003
4.6000+000	1.6184+002	1.9829+004	2.7742+003
4.8000+000	1.6048+002	2.0411+004	2.8961+003
5.0000+000	1.5904+002	2.1030+004	3.0205+003
5.2000+000	1.5752+002	2.1682+004	3.1467+003
6.0000+000	1.5139+002	2.4416+004	3.6350+003
7.0000+000	1.4266+002	2.8359+004	4.2418+003
8.0000+000	1.3278+002	3.2987+004	4.8594+003
9.0000+000	1.2188+002	3.8370+004	5.4938+003
1.0000+001	1.1038+002	4.4519+004	6.1438+003
1.2000+001	8.9066+001	5.8214+004	7.4074+003
1.4000+001	7.3835+001	7.1411+004	8.4403+003
1.6000+001	6.2869+001	8.4142+004	9.2917+003
1.8000+001	5.4661+001	9.6668+004	1.0030+004
2.0000+001	4.8395+001	1.0889+005	1.0674+004
2.2000+001	4.3474+001	1.2086+005	1.1244+004
2.4000+001	3.9511+001	1.3260+005	1.1755+004
2.6000+001	3.6248+001	1.4416+005	1.2218+004
2.8000+001	3.3512+001	1.5557+005	1.2641+004
3.0000+001	3.1183+001	1.6686+005	1.3030+004
4.0000+001	2.3277+001	2.2198+005	1.4617+004
5.0000+001	1.8660+001	2.7584+005	1.5819+004
6.0000+001	1.5605+001	3.2904+005	1.6789+004
7.0000+001	1.3424+001	3.8185+005	1.7603+004
8.0000+001	1.1785+001	4.3443+005	1.8305+004
9.0000+001	1.0506+001	4.8684+005	1.8923+004
1.0000+002	9.4802+000	5.3913+005	1.9474+004
1.2500+002	7.6230+000	6.6957+005	2.0638+004
1.5000+002	6.3762+000	7.9976+005	2.1588+004
1.7500+002	5.4807+000	9.2981+005	2.2389+004
2.0000+002	4.8060+000	1.0598+006	2.3084+004
2.2500+002	4.2793+000	1.1897+006	2.3696+004
2.5000+002	3.8568+000	1.3196+006	2.4243+004
2.7500+002	3.5102+000	1.4494+006	2.4738+004
3.0000+002	3.2208+000	1.5793+006	2.5190+004

C_p (J/kg · K)	C_v (J/kg · K)	Viscosity (kg/m · S)	k (W/m · K)
1.8306+002	1.8280+002		
4.9536+002	4.9320+002		
1.1865+003	1.1759+003		
2.2794+003	2.2370+003		
4.4857+003	4.3078+003		
1.0909+005	2.2644+004		
1.9660+003	1.9530+003	9.2537-006	1.5310-002
1.2903+003	1.2802+003	8.5335-006	1.6642-002
1.2666+003	1.2425+003	8.0758-006	1.7874-002
1.4047+003	1.3635+003	7.7493-006	1.9037-002
1.5537+003	1.4928+003	7.4981-006	2.0143-002
1.6868+003	1.5962+003	7.2917-006	2.1195-002
1.8183+003	1.6750+003	7.1139-006	2.2192-002
1.9396+003	1.7389+003	6.9553-006	2.3132-002
2.0710+003	1.7969+003	6.8097-006	2.4013-002
2.1953+003	1.8552+003	6.6732-006	2.4832-002
2.3352+003	1.9168+003	6.5438-006	2.5589-002
2.4853+003	1.9833+003	6.4198-006	2.6284-002
2.6620+003	2.0547+003	6.3001-006	2.6918-002
2.8266+003	2.1306+003	6.1838-006	2.7491-002
3.0395+003	2.2102+003	6.0700-006	2.8003-002
3.2258+003	2.2913+003	5.9582-006	2.8456-002
3.3717+003	2.3472+003	5.8483-006	2.8852-002
3.6255+003	2.3901+003	5.4852-006	3.0108-002
4.2496+003	2.5279+003	5.0860-006	3.0844-002
4.9536+003	2.6667+003	4.7374-006	3.0902-002
5.6851+003	2.7890+003	4.4408-006	3.0558-002
6.3283+003	2.8925+003	4.2039-006	3.0085-002
6.7458+003	3.0195+003	3.9427-006	2.9606-002
6.5043+003	3.0653+003	3.9179-006	3.0059-002
6.3420+003	3.1012+003	3.9956-006	3.0891-002
6.1852+003	3.1257+003	4.1195-006	3.1873-002
6.0439+003	3.1398+003	4.2673-006	3.2935-002
5.9236+003	3.1477+003	4.4270-006	3.4040-002
5.8234+003	3.1520+003	4.5920-006	3.5170-002
5.7404+003	3.1540+003	4.7589-006	3.6313-002
5.6715+003	3.1548+003	4.9257-006	3.7463-002
5.6141+003	3.1548+003	5.0913-006	3.8617-002
5.4346+003	3.1502+003	5.8873-006	4.4359-002
5.3466+003	3.1449+003	6.6262-006	4.9976-002
5.2974+003	3.1405+003	7.3175-006	5.5434-002
5.2675+003	3.1370+003	7.9718-006	6.0734-002
5.2481+003	3.1343+003	8.5971-006	6.5884-002
5.2348+003	3.1322+003	9.1998-006	7.0896-002
5.2254+003	3.1305+003	9.7843-006	7.5782-002
5.2114+003	3.1274+003	1.1187-005	8.7507-002
5.2042+003	3.1254+003	1.2533-005	9.8642-002
5.2000+003	3.1240+003	1.3842-005	1.0929-001
5.1975+003	3.1230+003	1.5143-005	1.1953-001
5.1959+003	3.1222+003	1.6409-005	1.2941-001
5.1948+003	3.1216+003	1.7632-005	1.3899-001
5.1941+003	3.1211+003	1.8820-005	1.4830-001
5.1936+003	3.1208+003	1.9976-005	1.5737-001

Helium—Pressure is 25 atm = 25.325 × 10^5 Pa

Temperature (K)	Density (kg/m³)	Enthalpy (J/kg)	Entropy (J/kg · K)
1.0000+000	1.7264+002	1.5815+004	3.0485+001
1.2000+000	1.7261+002	1.5902+004	9.7553+001
1.4000+000	1.7290+002	1.6084+004	2.3619+002
1.6000+000	1.7355+002	1.6473+004	4.9506+002
1.8000+000	1.7512+002	1.7300+004	9.7763+002
1.8396 T-Lambda	1.7688+002	1.7588+004	1.1527+003
2.0000+000	1.7623+002	1.7866+004	1.2989+003
2.2000+000	1.7523+002	1.8107+004	1.4148+003
2.4000+000	1.7494+002	1.8326+004	1.5116+003
2.6000+000	1.7461+002	1.8561+004	1.6071+003
2.8000+000	1.7421+002	1.8825+004	1.7059+003
3.0000+000	1.7373+002	1.9117+004	1.8073+003
3.2000+000	1.7316+002	1.9433+004	1.9099+003
3.4000+000	1.7250+002	1.9773+004	2.0132+003
3.6000+000	1.7176+002	2.0139+004	2.1174+003
3.8000+000	1.7092+002	2.0530+004	2.2227+003
4.0000+000	1.7001+002	2.0951+004	2.3296+003
4.2000+000	1.6903+002	2.1401+004	2.4383+003
4.4000+000	1.6797+002	2.1884+004	2.5492+003
4.6000+000	1.6685+002	2.2400+004	2.6623+003
4.8000+000	1.6565+002	2.2952+004	2.7778+003
5.0000+000	1.6439+002	2.3538+004	2.8954+003
5.2000+000	1.6307+002	2.4154+004	3.0145+003
6.0000+000	1.5775+002	2.6729+004	3.4746+003
7.0000+000	1.5032+002	3.0394+004	4.0386+003
8.0000+000	1.4208+002	3.4617+004	4.6022+003
9.0000+000	1.3310+002	3.9433+004	5.1696+003
1.0000+001	1.2355+002	4.4872+004	5.7442+003
1.2000+001	1.0428+002	5.7391+004	6.8983+003
1.4000+001	8.8344+001	7.0330+004	7.9151+003
1.6000+001	7.6193+001	8.2928+004	8.7600+003
1.8000+001	6.6740+001	9.5420+004	9.4957+003
2.0000+001	5.9337+001	1.0771+005	1.0143+004
2.2000+001	5.3432+001	1.1978+005	1.0719+004
2.4000+001	4.8628+001	1.3166+005	1.1235+004
2.6000+001	4.4652+001	1.4336+005	1.1704+004
2.8000+001	4.1306+001	1.5492+005	1.2132+004
3.0000+001	3.8451+001	1.6634+005	1.2526+004
4.0000+001	2.8742+001	2.2204+005	1.4130+004
5.0000+001	2.3066+001	2.7629+005	1.5341+004
6.0000+001	1.9310+001	3.2977+005	1.6316+004
7.0000+001	1.6627+001	3.8278+005	1.7133+004
8.0000+001	1.4609+001	4.3551+005	1.7837+004
9.0000+001	1.3033+001	4.8803+005	1.8456+004
1.0000+002	1.1768+001	5.4042+005	1.9008+004
1.2500+002	9.4740+000	6.7101+005	2.0174+004
1.5000+002	7.9316+000	8.0129+005	2.1124+004
1.7500+002	6.8222+000	9.3139+005	2.1926+004
2.0000+002	5.9855+000	1.0614+006	2.2620+004
2.2500+002	5.3319+000	1.1913+006	2.3233+004
2.5000+002	4.8070+000	1.3212+006	2.3780+004
2.7500+002	4.3763+000	1.4511+006	2.4275+004
3.0000+002	4.0164+000	1.5809+006	2.4727+004

C_p (J/kg·K)	C_v (J/kg·K)	Viscosity (kg/m·S)	k (W/m·K)
2.1431+002	2.1378+002		
5.6735+002	5.6335+002		
1.3539+003	1.3340+003		
2.6501+003	2.5607+003		
5.8404+003	5.2783+003		
1.0466+005	1.5771+004		
1.6190+003	1.5870+003	1.2149-005	1.5751-002
1.1482+003	1.1338+003	1.0618-005	1.7146-002
1.1824+003	1.1616+003	9.7196-006	1.8445-002
1.3289+003	1.2954+003	9.1190-006	1.9677-002
1.4789+003	1.4222+003	8.6828-006	2.0853-002
1.6014+003	1.5208+003	8.3434-006	2.1976-002
1.7328+003	1.5967+003	8.0664-006	2.3043-002
1.8306+003	1.6603+003	7.8310-006	2.4053-002
1.9569+003	1.7201+003	7.6251-006	2.5004-002
2.0783+003	1.7816+003	7.4397-006	2.5893-002
2.2134+003	1.8473+003	7.2708-006	2.6721-002
2.3433+003	1.9183+003	7.1140-006	2.7486-002
2.5214+003	1.9943+003	6.9671-006	2.8191-002
2.6812+003	2.0746+003	6.8282-006	2.8835-002
2.8747+003	2.1584+003	6.6954-006	2.9420-002
3.0356+003	2.2431+003	6.5676-006	2.9947-002
3.1656+003	2.3023+003	6.4444-006	3.0417-002
3.3988+003	2.3551+003	6.0437-006	3.1958-002
3.9211+003	2.4983+003	5.6206-006	3.3044-002
4.4853+003	2.6374+003	5.2603-006	3.3426-002
5.0651+003	2.7570+003	4.9543-006	3.3337-002
5.6231+003	2.8577+003	4.7003-006	3.2995-002
6.3380+003	2.9927+003	4.3586-006	3.2300-002
6.3642+003	3.0497+003	4.2426-006	3.2343-002
6.2909+003	3.0906+003	4.2600-006	3.2924-002
6.1964+003	3.1201+003	4.3425-006	3.3744-002
6.0897+003	3.1382+003	4.4611-006	3.4692-002
5.9868+003	3.1491+003	4.5994-006	3.5706-002
5.8938+003	3.1554+003	4.7484-006	3.6759-002
5.8123+003	3.1589+003	4.9029-006	3.7835-002
5.7418+003	3.1606+003	5.0600-006	3.8925-002
5.6812+003	3.1611+003	5.2177-006	4.0023-002
5.4820+003	3.1572+003	5.9898-006	4.5539-002
5.3795+003	3.1513+003	6.7164-006	5.0991-002
5.3209+003	3.1462+003	7.3999-006	5.6326-002
5.2848+003	3.1421+003	8.0484-006	6.1531-002
5.2611+003	3.1388+003	8.6692-006	6.6607-002
5.2449+003	3.1362+003	9.2678-006	7.1559-002
5.2333+003	3.1341+003	9.8487-006	7.6396-002
5.2160+003	3.1303+003	1.1244-005	8.8035-002
5.2070+003	3.1278+003	1.2582-005	9.9113-002
5.2018+003	3.1261+003	1.3886-005	1.0972-001
5.1987+003	3.1248+003	1.5180-005	1.1993-001
5.1966+003	3.1238+003	1.6440-005	1.2979-001
5.1953+003	3.1231+003	1.7658-005	1.3936-001
5.1944+003	3.1225+003	1.8840-005	1.4865-001
5.1937+003	3.1220+003	1.9991-005	1.5771-001

Appendix 3

Compressibility Factor for Cryogenic Fluids

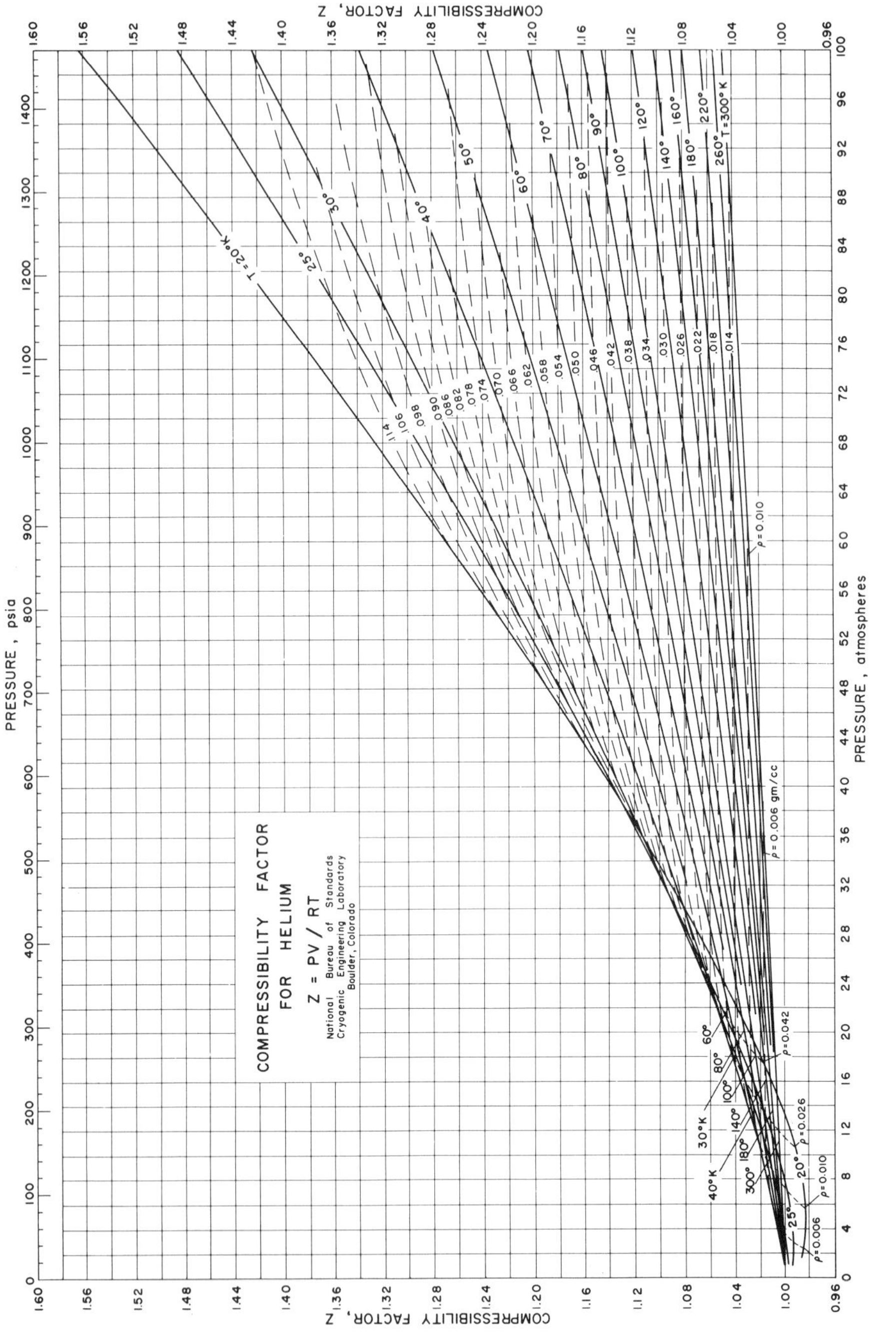

Compressibility factor for helium. $Z = PV/RT$.

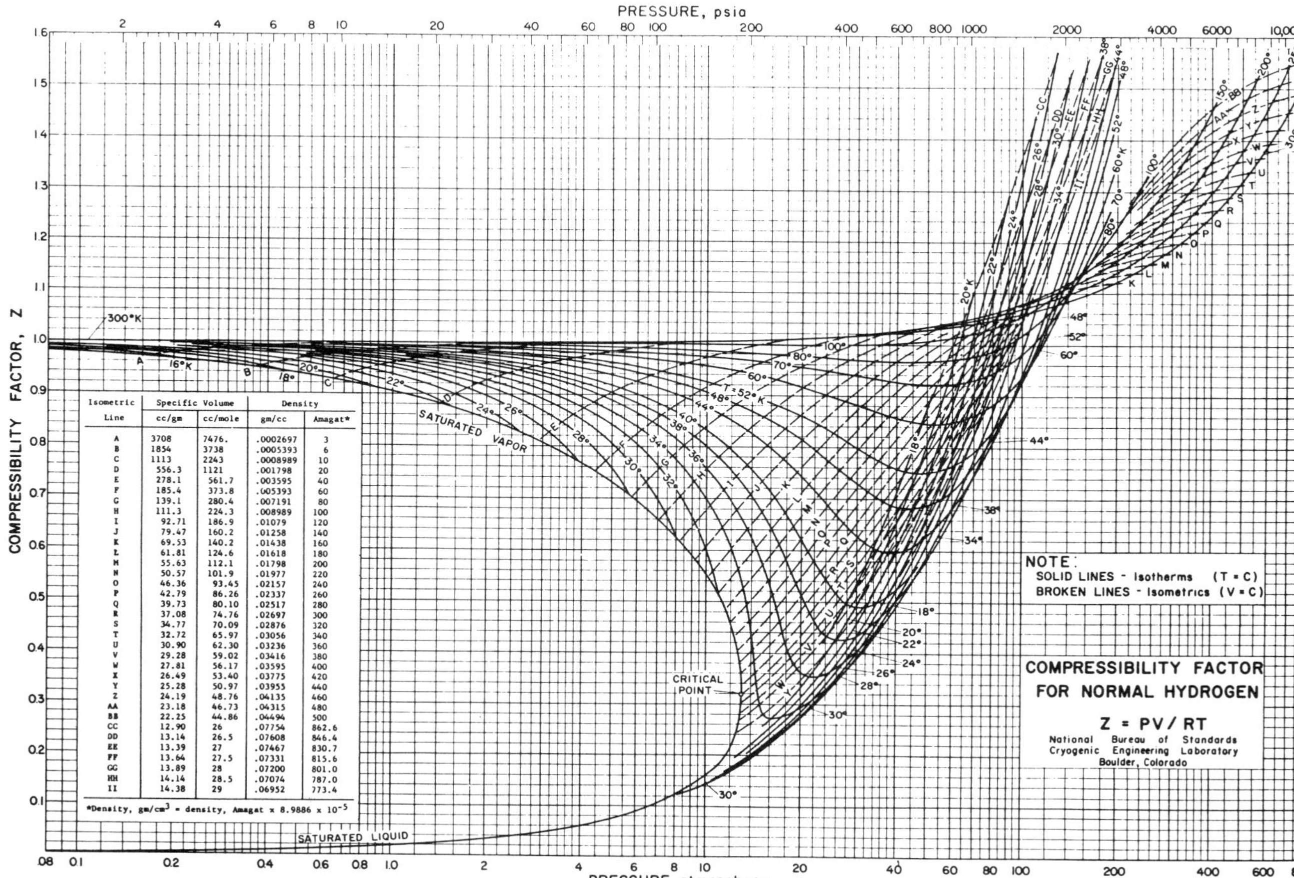

Isometric Line	Specific Volume cc/gm	Specific Volume cc/mole	Density gm/cc	Density Amagat*
A	3708	7476.	.0002697	3
B	1854	3738	.0005393	6
C	1113	2243	.0008989	10
D	556.3	1121	.001798	20
E	278.1	561.7	.003595	40
F	185.4	373.8	.005393	60
G	139.1	280.4	.007191	80
H	111.3	224.3	.008989	100
I	92.71	186.9	.01079	120
J	79.47	160.2	.01258	140
K	69.53	140.2	.01438	160
L	61.81	124.6	.01618	180
M	55.63	112.1	.01798	200
N	50.57	101.9	.01977	220
O	46.36	93.45	.02157	240
P	42.79	86.26	.02337	260
Q	39.73	80.10	.02517	280
R	37.08	74.76	.02697	300
S	34.77	70.09	.02876	320
T	32.72	65.97	.03056	340
U	30.90	62.30	.03236	360
V	29.28	59.02	.03416	380
W	27.81	56.17	.03595	400
X	26.49	53.40	.03775	420
Y	25.28	50.97	.03955	440
Z	24.19	48.76	.04135	460
AA	23.18	46.73	.04315	480
BB	22.25	44.86	.04494	500
CC	12.90	26	.07754	862.6
DD	13.14	26.5	.07608	846.4
EE	13.39	27	.07467	830.7
FF	13.64	27.5	.07331	815.6
GG	13.89	28	.07200	801.0
HH	14.14	28.5	.07074	787.0
II	14.38	29	.06952	773.4

*Density, gm/cm^3 = density, Amagat × 8.9886 × 10^{-5}

Compressibility factor for normal hydrogen. $Z = PV/RT$.

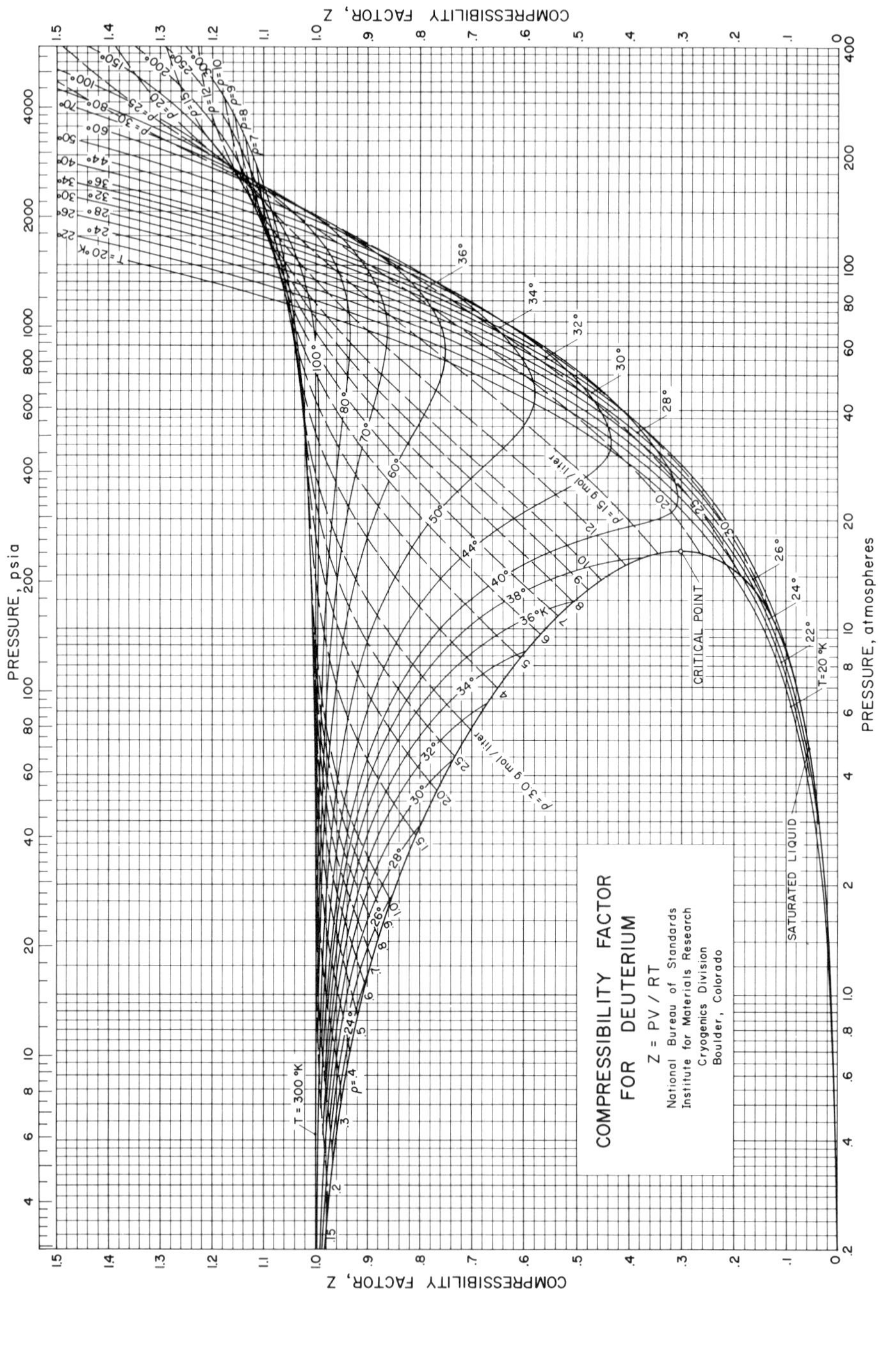

Compressibility factor for deuterium. $Z = PV/RT$.

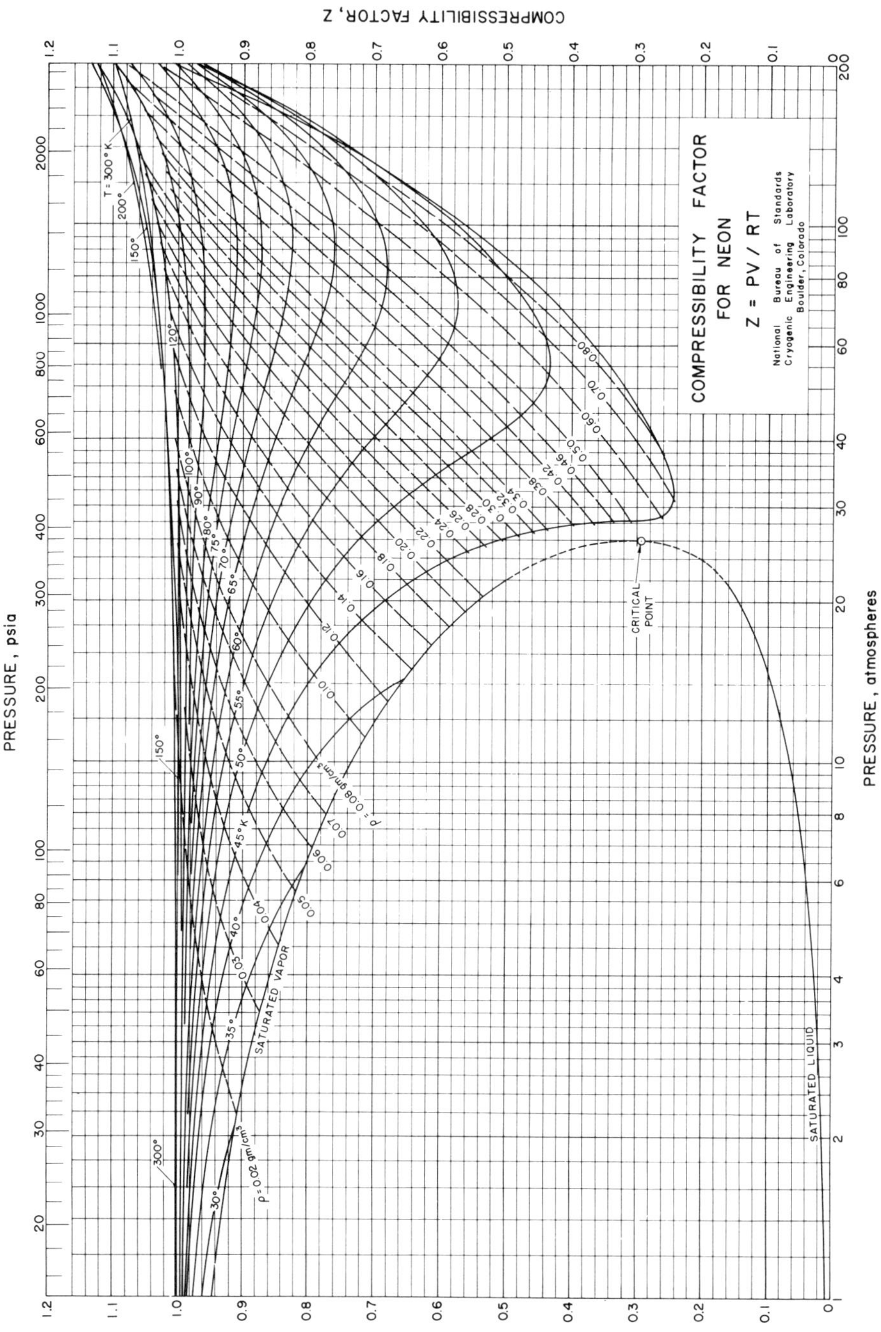

Compressibility factor for neon. $Z = PV/RT$.

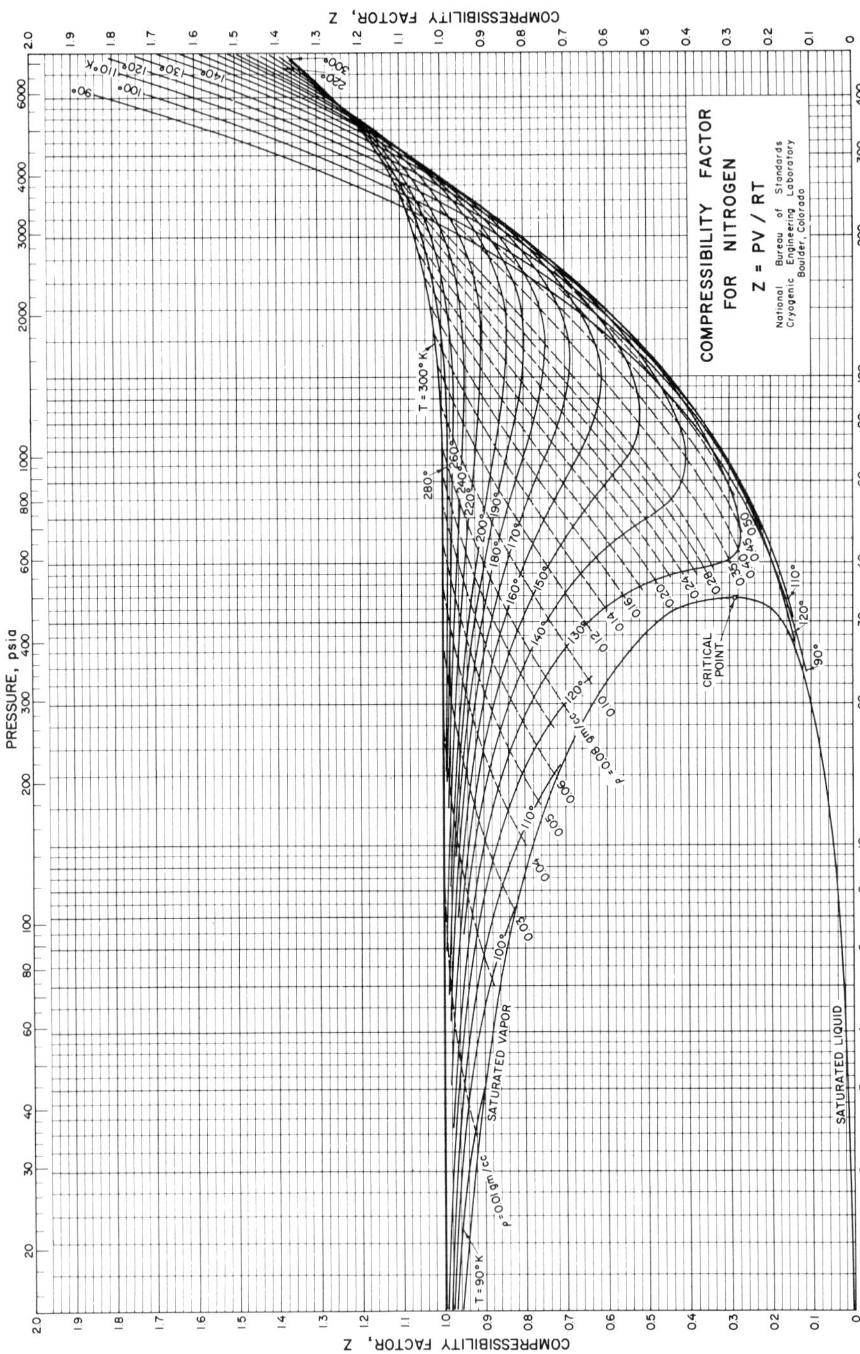

Compressibility factor for nitrogen. $Z = PV/RT$.

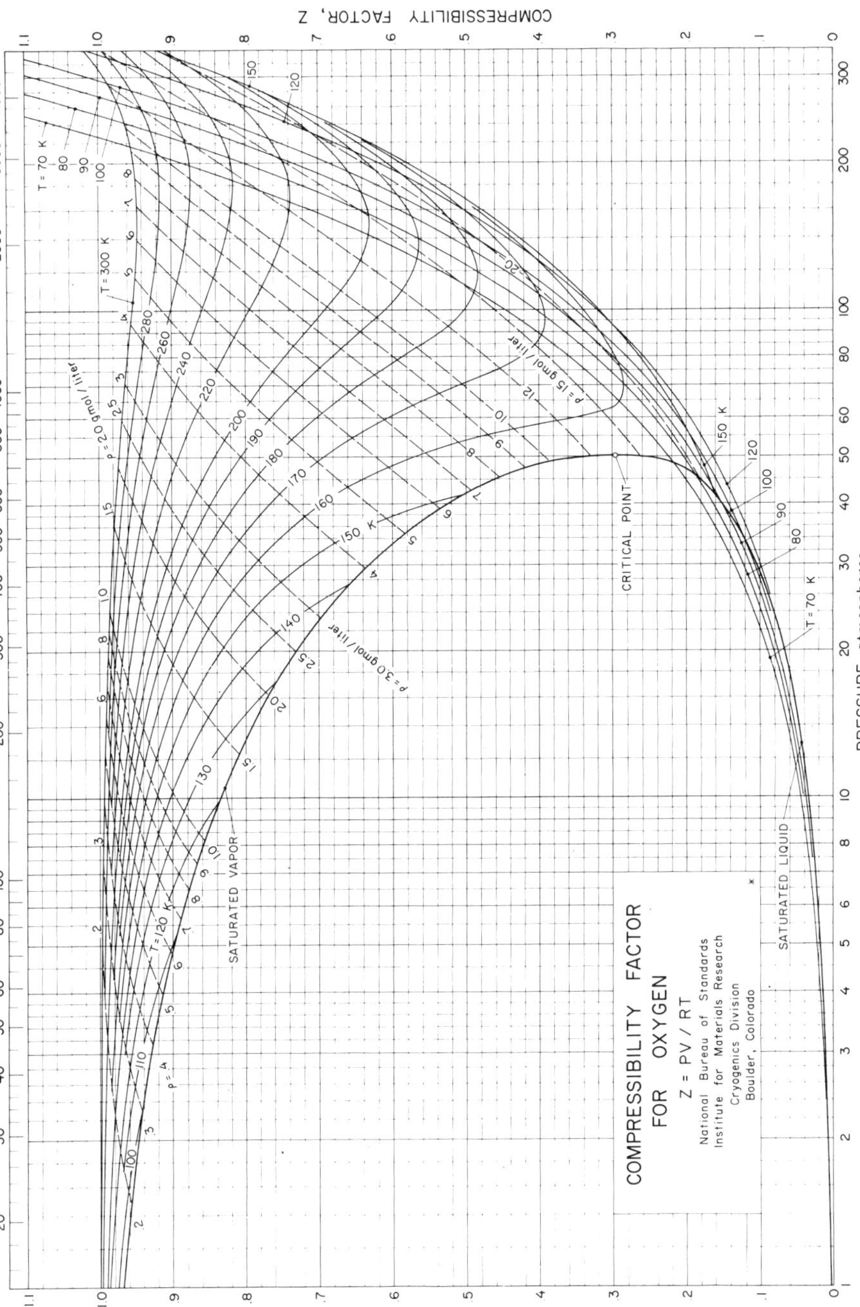

Reprinted from: "The Thermodynamic Properties of Oxygen" by Richard B. Stewart, Ph.D. Dissertation, University of Iowa, Iowa City (June, 1966)
NBS File No. D 6100

Compressibility factor for oxygen. $Z = PV/RT$.

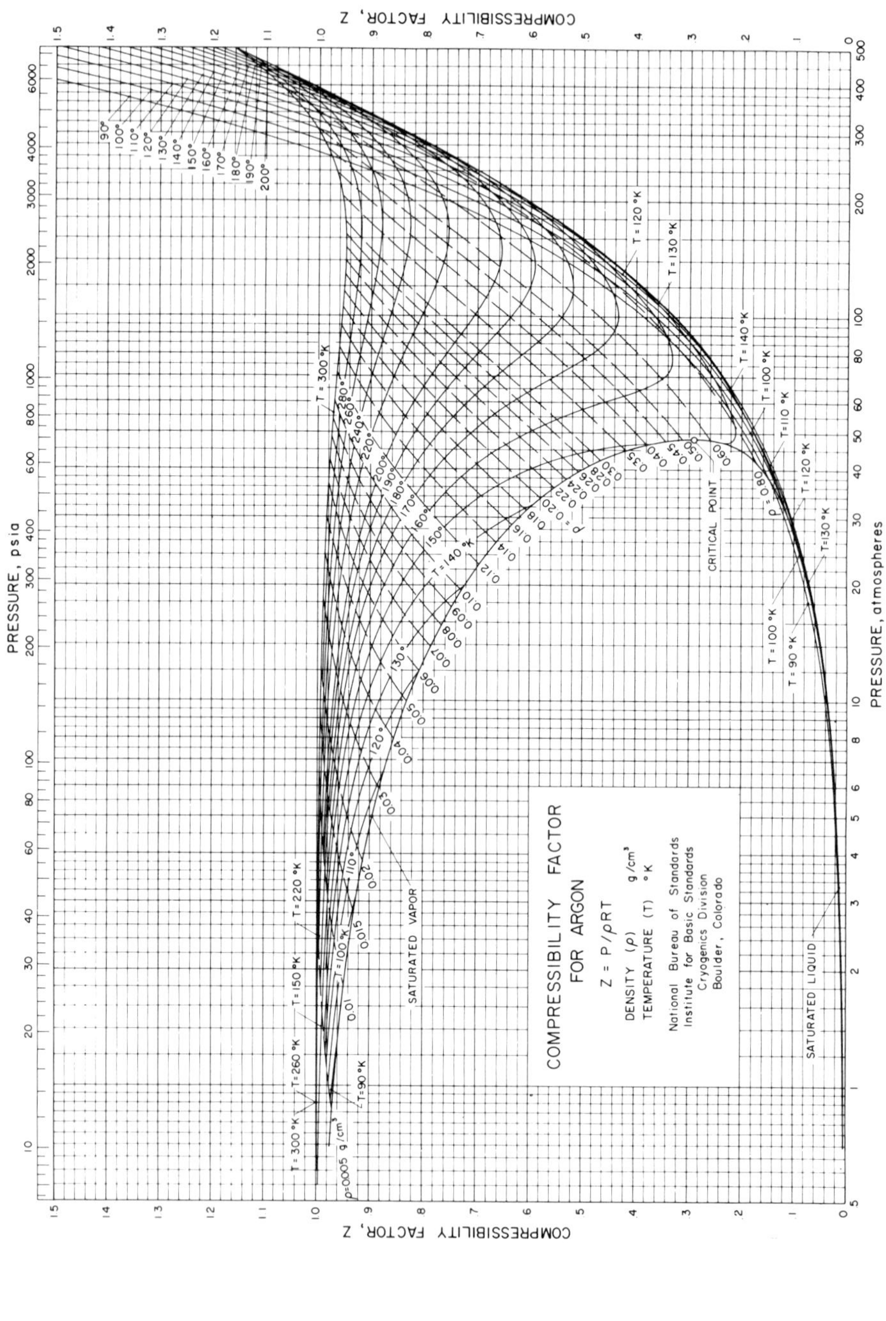

Compressibility factor for argon. $Z = P/\rho RT$. Density (ρ) is in g/cm^3. Temperature (T) is in K.

Appendix 4

Temperature–Entropy Diagrams for Cryogenic Fluids

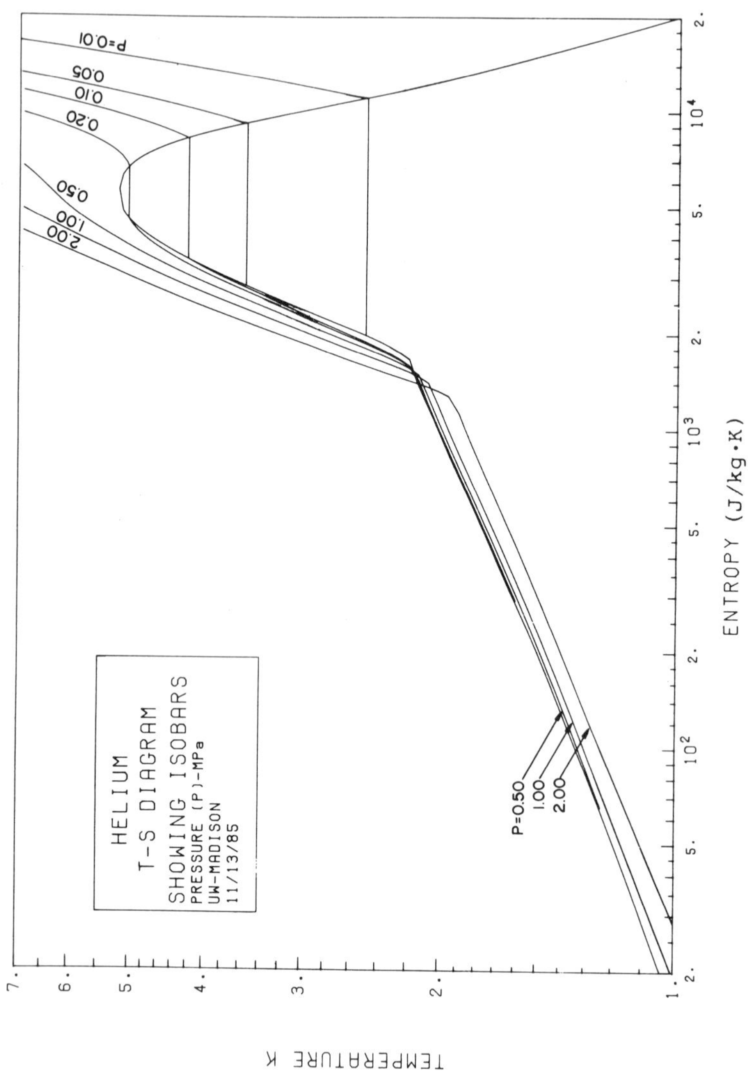

Helium *T–S* diagram showing isobars. Pressure *P* is in MPa.

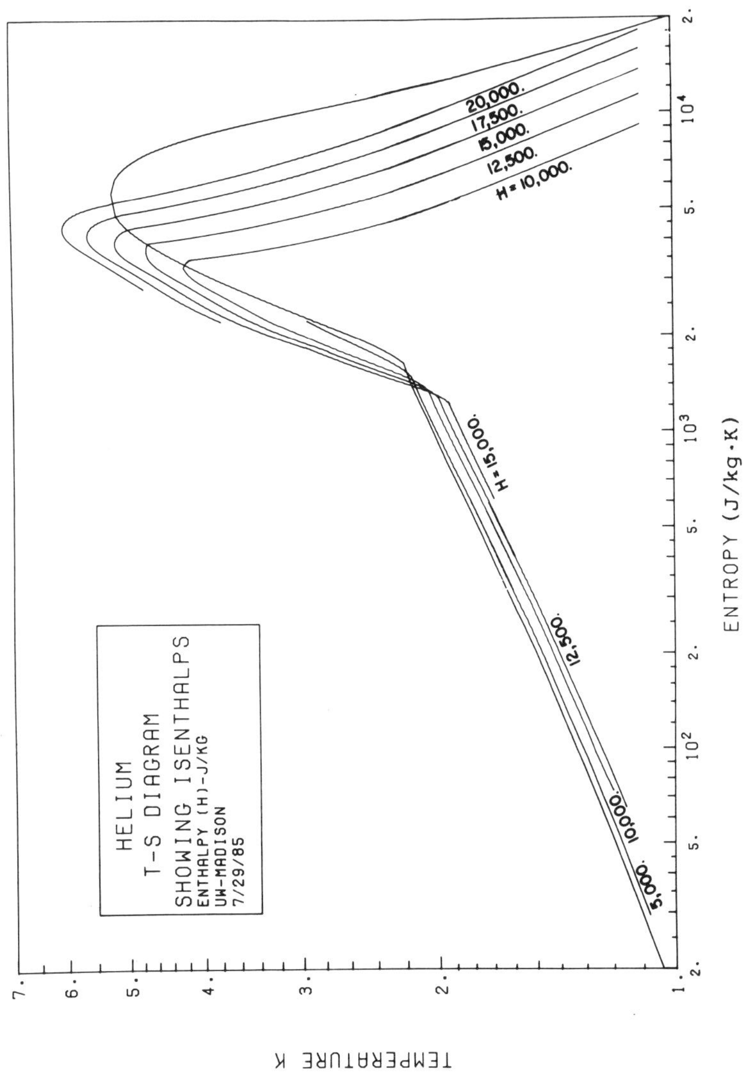

Helium T–S diagram showing isenthalps. Enthalpy H is in J/kg.

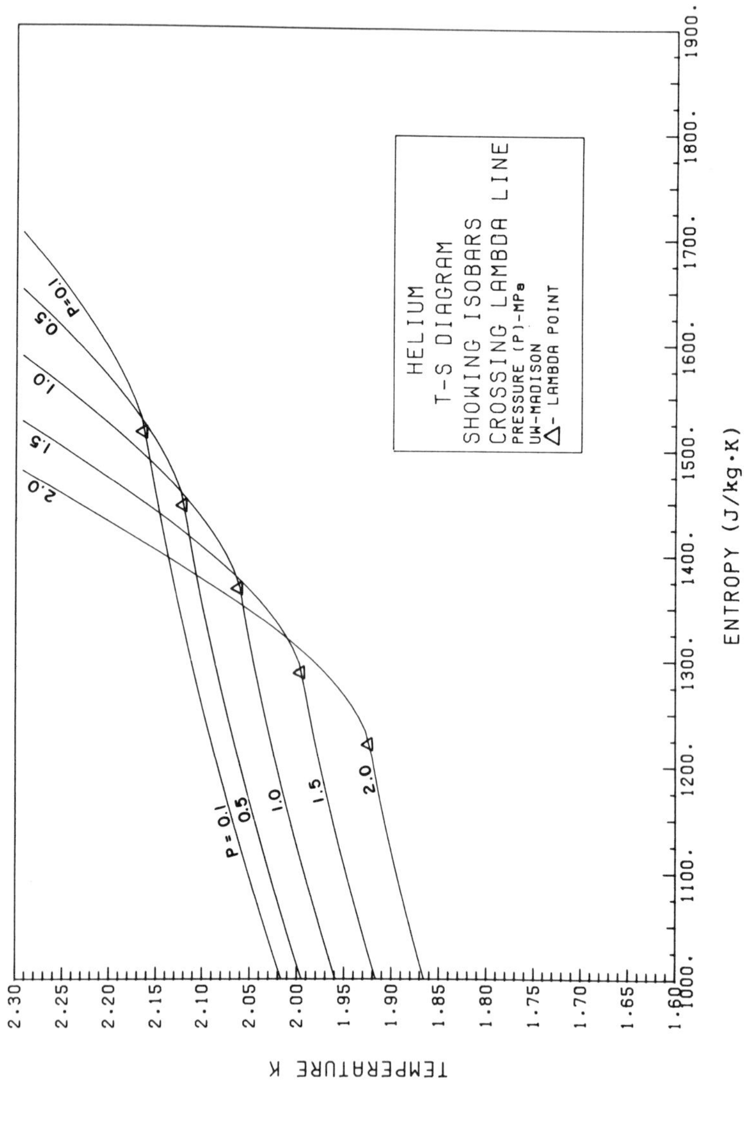

Helium *T–S* diagram showing isobars crossing λ-line: Δ represents λ-point and pressure *P* is in MPa.

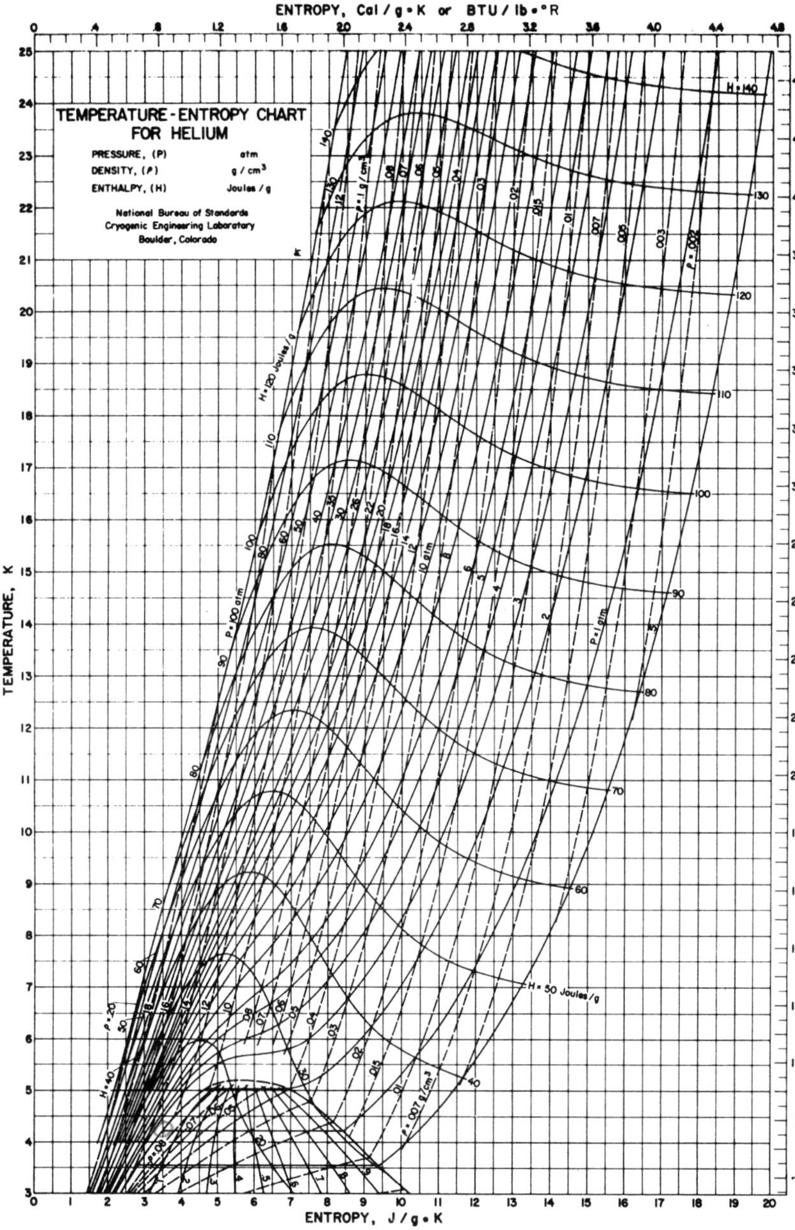

T–S chart for helium. Pressure P is in atm, density ρ is in g/cm^3, temperature T is in K or °R, enthalpy H is in J/g, and entroy S is in Cal/g · K or Btu/lb · °R. From National Bureau of Standards Cryogenic Engineering Laboratory, Boulder, CO.

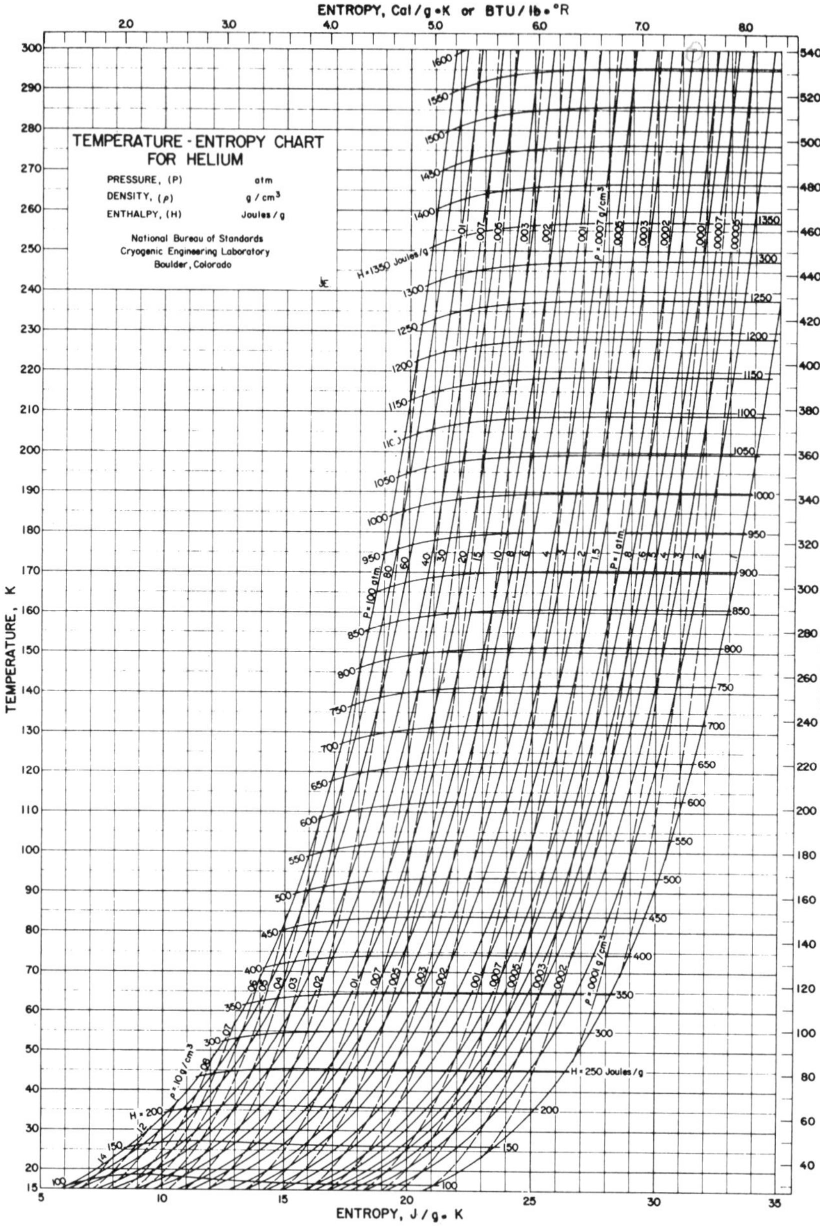

T–S chart for helium. Pressure *P* is in atm, density *ρ* is in g/cm^3, temperature *T* is in K or °R, enthalpy *H* is in J/g, and entropy *S* is in Cal/g · K or Btu/lb · °R. From National Bureau of Standards Cryogenic Engineering Laboratory, Boulder, CO.

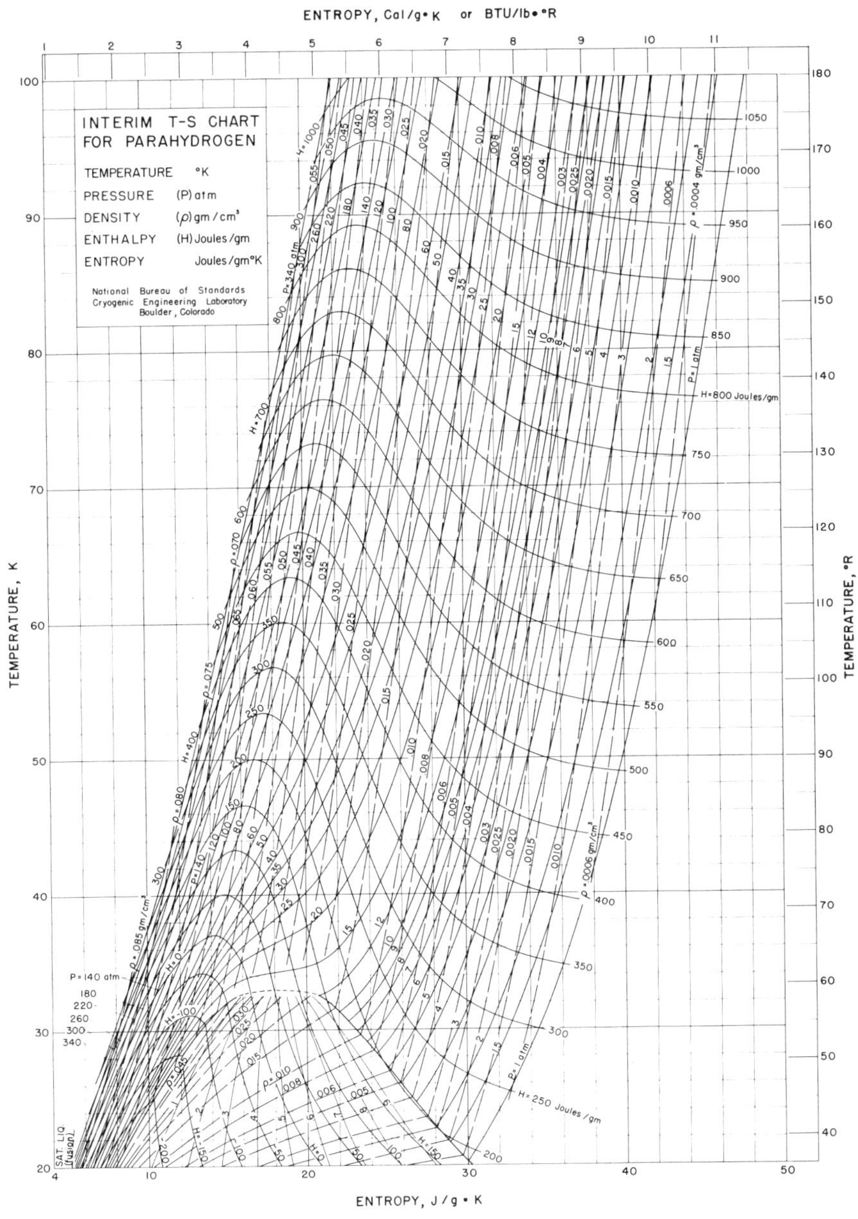

T–S chart for parahydrogen. Pressure *P* is in psia, density ρ is in lb/ft^3, temperature *T* is in K or °R, enthalpy *H* is in Btu/lb, and entropy *S* is in J/g · K or Btu/lb · °R. From National Bureau of Standards Cryogenic Engineering Laboratory, Boulder, CO.

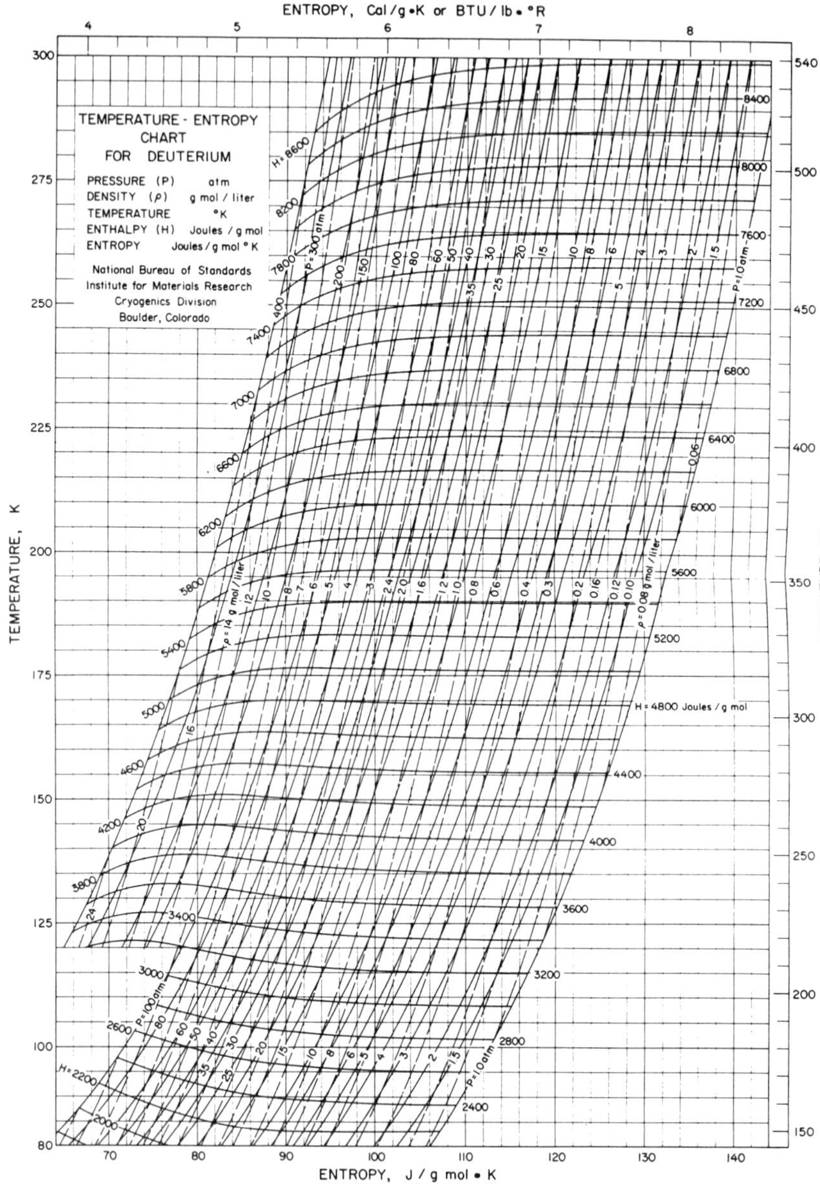

T–S chart for deuterium. Pressure *P* is in atm, density *ρ* is in g/L, temperature *T* is in K or °R, enthalpy *H* is in J/g, and entropy *S* is in Cal/g · K or Btu/lb · °R or J/g · K. From National Bureau of Standards Institute for Materials Research, Cryogenics Division, Boulder, CO.

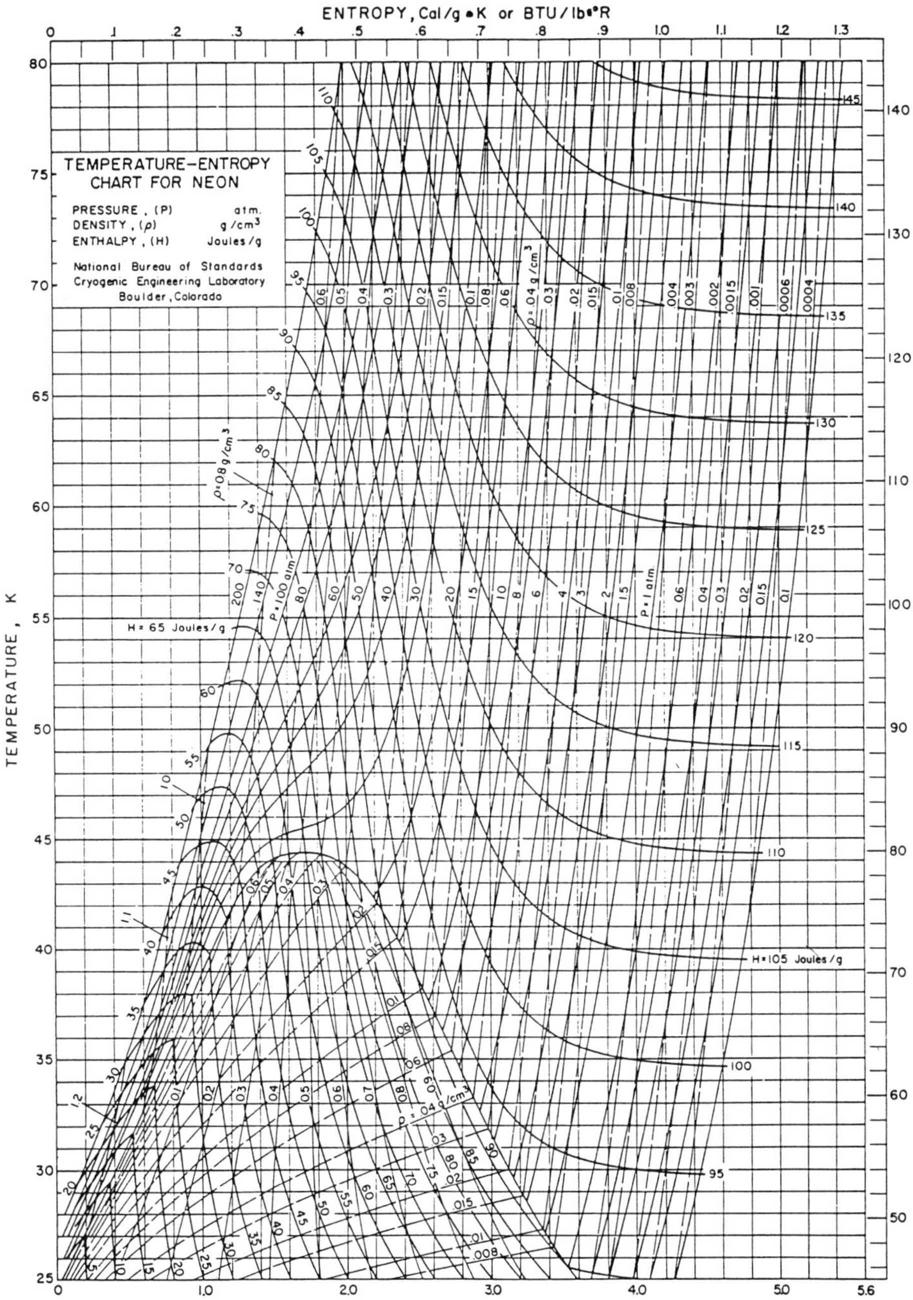

T–S chart for neon. Pressure P is in atm, density ρ is in g/L, temperature T is in K or °R, enthalpy H is in J/g, and entropy S is in Cal/g · K or Btu/lb · °R or J/g · K. From National Bureau of Standards Institute for Materials Research, Cryogenics Division, Boulder, CO.

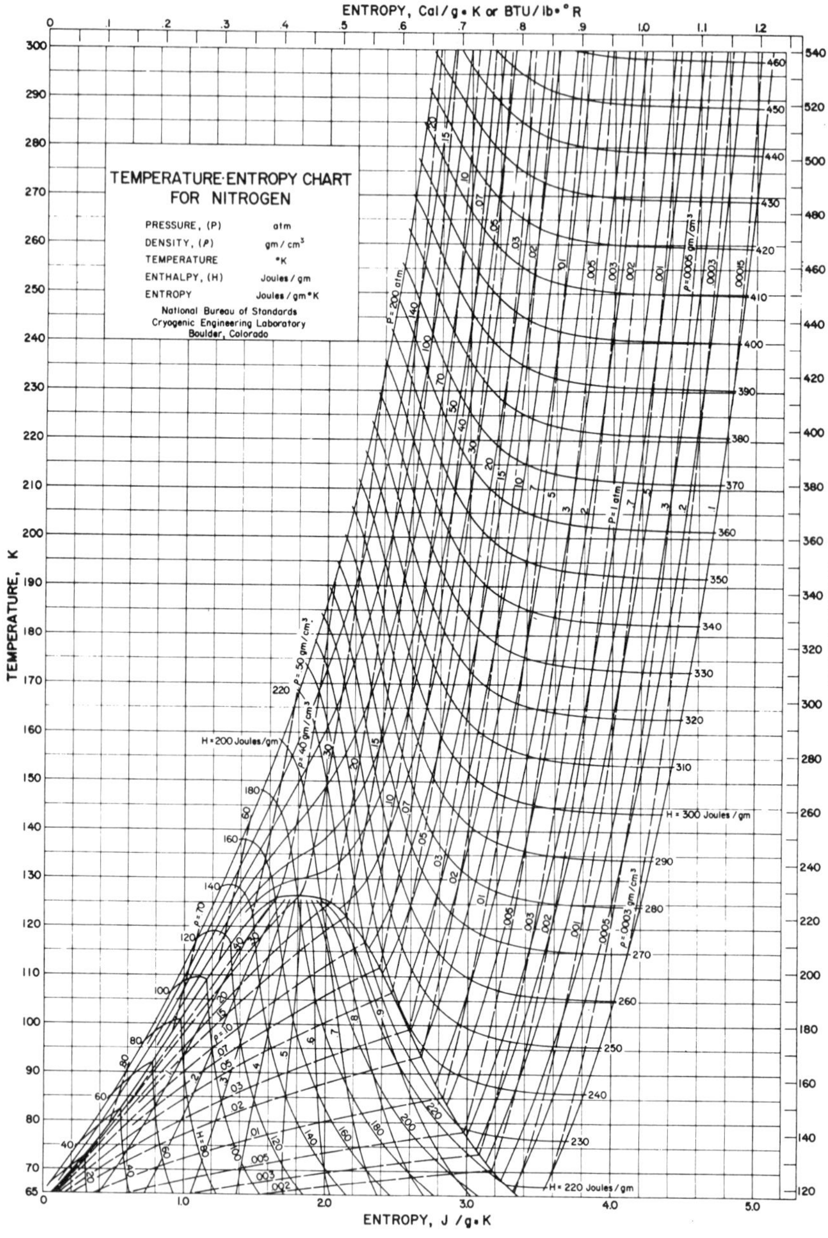

T–S chart for nitrogen. Pressure *P* is in atm, density *ρ* is in g/cm³, temperature *T* is in K or °R, enthalpy *H* is in J/g, and entropy *S* is in Cal/g · K or Btu/lb · °R or J/g · K. From National Bureau of Standards Cryogenic Engineering Laboratory, Boulder, CO.

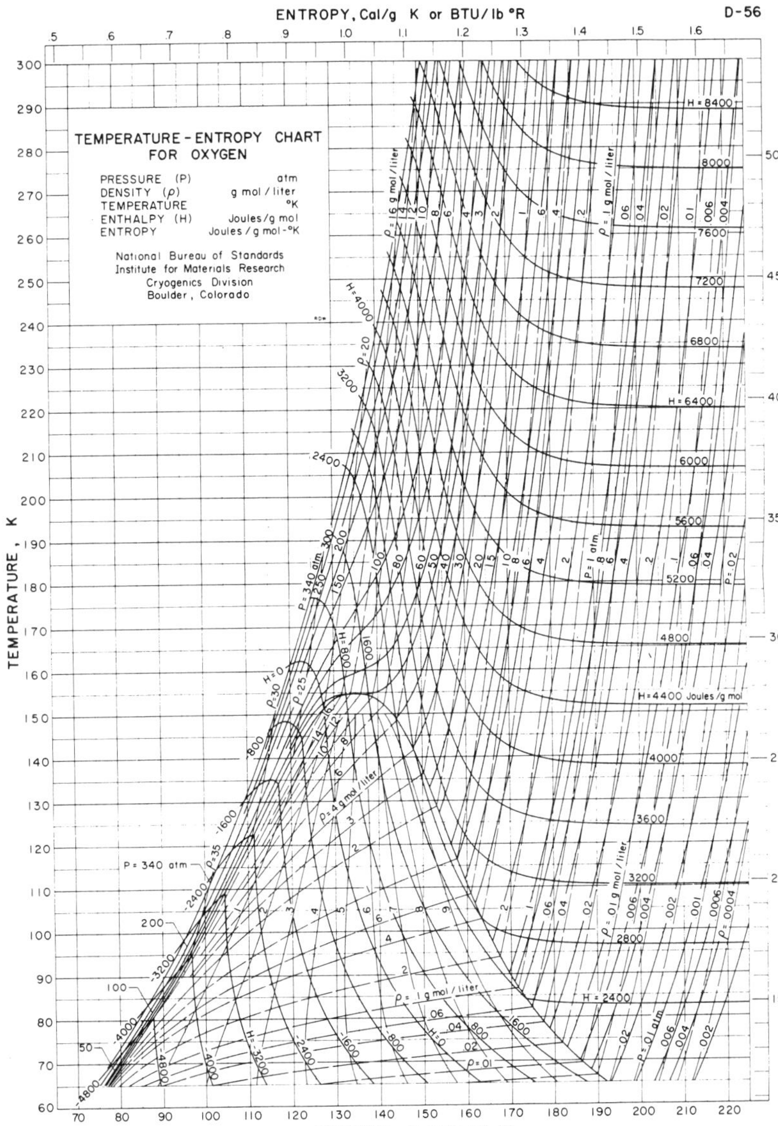

T–S chart for oxygen. Pressure *P* is in psia, density *ρ* is in lb/ft^3, temperature *T* is in K or °R, enthalpy *H* is in Btu/lb, and entropy *S* is in J/g · K or Btu/lb · °R. From National Bureau of Standards Institute for Basic Standards, Cryogenics Division, Boulder, CO.

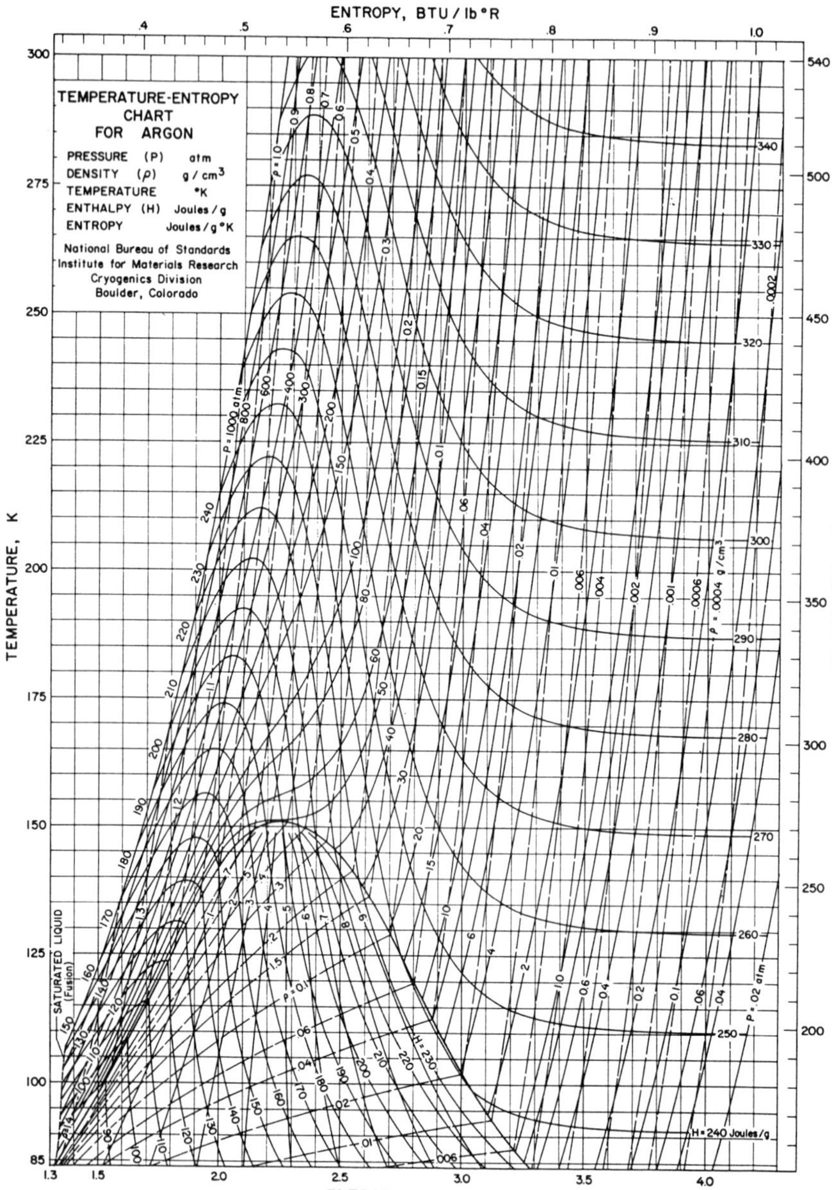

T–S chart for argon. Pressure *P* is in atm, volume *V* is in cm^3/mol, temperature *T* is in K, enthalpy *H* is in J/mol, and entropy is in J/mol · K. Entropy and enthalpy would be zero for a perfect crystal at absolute zero. Reproduced with permission of Butterworths Scientific Publications, London.

Further Reading

LOW-TEMPERATURE PROPERTIES OF MATERIALS

Berman, R., *Thermal Conduction in Solids*, Oxford Studies in Physics, Clarendon Press, Oxford, 1976.

Blakemore, J. S., *Solid State Physics*, W. B. Saunders, Philadelphia, 1974.

Gopal, E. S. R., *Specific Heats at Low Temperatures*, Plenum Press, New York, 1966.

Kittel, C., *Introduction to Solid State Physics*, 5th ed., Wiley, New York, 1976.

Krishnan, R. S., Srinovasan, R., and Devanarayanan, S., *Thermal Expansion of Crystals*, Pergamon Press, Oxford, 1979.

Lynton, E. A., *Superconductivity*, 3rd ed., Chapman and Hall, London, 1969.

McClintock, P. V. E., Meredith, D. J., and Wigmore, J. K., *Matter at Low Temperatures*, Blackie and Sons, Glasgow, 1984.

Meaden, G. T., *Electrical Resistance of Metals*, Plenum Press, New York, 1965.

Parrott, J. E. and Struckes, A. D., *Thermal Conductivity of Solids*, Pion Ltd., London, 1975.

Pollock, D. D., *Physical Properties of Materials for Engineers*, Vol. I, CRC Press, Boca Raton, FL, 1982.

Rose–Innes, A. C. and Rhoderick, E. H., *Introduction to Superconductivity*, Pergamon Press, Oxford, 1978.

Ziman, J. M., *Principles of the Theory of Solids*, 2nd ed., Cambridge University Press, Cambridge, 1972.

LOW-TEMPERATURE TECHNIQUES

Croft, A. J., *Cryogenic Laboratory Equipment*, Plenum Press, New York, 1970.

Hoare, F. E., Jackson, L. C., and Kurti, N., *Experimental Cryophysics*, Butterworth, London, 1961.

Hudson, R. P., *Principles and Applications of Magnetic Cooling*, North Holland Publishing, Amsterdam, 1972.

Lounasmaa, O. V., *Experimental Principles and Methods Below 1 K*, Academic Press, New York, 1974.

Rose–Innes, A. C., *Low Temperature Laboratory Techniques*, 2nd ed., English Universities Press, London, 1973.

White G. K., *Experimental Techniques in Low Temperature Physics*, 3rd ed., Clarendon Press, Oxford, 1979.

CRYOGENIC ENGINEERING

Bailey, C. A., *Advanced Cryogenics*, Plenum Press, London, 1971.
Barron, R. F., *Cryogenic Systems*, 2nd ed., Clarendon Press, Oxford, 1985.
Bell, J. H., *Cryogenic Engineering*, Prentice–Hall, Englewood Cliffs, NJ, 1963.
Brechna, H., *Superconducting Magnet Systems*, Chap. 5, Springer–Verlag, New York, 1973.
Conte, R. R., *Elements de Cryogenie*, Masson, Paris, 1970.
Haselden, G. G., *Cryogenic Fundamentals*, Academic Press, New York, 1971.
Kropschot, R. H., Birmingham, B. W., and Mann, D. B., *Technology of Liquid Helium*, National Bureau of Standards Monograph 111, U.S. Government Printing Office, Washington, DC, 1968.
Lady, E. R., *Cryogenic Engineering*, unpublished short course, University of Michigan, 1965.
Scott, R. B., *Cryogenic Engineering*, Van Nostrand, Princeton, NJ, 1959.
Vance, R. W. and Duke, W. M., *Applied Cryogenic Engineering*, Wiley, New York, 1962.
Walker, G., *Cryocoolers Parts I and II*, Plenum Press, New York, 1983.
Williamson, K. D. Jr. and Edeskuty, F. J., *Liquid Cryogens*, Vols. I and II, CRC Press, Boca Raton, FL, 1983.

STATISTICS AND THERMODYNAMICS

Guggenheim, E. A., *Thermodynamics*, North Holland Publishing, Amsterdam, 1977.
Huang, K. *Statistical Mechanics*, Wiley, New York, 1963.
Kittel, C., *Elementary Statistical Physics*, Wiley, New York, 1958.
Landau, L. D. and Lifshitz, E. M., *Statistical Physics*, 2nd ed., Addison–Wesley, Reading, MA, 1969.
Reif, F., *Fundamentals of Statistical and Thermal Physics*, McGraw–Hill, New York, 1965.
Walton, A. J., *Three Phases of Matter*, Clarendon Press, Oxford, 1983.
Zemansky, M. W., *Heat and Thermodynamics*, 5th ed., McGraw–Hill, New York, 1968.

HELIUM PHYSICS

Armitage, J. G. M. and Farquhar, J. E., *The Helium Liquids*, Academic Press, London, 1975.
Atkins, K. R., *Liquid Helium*, Cambridge University Press, Cambridge, 1959.
Benneman, K. H. and Kitterson, J. B., *The Physics of Liquid and Solid Helium*, Parts I and II, Wiley, New York, 1976.
Dash, J. G., *Films and Solid Surfaces*, Academic Press, New York, 1975.
Keller, W. E., *Helium-3 and Helium-4*, Plenum Press, New York, 1969.
Khalatnikov, I. M., *Introduction to the Theory of Superfluidity*, W. A. Benjamin, New York, 1965.
London, F., *Superfluids*, Vol. II, *Macroscopic Theory of Superfluid Helium*, Dover Publications, New York, 1954.
Putterman, S. J., *Superfluid Hydrodynamics*, North Holland Publishing, Amsterdam, 1974.
Wilks, J., *The Properties of Liquid and Solid Helium*, Clarendon Press, Oxford, 1967.

FLUIDS AND HEAT TRANSFER

Bejan, A., *Convection Heat Transfer*, Wiley, New York, 1984.
Frenkel, J., *Kinetic Theory of Liquids*, Oxford University Press, Oxford, 1946.
Frost, W., *Heat Transfer at Low Temperatures*, Plenum Press, New York, 1975.
Incorpera, F. P. and DeWitt, D. P., *Fundamentals of Heat Transfer*, Wiley, New York, 1981.
Hirschfelder, J. O., Curtiss, C. F., and Bird, R. B., *Molecular Theory of Gases and Liquids*, Wiley, New York, 1954.
Kays, W. M. and Crawford, M. E., *Convection Heat and Mass Transfer*, McGraw–Hill, New York, 1980.
Kennard, E. H., *Kinetic Theory of Gases*, McGraw–Hill, New York, 1954.
Rowlinson, J. S., *Liquids and Liquid Mixtures*, Plenum Press, New York, 1969.

Conversion Factors

	SI	CGS	English
Length	1 m	100 cm	3.281 ft
Area	1 m^2	10^4 cm^2	10.76 ft^2
Volume	1 m^3	10^6 cm^3 (10^3 L)	35.31 ft^3
Mass	1 kg	1000 g	2.205 lb_m
Density	1 kg/m^3	10^{-3} g/cm^3	6.243×10^{-2} lb_m/ft^3
Velocity	1 m/s	100 cm/s	3.281 ft/s
Force	1 N	10^5 dynes	0.2248 lb_f
Pressure	1 Pa	10 $dynes/cm^2$ (9.869×10^{-6} atm) (7.501×10^{-3} torr) (10^{-5} bar)	1.45×10^{-4} $lb_f/in.^2$
Temperature	1 K	1K	9/5°R
Energy	1 J	10^7 erg	9.479×10^{-4} Btu
Heat transfer rate	1 W	10^7 erg/s	3.412 Btu/h
Heat transfer coefficient	1 $W/m^2 \cdot K$	10^{-4} $W/cm^2 \cdot K$	0.176 $Btu/h \cdot ft^2 \cdot °F$
Viscosity	1 $N \cdot s/m^2$	10 poise	5.8×10^{-6} $lb_f \cdot h/ft^2$
Thermal conductivity	1 $W/m \cdot K$	10^{-2} $W/cm \cdot K$	0.578 $Btu/h \cdot ft \cdot °F$

Physical Constants

Universal gas constant	$R = 0.0823\ \mathrm{L \cdot atm/mol \cdot K}$, $8.31\ \mathrm{J/mol \cdot K}$
Speed of light in vacuum	$c = 2.998 \times 10^{8}\ \mathrm{m/s}$
Avogadro's number	$N_0 = 6.024 \times 10^{23}$ molecules/mol
Boltzmann constant	$k_B = 1.38 \times 10^{-23}\ \mathrm{J/K}$ · molecule
Planck's constant	$h = 6.625 \times 10^{-34}\ \mathrm{J \cdot s}$/molecule
Stefan–Boltzmann constant	$\sigma = 5.67 \times 10^{-8}\ \mathrm{W/m^2 \cdot K^4}$
Electron mass	$m_e = 9.11 \times 10^{-31}\ \mathrm{kg}$
Proton mass	$m_p = 1.67 \times 10^{-27}\ \mathrm{kg}$
Permeability constant	$\mu_0 = 1.26 \times 10^{-6}\ \mathrm{H/m}$
Permittivity constant	$\varepsilon_0 = 8.85 \times 10^{-12}\ \mathrm{F/m}$
Bohr magneton	$\mu_e = 0.927 \times 10^{-23}\ \mathrm{J/T}$
Elementary charge	$e = 1.60 \times 10^{-19}\ \mathrm{C}$
Gravitational acceleration	$g = 9.807\ \mathrm{m/s^2}$

Index